Student Solutions Manual

for

McKeague's

Prealgebra

Fifth Edition

AF698994

Sheryl W. Sippel
Hillsborough Community College

THOMSON
BROOKS/COLE

Australia • Canada • Mexico • Singapore • Spain • United Kingdom • United States

© 2005 Thomson Brooks/Cole, a part of The Thomson Corporation. Thomson, the Star logo, and Brooks/Cole are trademarks used herein under license.

ALL RIGHTS RESERVED. No part of this work covered by the copyright hereon may be reproduced or used in any form or by any means—graphic, electronic, or mechanical, including photocopying, recording, taping, Web distribution, information storage and retrieval systems, or in any other manner—without the written permission of the publisher.

Printed in the United States of America
1 2 3 4 5 6 7 08 07 06 05 04

Printer: Malloy Incorporated

ISBN: 0-534-46407-6

For more information about our products,
contact us at:
Thomson Learning Academic Resource Center
1-800-423-0563

For permission to use material from this text or product, submit a request online at
http://www.thomsonrights.com.
Any additional questions about permissions can be submitted by email to **thomsonrights@thomson.com.**

Thomson Higher Education
10 Davis Drive
Belmont, CA 94002-3098
USA

Asia (including India)
Thomson Learning
5 Shenton Way
#01-01 UIC Building
Singapore 068808

Australia/New Zealand
Thomson Learning Australia
102 Dodds Street
Southbank, Victoria 3006
Australia

Canada
Thomson Nelson
1120 Birchmount Road
Toronto, Ontario M1K 5G4
Canada

UK/Europe/Middle East/Africa
Thomson Learning
High Holborn House
50–51 Bedford Road
London WC1R 4LR
United Kingdom

Latin America
Thomson Learning
Seneca, 53
Colonia Polanco
11560 Mexico
D.F. Mexico

Spain (including Portugal)
Thomson Paraninfo
Calle Magallanes, 25
28015 Madrid, Spain

TABLE OF CONTENTS

Chapter 1	**Whole Numbers**	
	Chapter 1 Pretest	1
Section 1.1	Place Value and Names for Numbers	2
Section 1.2	Addition with Whole Numbers and Perimeter	4
Section 1.3	Rounding Numbers, Estimating Answers, and Displaying Information	7
Section 1.4	Subtraction with Whole Numbers	11
Section 1.5	Multiplication with Whole Numbers	14
Section 1.6	Division with Whole Numbers	19
Section 1.7	Exponents, Order of Operations, and Averages	24
Section 1.8	Area and Volume	28
	Chapter 1 Review	33
	Chapter 1 Test	36
Chapter 2	**Introduction to Algebra**	
	Chapter 2 Preview	38
	Chapter 2 Pretest	39
Section 2.1	Positive and Negative Numbers	40
Section 2.2	Addition with Negative Numbers	43
Section 2.3	Subtraction with Negative Numbers	47
Section 2.4	Multiplication with Negative Numbers	50
Section 2.5	Division with Negative Numbers	54
Section 2.6	Simplifying Algebraic Expressions	58
	Chapter 2 Review	61
	Chapters 1-2 Cumulative Review	63
	Chapter 2 Test	65

Chapter 3	**Fractions and Mixed Numbers**	
	Chapter 3 Preview	66
	Chapter 3 Pretest	67
Section 3.1	The Meaning and Properties of Fractions	68
Section 3.2	Prime Numbers, Factors, and Reducing to Lowest Terms	72
Section 3.3	Multiplication with Fractions and the Area of a Triangle	76
Section 3.4	Division with Fractions	82
Section 3.5	Addition and Subtraction with Fractions	85
Section 3.6	Mixed-Number Notation	91
Section 3.7	Multiplication and Division with Mixed Numbers	94
Section 3.8	Addition and Subtraction with Mixed Numbers	99
Section 3.9	Combinations of Operations and Complex Fractions	104
	Chapter 3 Review	110
	Chapters 1-3 Cumulative Review	112
	Chapter 3 Test	114
Chapter 4	**Solving Equations**	
	Chapter 4 Preview	116
	Chapter 4 Pretest	117
Section 4.1	The Distributive Property and Algebraic Expressions	119
Section 4.2	The Addition Property of Equality	122
Section 4.3	The Multiplication Property of Equality	125
Section 4.4	Linear Equations in One Variable	129
Section 4.5	Applications	134
Section 4.6	Evaluating Formulas	142
Section 4.7	Paired Data and Equations in Two Variables	147
Section 4.8	The Rectangular Coordinate System	153
Section 4.9	Graphing Straight Lines	156
	Chapter 4 Review	162
	Chapters 1-4 Cumulative Review	166
	Chapter 4 Test	168

Chapter 5	**Decimals**	
	Chapter 5 Preview	170
	Chapter 5 Pretest	171
Section 5.1	Decimal Notation and Place Value	172
Section 5.2	Addition and Subtraction with Decimals	176
Section 5.3	Multiplication with Decimals; Circumference and Area of a Circle	180
Section 5.4	Division with Decimals	184
Section 5.5	Fractions and Decimals and the Volume of a Sphere	189
Section 5.6	Equations Containing Decimals	195
Section 5.7	Square Roots and the Pythagorean Theorem	198
Section 5.8	Simplifying Square Roots	202
Section 5.9	Adding and Subtracting Square Roots	205
	Chapter 5 Review	208
	Chapters 1-5 Cumulative Review	211
	Chapter 5 Test	213
Chapter 6	**Ratio and Proportion**	
	Chapter 6 Preview	214
	Chapter 6 Pretest	215
Section 6.1	Ratios	216
Section 6.2	Rates and Unit Pricing	218
Section 6.3	Proportions	219
Section 6.4	Applications of Proportions	222
Section 6.5	Similar Figures	224
	Chapter 6 Review	227
	Chapters 1-6 Cumulative Review	228
	Chapter 6 Test	230

Chapter 7	**Percent**	
	Chapter 7 Preview	231
	Chapter 7 Pretest	232
Section 7.1	Percents, Decimals, and Fractions	233
Section 7.2	Basic Percent Problems	237
Section 7.3	General Applications of Percent	240
Section 7.4	Sales Tax and Commission	242
Section 7.5	Percent Increase or Decrease and Discount	245
Section 7.6	Interest	248
Section 7.7	Pie Charts	251
	Chapter 7 Review	253
	Chapters 1-7 Cumulative Review	255
	Chapter 7 Test	257
Chapter 8	**Measurement**	
	Chapter 8 Preview	258
	Chapter 8 Pretest	259
Section 8.1	Unit Analysis I: Length	260
Section 8.2	Unit Analysis II: Area and Volume	264
Section 8.3	Unit Analysis III: Weight	268
Section 8.4	Converting Between the Two Systems and Temperature	271
Section 8.5	Time and Mixed Units	274
	Chapter 8 Review	277
	Chapters 1-8 Cumulative Review	279
	Chapter 8 Test	281

Chapter 9	**Exponents and Polynomials**	
	Chapter 9 Preview	282
	Chapter 9 Pretest	283
Section 9.1	Multiplication Properties of Exponents	284
Section 9.2	Adding and Subtracting Polynomials	287
Section 9.3	Multiplying Polynomials	290
Section 9.4	Negative Exponents	294
Section 9.5	Scientific Notation	296
Section 9.6	More About Scientific Notation	298
	Chapter 9 Review	300
	Chapters 1-9 Cumulative Review	302
	Chapter 9 Test	304

Chapter 1 Pretest

1. $7{,}062 = 7{,}000 + 60 + 2$

3. Eighteen thousand, five hundred seven = 18,507

5.
$$\begin{array}{r} \scriptstyle 1\,1 \\ 1{,}029 \\ +\ 4{,}381 \\ \hline 5{,}410 \end{array}$$

7. Round 6,798 to the nearest hundred:
locate the digit to the right of the hundreds: 6,7<u>9</u>8: *9 ≥ 5 so replace it and all digits to the right with a 0 and add one to the hundreds place.* 6,800

9.
$$\begin{array}{r} \scriptstyle 6\ 9\ 10 \\ 1{,}\not 7\,\not 0\,\not 0 \\ -1{,}4\ 3\ 6 \\ \hline 2\ 6\ 4 \end{array}$$

11.
$$\begin{array}{rl} 536 & \\ \times\ 40 & \\ \hline 000 & \leftarrow 0(536) \\ \underline{21{,}440} & \leftarrow 40(536) \\ 21{,}440 & \text{Add} \end{array}$$

13.
$$\begin{array}{r} 174\ \text{R}16 \\ 23\overline{)4{,}018} \\ 23\downarrow \\ \hline 171 \\ 161\downarrow \\ \hline 108 \\ 92 \\ \hline 16 \end{array}$$

15.
$$\begin{aligned} P &= 2l + 2w \\ &= 2(36\text{ in}) + 2(27\text{ in}) \\ &= 72\text{ in} + 54\text{ in} \\ &= 126\text{ in} \end{aligned}$$

$$\begin{aligned} A &= lw \\ &= (36\text{ in})(27\text{ in}) \\ &= 972\text{ in}^2 \end{aligned}$$

17.
$$\begin{aligned} &7 + 3\cdot 2^3 && \textit{exponentiate} \\ &= 7 + 3\cdot 8 && \textit{multiply} \\ &= 7 + 24 && \textit{add} \\ &= 31 \end{aligned}$$

19.
$$\begin{aligned} \text{Mean} &= \frac{\text{sum of values}}{\text{number of values}} \\ &= \frac{4+5+7+9+15}{5} \\ &= \frac{40}{5} \\ &= 8 \end{aligned}$$

Because there is an odd number of values, the median is the middle value: Median = 7

Range = high − low = 15 − 4 = 11

Section 1.1

1. 78: 8 ones, 7 tens

3. 45: 5 ones, 4 tens

5. 348: 8 ones, 4 tens, 3 hundreds

7. 608: 8 ones, 0 tens, 6 hundreds

9. 2,378: 8 ones, 7 tens, 3 hundreds, 2 thousands

11. 273,569: 9 ones, 6 tens, 5 hundreds, 3 thousands, 7 ten thousands, 2 hundred thousands

13. 458,992
↑
ten thousands

15. 507,994,787
↑
hundred millions

17. 267,894,335
↑
ones

19. 4,569,000
↑
hundred thousands

21. $658 = 600 + 50 + 8$

23. $68 = 60 + 8$

25. $4,587 = 4,000 + 500 + 80 + 7$

27. $32,674 = 30,000 + 2,000 + 600 + 70 + 4$

29. $3,462,577 = 3,000,000 + 400,000 + 60,000 + 2,000 + 500 + 70 + 7$

31. $407 = 400 + 7$

33. $30,068 = 30,000 + 60 + 8$

35. $3,004,008 = 3,000,000 + 4,000 + 8$

37. 29 = twenty-nine

39. 40 = forty

41. 573 = five hundred seventy-three

43. 707 = seven hundred seven

45. 770 = seven hundred seventy

47. 23,540 = twenty-three thousand, five hundred forty

49. 3,004 = three thousand, four

51. 3,040 = three thousand, forty

53. 104,065,780 = one hundred four million, sixty-five thousand, seven hundred eighty

55. 5,003,040,008 = five billion, three million, forty thousand, eight

57. 2,546,731 = two million, five hundred forty-six thousand, seven hundred thirty-one

59. Three hundred twenty-five = 325

61. Five thousand, four hundred thirty-two = 5,432

63. Eighty-six thousand, seven hundred sixty-two = 86,762

65. Two million, two hundred = 2,000,200

67. Two million, two thousand, two hundred = 2,002,200

Applying the Concepts

69. 3,100 miles
↑
thousands

71. $106{,}721 = 100{,}000 + 6{,}000 + 700 + 20 + 1$

73. 2,555,476 = two million, five hundred fifty-five thousand, four hundred seventy-six

75. Two hundred seventy-five million = 275,000,000

77. 127,000,000 = one hundred twenty-seven million

79. Thirty million = 30,000,000

81. 8,800,000 = eight million, eight hundred thousand

Extending the Concepts

83.

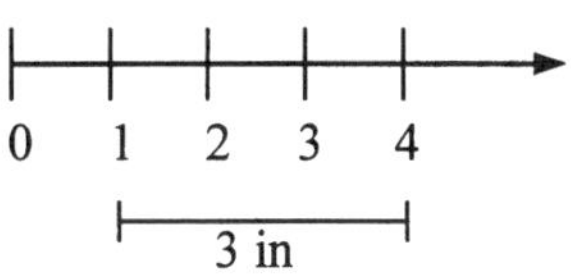

3 inches.

85.

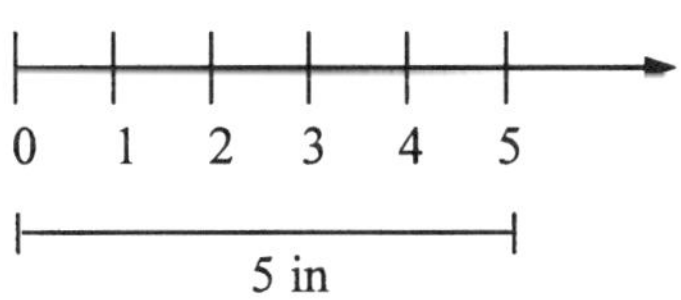

5 inches.

87.

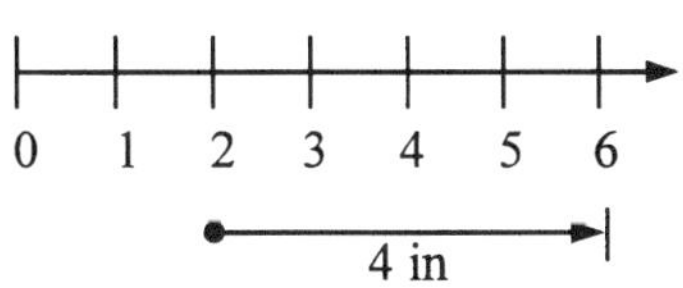

The number 6.

89.

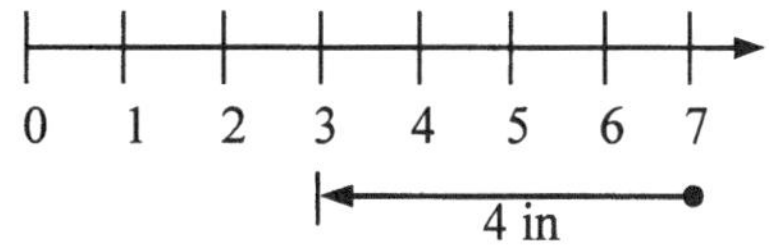

The number 3.

91.

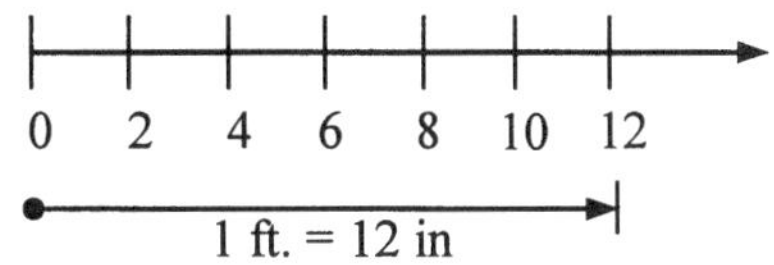

The number 12.

Section 1.2

1. $3+5+7=15$

3. $1+4+9=14$

5. $5+9+4+6=24$

7. $1+2+3+4+5=15$

9. $9+1+8+2=20$

11. $$\begin{array}{r} 43 \\ +\ 25 \\ \hline 68 \end{array}$$

13. $$\begin{array}{r} 81 \\ +\ 17 \\ \hline 98 \end{array}$$

15. $$\begin{array}{r} 4{,}281 \\ +\ 3{,}016 \\ \hline 7{,}297 \end{array}$$

17. $$\begin{array}{r} 3{,}482 \\ +\ 3{,}005 \\ \hline 6{,}487 \end{array}$$

19. $$\begin{array}{r} 32 \\ 21 \\ +\ 43 \\ \hline 96 \end{array}$$

21. $$\begin{array}{r} 6{,}245 \\ 203 \\ +\ 1{,}001 \\ \hline 7{,}449 \end{array}$$

23. $$\begin{array}{r} \scriptstyle 1 \\ 49 \\ +\ 16 \\ \hline 65 \end{array}$$

25. $$\begin{array}{r} \scriptstyle 1 \\ 74 \\ +\ 28 \\ \hline 102 \end{array}$$

27. $$\begin{array}{r} \scriptstyle 1 \\ 682 \\ +\ 193 \\ \hline 875 \end{array}$$

29. $$\begin{array}{r} \scriptstyle 1 \\ 638 \\ +\ 191 \\ \hline 829 \end{array}$$

31. $$\begin{array}{r} \scriptstyle 1\ \ 1 \\ 4{,}963 \\ +\ 5{,}428 \\ \hline 10{,}391 \end{array}$$

33. $$\begin{array}{r} \scriptstyle 1\ 1\ 1 \\ 6{,}205 \\ +\ 9{,}999 \\ \hline 16{,}204 \end{array}$$

35. $$\begin{array}{r} \scriptstyle 1\ 1\ 1\ 1 \\ 56{,}789 \\ +\ 98{,}765 \\ \hline 155{,}554 \end{array}$$

37. $$\begin{array}{r} \scriptstyle 1\ 1\ 1\ 1 \\ 52{,}468 \\ +\ 58{,}642 \\ \hline 111{,}110 \end{array}$$

39.
$$\begin{array}{r} \scriptstyle 1\ 11 \\ 4{,}296 \\ 8{,}720 \\ +\ 4{,}375 \\ \hline 17{,}391 \end{array}$$

41.
$$\begin{array}{r} \scriptstyle 1\ 12 \\ 4{,}994 \\ 449 \\ +\ 9{,}449 \\ \hline 14{,}892 \end{array}$$

43.
$$\begin{array}{r} \scriptstyle 2 \\ 12 \\ 34 \\ 56 \\ +\ 78 \\ \hline 180 \end{array}$$

45.
$$\begin{array}{r} \scriptstyle 22 \\ 999 \\ 444 \\ 555 \\ +\ 222 \\ \hline 2{,}220 \end{array}$$

47.
$$\begin{array}{r} \scriptstyle 1\ 11 \\ 9{,}245 \\ 672 \\ 8{,}341 \\ +\ 27 \\ \hline 18{,}285 \end{array}$$

49.
$$\begin{array}{r} 61 \\ +\ 38 \\ \hline 99 \end{array} \quad \begin{array}{r} 63 \\ +\ 36 \\ \hline 99 \end{array} \quad \begin{array}{r} 65 \\ +\ 34 \\ \hline 99 \end{array} \quad \begin{array}{r} 67 \\ +\ 32 \\ \hline 99 \end{array}$$

Note the pattern: Each of the 1st numbers is increased by 2, while each of the 2nd numbers is decreased by 2. The sum remains the same.

51.
$$\begin{array}{r} 9 \\ +\ 16 \\ \hline 25 \end{array} \quad \begin{array}{r} 36 \\ +\ 64 \\ \hline 100 \end{array} \quad \begin{array}{r} 81 \\ +\ 144 \\ \hline 225 \end{array} \quad \begin{array}{r} 144 \\ +\ 256 \\ \hline 400 \end{array}$$

Note: All of these numbers are "perfect squares". For example, $9+16=25$ is the same as $3^2+4^2=5^2$.

53. $5+9=9+5$

55. $3+8=8+3$

57. $6+4=4+6$

59. $(1+2)+3=1+(2+3)$

61. $(2+1)+6=2+(1+6)$

63. $1+(9+1)=(1+9)+1$

65. $(4+n)+1=4+(n+1)$

67. The solution to $n+6=10$ is $n=4$, because $4+6=10$.

69. The solution to $n+8=13$ is $n=5$, because $5+8=13$.

71. The solution to $4+n=12$ is $n=8$, because $4+8=12$.

73. The solution to $17=n+9$ is $n=8$, because $17=8+9$.

75. $4\ +\ 9$

$\downarrow$

The sum of 4 and 9.

77. $8\ +\ 1$

$\downarrow$

The sum of 8 and 1.

79. 2 + 3 = 5

↓ ↓

The sum of 2 and 3 is 5.

81. Perimeter = sum of all sides

$P = 3 \text{ in.} + 3 \text{ in.} + 3 \text{ in.} + 3 \text{ in.}$

$P = 12 \text{ in.}$

83. Perimeter = sum of all sides

$P = 4 \text{ ft.} + 4 \text{ ft.} + 4 \text{ ft.} + 4 \text{ ft.}$

$P = 16 \text{ ft.}$

85. Perimeter = sum of all sides

$P = 10 \text{ yd.} + 3 \text{ yd.} + 10 \text{ yd.} + 3 \text{ yd.}$

$P = 26 \text{ yd.}$

87. Perimeter = sum of all sides

$P = 5 \text{ in.} + 6 \text{ in.} + 7 \text{ in.}$

$P = 18 \text{ in.}$

89.

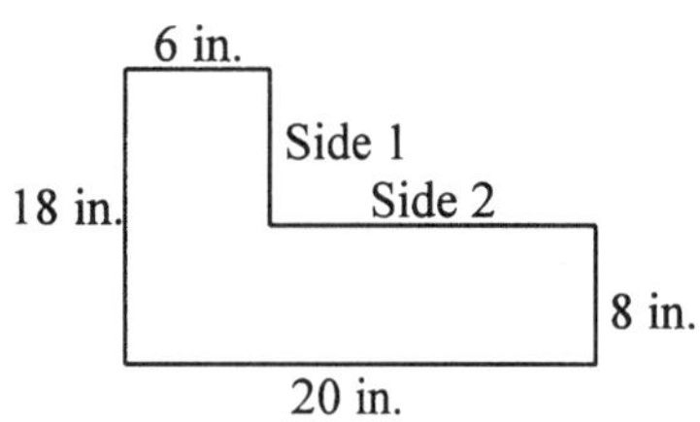

Find lengths of missing sides:
Side 1: $18 \text{ in.} - 8 \text{ in.} = 10 \text{ in.}$
Side 2: $20 \text{ in.} - 6 \text{ in.} = 14 \text{ in.}$

$P = 18 \text{ in.} + 6 \text{ in.} + 10 \text{ in.} + 14 \text{ in.} + 8 \text{ in.} + 20 \text{ in.}$

$P = 76 \text{ in.}$

91.

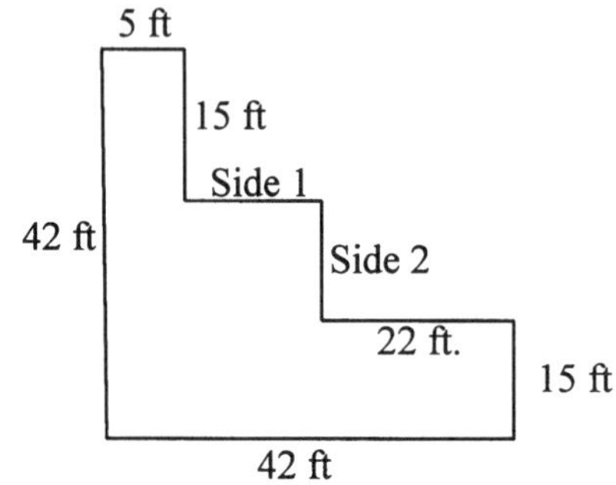

Find lengths of missing sides:
Side 1: 42 ft. – 22 ft. – 5 ft. = 15 ft.
Side 2: 42 ft. – 15 ft. – 15 ft. = 12 ft.

$P = 42 \text{ ft.} + 5 \text{ ft.} + 15 \text{ ft.} + 15 \text{ ft.} + 12 \text{ ft.} + 22 \text{ ft.}$
$+ 15 \text{ ft.} + 42 \text{ ft.}$

$P = 168 \text{ ft.}$

Applying the Concepts

93.

	1st gas purchased:	18 gallons
+	2nd gas purchased:	+ 16 gallons
	Total gas purchased:	34 gallons

95.

	Monday's balance:	$241
+	Tuesday's deposit:	+ 108
	Wednesday's balance:	$349

97. a.

	Ages 15-19:	425,493
+	Under 15:	+ 7,315
	Total babies:	432,808 babies

b.

	Ages 20-29:	2,082,497
	Ages 15-19:	425,493
+	Under 15:	+ 7,315
	Total babies:	2,515,305 babies

c.

	Ages 30-39:	1,405,146
	Ages 20-29:	2,082,497
	Ages 15-19:	425,493
+	Under 15:	+ 7,315
	Total babies:	3,920,451 babies

Section 1.3

1. Round 42 to the nearest ten:
locate the digit to the right of the tens:
42: *$2 < 5$ so replace it with a 0:* 40

3. Round 46 to the nearest ten:
locate the digit to the right of the tens:
46: *$6 \geq 5$ so replace it with a 0 and add one to the tens place:* 50

5. Round 45 to the nearest ten:
locate the digit to the right of the tens:
45: *$5 \geq 5$ so replace it with a 0 and add one to the tens place:* 50

7. Round 77 to the nearest ten:
locate the digit to the right of the tens:
77: *$7 \geq 5$ so replace it with a 0 and add one to the tens place:* 80

9. Round 458 to the nearest ten:
locate the digit to the right of the tens:
458: *$8 \geq 5$ so replace it with a 0 and add one to the tens place:* 460

11. Round 471 to the nearest ten:
locate the digit to the right of the tens:
471: *$1 < 5$ so replace it with a 0:* 470

13. Round 56,782 to the nearest ten:
locate the digit to the right of the tens:
56,782: *$2 < 5$ so replace it with a 0:* 56,780

15. Round 4,504 to the nearest ten:
locate the digit to the right of the tens:
4,504: *$4 < 5$ so replace it with a 0:* 4,500

17. Round 549 to the nearest hundred:
locate the digit to the right of the hundreds:
549: *$4 < 5$ so replace it and all digits to the right with a 0:* 500

19. Round 833 to the nearest hundred:
locate the digit to the right of the hundreds:
833: *$3 < 5$ so replace it and all digits to the right with a 0:* 800

21. Round 899 to the nearest hundred:
locate the digit to the right of the hundreds:
899: *$9 \geq 5$ so replace it and all digits to the right with a 0 and add one to the hundreds place.* 900

23. Round 1,090 to the nearest hundred:
locate the digit to the right of the hundreds:
1090: *$9 \geq 5$ so replace it and all digits to the right with a 0 and add one to the hundreds place.* 1,100

25. Round 5,044 to the nearest hundred:
locate the digit to the right of the hundreds:
5,044: *$4 < 5$ so replace it and all digits to the right with a 0:* 5,000

27. Round 39,603 to the nearest hundred:
locate the digit to the right of the hundreds:
39,603: *$0 < 5$ so replace it and all digits to the right with a 0:* 39,600

29. Round 4,670 to the nearest thousand:
locate the digit to the right of the thousands:
4,670: *$6 \geq 5$ so replace it and all digits to the right with a 0 and add one to the thousands place.* 5,000

31. Round 9,760 to the nearest thousand:
locate the digit to the right of the thousands:
9,760: *$7 \geq 5$ so replace it and all digits to the right with a 0 and add one to the thousands place.* 10,000
(Note: you must carry when you add one to the thousands place).

33. Round 978 to the nearest thousand:
locate the digit to the right of the thousands:
978: *$9 \geq 5$ so replace it and all digits to the right with a 0 and add one to the thousands place.* 1,000
(Note: The thousands place was previously 0).

35. Round 657,892 to the nearest thousand:
locate the digit to the right of the thousands:
657,<u>**8**</u>92: *8 ≥ 5 so replace it and all digits to the right with a 0 and add one to the thousands place.* 658,000

37. Round 509,905 to the nearest thousand:
locate the digit to the right of the thousands:
509,<u>**9**</u>05: *9≥ 5 so replace it and all digits to the right with a 0 and add one to the thousands place.* 510,000
(Note: you must carry when you add one to the thousands place).

39. Round 3,789,345 to the nearest thousand:
locate the digit to the right of the thousands:
3,789,<u>**3**</u>45: 3 < *5 so replace it and all digits to the right with a 0:* 3,789,000

41. *Locate the digit to the right of the given place value and compare to 5:*

Tens	Hundred	Thousand
7,82<u>**1**</u> 1 < 5 7,820	7,8<u>**2**</u>1 2 < 5 7,800	7,<u>**8**</u>21 8 > 5 8,000

43. *Locate the digit to the right of the given place value and compare to 5:*

Tens	Hundred	Thousand
5,99<u>**9**</u> 9≥ 5 6,000	5,9<u>**9**</u>9 9≥ 5 6,000	5,<u>**9**</u>99 9≥ 5 6,000

Note: Adding one to 9 causes carrying.

45. *Locate the digit to the right of the given place value and compare to 5:*

Tens	Hundred	Thousand
10,98<u>**5**</u> 5 ≥ 5 10,990	10,9<u>**8**</u>5 8 ≥ 5 11,000	10,<u>**9**</u>85 9 ≥ 5 11,000

Note: Adding one to 9 causes carrying.

47. *Locate the digit to the right of the given place value and compare to 5:*

Tens	Hundreds	Thousands
99,99<u>**9**</u> 9 ≥ 5 100,000	99,9<u>**9**</u>9 9 ≥ 5 100,000	99,<u>**9**</u>99 9 ≥ 5 100,000

Note: Adding one to 9 causes carrying.

Applying the Concepts

49. Round \$2,486,609 to the nearest hundred thousand:
2,4<u>**8**</u>6,609: *8≥ 5 so replace it and all digits to the right with 0 and add 1 to the hundred thousands place:* \$2,500,000

51. $7{,}315+425{,}493+2{,}082{,}497+1{,}405{,}146$
$+95{,}788+5{,}224+263$
$=4{,}021{,}726$ babies

53. Round 2,082,497 to the nearest hundred thousand
2,0<u>**8**</u>2,497: $8\geq 5$ *so add 1 to the hundred thousands place and replace all digits to the right with 0:* 2,100,000 babies

55. From the pie chart:

Supply Expenses	\$11,456
+ Postage Expenses	+ 3,792
Postage & Supply Expenses	\$15,248

To the nearest hundred, \$15,248 rounds to \$15,200.

57.

Supply Expenses	\$11,456
Postage Expenses	3,792
Telephone Expenses	3,652
Car Expenses	3,205
+ All Other Expenses	+ 8,496
Total	\$30,601

To the nearest thousand, \$30,601 rounds to \$31,000.

Estimating

59.

750	rounds to	800
275	rounds to	300
+ 120	rounds to	+ 100
		1,200

61.

472	rounds to	500
422	rounds to	400
536	rounds to	500
+ 511	rounds to	+ 500
		1,900

63.

25,399	rounds to	25,000
7,601	rounds to	8,000
18,744	rounds to	19,000
+ 6,298	rounds to	+ 6,000
		58,000

65. Draw a horizontal line from the top of the bar above 3 seconds over to the vertical axis. The line is closer to 160 mph.

67. Draw a horizontal line from the top of the bar above 1 second over to the vertical axis. Estimate the speed to be about 75 mph.

69. On the vertical axis, mark the scale at equal intervals of 20, starting at 0 and ending at 100 (the highest level of caffeine). On the horizontal axis, draw a bar of the appropriate height for each drink. Label each bar, and the vertical axis.

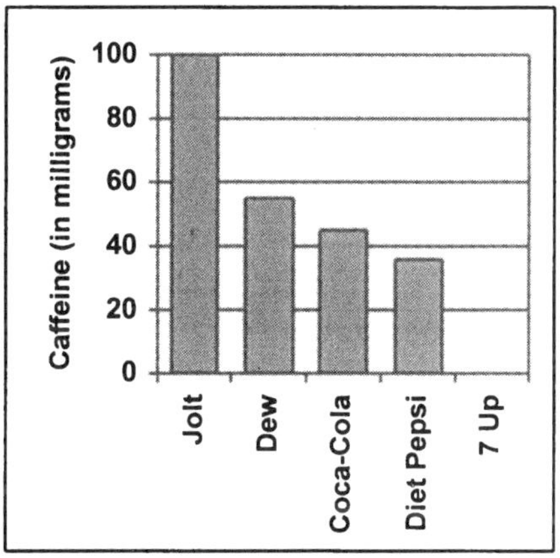

71. On the vertical axis, mark the scale at equal intervals of 100, starting at 0 and ending at 700, (just above the highest level of calories). On the horizontal axis, draw a bar of the appropriate height for each activity. Label each bar, and the vertical axis.

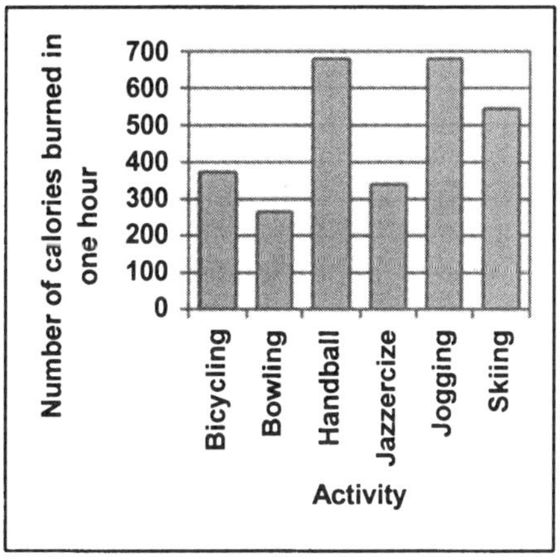

6

Extending the Concepts

73. 1,2,3,4,... is the sequence of counting numbers. The next number in the sequence is 5.
This is an *arithmetic* sequence because the sequence begins with 1, and each number thereafter comes from *adding* 1 to the number before it.

75. 2, 4, 6, 8,... is the sequence of even numbers.
The next number in the sequence is 10.
This is an *arithmetic* sequence because the sequence begins with 2, and each number thereafter comes from *adding* 2 to the number before it.

77. 5,8,11,...
The next number in the sequence is 14.
This is an *arithmetic* sequence because the sequence begins with 5, and each number thereafter comes from *adding* 3 to the previous number.

79. 10,35,60, ...
The next number in the sequence is 85.
This is an *arithmetic* sequence because the sequence begins with 10, and each number thereafter comes from *adding* 25 to the previous number.

Improving Your Quantitative Literacy

81. Answers will vary.

Section 1.4

1. Subtract 24 from 56:

$$\begin{array}{r} 56 \\ -\ 24 \\ \hline 32 \end{array}$$

3. Subtract 23 from 45:

$$\begin{array}{r} 45 \\ -\ 23 \\ \hline 22 \end{array}$$

5. Find the difference of 29 and 19:

$$\begin{array}{r} 29 \\ -\ 19 \\ \hline 10 \end{array}$$

7. Find the difference of 126 and 15:

$$\begin{array}{r} 126 \\ -\ 15 \\ \hline 111 \end{array}$$

9.

$$\begin{array}{r} 975 \\ -\ 663 \\ \hline 312 \end{array}$$

11.

$$\begin{array}{r} 904 \\ -\ 501 \\ \hline 403 \end{array}$$

13.

$$\begin{array}{r} 9{,}876 \\ -\ 8{,}765 \\ \hline 1{,}111 \end{array}$$

15.

$$\begin{array}{r} 7{,}976 \\ -\ 3{,}432 \\ \hline 4{,}544 \end{array}$$

17.

$$\begin{array}{r} \overset{4}{\cancel{5}}\ \overset{12}{\cancel{2}} \\ -3\ 7 \\ \hline 1\ 5 \end{array}$$

19.

$$\begin{array}{r} \overset{6}{\cancel{7}}\ \overset{10}{\cancel{0}} \\ -3\ 7 \\ \hline 3\ 3 \end{array}$$

21.

$$\begin{array}{r} \overset{6}{\cancel{7}}\ \overset{14}{\cancel{4}} \\ -6\ 9 \\ \hline 5 \end{array}$$

23.

$$\begin{array}{r} \overset{4}{\cancel{5}}\ \overset{11}{\cancel{1}} \\ -1\ 8 \\ \hline 3\ 3 \end{array}$$

25.

$$\begin{array}{r} \overset{2}{\cancel{3}}\ \overset{12}{\cancel{2}}\ 9 \\ -2\ 3\ 4 \\ \hline 9\ 5 \end{array}$$

27.

$$\begin{array}{r} \overset{2}{\cancel{3}}\ \overset{14}{\cancel{4}}\ 8 \\ -1\ 9\ 6 \\ \hline 1\ 5\ 2 \end{array}$$

29.

$$\begin{array}{r} \overset{8}{\cancel{9}}\ \overset{12}{\cancel{3}}\ \overset{12}{\cancel{2}} \\ -6\ 5\ 8 \\ \hline 2\ 7\ 4 \end{array}$$

31.

$$\begin{array}{r} \overset{5}{\cancel{6}}\ \overset{13}{\cancel{4}}\ \overset{17}{\cancel{7}} \\ -1\ 5\ 9 \\ \hline 4\ 8\ 8 \end{array}$$

33.
$$\begin{array}{r} {\scriptstyle 8\;9\;15} \\ \not{9}\,\not{0}\,\not{5} \\ -3\,6\,7 \\ \hline 5\,3\,8 \end{array}$$

35.
$$\begin{array}{r} {\scriptstyle 5\;9\;10} \\ \not{6}\,\not{0}\,\not{0} \\ -4\,3\,7 \\ \hline 1\,6\,3 \end{array}$$

37.
$$\begin{array}{r} {\scriptstyle 3\;\;15\;\;\;\;} \\ \not{4},\not{5}\,8\,3 \\ -2,9\,7\,3 \\ \hline 1,6\,1\,0 \end{array}$$

39.
$$\begin{array}{r} {\scriptstyle 8\;\;9\;13\;10} \\ 7\,\not{9},\not{0}\,\not{4}\,\not{0} \\ -3\,2,9\,5\,7 \\ \hline 4\,6,0\,8\,3 \end{array}$$

41.
$$\begin{array}{rrrr} {\scriptstyle 1\;15} & {\scriptstyle 1\;14} & {\scriptstyle 1\;13} & {\scriptstyle 1\;12} \\ \not{2}\,\not{5} & \not{2}\,\not{4} & \not{2}\,\not{3} & \not{2}\,\not{2} \\ -\;1\,5 & -\;1\,6 & -\;1\,7 & -\;1\,8 \\ \hline 1\,0 & 8 & 6 & 4 \end{array}$$

43.
$$\begin{array}{rrrr} {\scriptstyle 3\;9\;10} & {\scriptstyle 3\;9\;10} & {\scriptstyle 1\;12\;\;} & {\scriptstyle 1\;12\;\;} \\ \not{4}\,\not{0}\,\not{0} & \not{4}\,\not{0}\,\not{0} & \not{2}\,\not{2}\,5 & \not{2}\,\not{2}\,5 \\ -2\,5\,6 & -1\,4\,4 & -1\,4\,4 & -\;8\,1 \\ \hline 1\,4\,4 & 2\,5\,6 & 8\,1 & 1\,4\,4 \end{array}$$

Applying the Concepts

45.

Opening balance:	\$504
− Total checks:	− 249
Amount left in account:	\$255

47.

Height in 1954:	29,028 ft.
+ 7 feet:	+ 7 ft.
Height in 1999:	29,035 ft.

Height in 1999:	29,035 ft.
− Height in 1847:	− 29,002 ft.
Difference:	33 ft.

49.

Past enrollment:	567 students
− Current enrollment:	− 399 students
Decrease:	168 students

51.

Opening balance:	\$425
+ Deposit:	+ 149
New balance:	\$574

53. **53.** Total Checks:

Market:	\$37
+ Credit Card:	+ 188
Total:	\$225

Balance after deposit:	\$574
− Total checks:	− 225
New balance:	\$349 ≈ \$350

55. a. For each year, enter the sales:

Year	Sales (in millions)
1996	386
1997	518
1998	573
1999	819

b.

1999 sales:	\$819 million
− 1997 sales:	− 518 million
Increase:	\$301 million

Improving Your Quantitative Literacy

57. a. It appears they have been rounded to the nearest hundred.

b.

2003 participants:	19,100
− 2000 participants:	− 13,000
Increase:	6,100

c. There was an increase of 6,100 participants from 2000-2003. If the same increase occurs over the next three years:

2003 participants:	19,100
+ increase:	+ 6,100
2006 participants:	25,200

Section 1.5

1. $3 \cdot 100 = 300$
$3 \cdot 1 = 3$, *attach two 0's*

3. $3 \cdot 200 = 600$
$3 \cdot 2 = 6$, *attach two 0's*

5. $6 \cdot 500 = 3{,}000$
$6 \cdot 5 = 30$, *attach two 0's*

7. $5 \cdot 1{,}000 = 5{,}000$
$5 \cdot 1 = 5$, *attach three 0's*

9. $3 \cdot 7{,}000 = 21{,}000$
$3 \cdot 7 = 21$, *attach three 0's*

11. $9 \cdot 9{,}000 = 81{,}000$
$9 \cdot 9 = 81$, *attach three 0's*

13.
$$\begin{array}{r} \overset{2}{2}5 \\ \times\ \ 4 \\ \hline 100 \end{array}$$

15.
$$\begin{array}{r} \overset{4}{3}8 \\ \times\ \ 6 \\ \hline 228 \end{array}$$

17.
$$\begin{array}{r} \overset{1}{1}8 \\ \times\ \ 2 \\ \hline 36 \end{array}$$

19.
$$\begin{array}{rl} 72 & \\ \times\ 20 & \\ \hline 00 & \leftarrow 0(72) \\ 1{,}440 & \leftarrow 20(72) \\ \hline 1{,}440 & \text{Add} \end{array}$$

21.
$$\begin{array}{rl} 19 & \\ \times\ 50 & \\ \hline 00 & \leftarrow 0(19) \\ 950 & \leftarrow 50(19) \\ \hline 950 & \text{Add} \end{array}$$

23.
$$\begin{array}{rl} 69 & \\ \times\ 25 & \\ \hline 345 & \leftarrow 5(69) \\ 1{,}380 & \leftarrow 20(69) \\ \hline 1{,}725 & \text{Add} \end{array}$$

25.
$$\begin{array}{rl} 11 & \\ \times\ 11 & \\ \hline 11 & \leftarrow 1(11) \\ 110 & \leftarrow 10(11) \\ \hline 121 & \text{Add} \end{array}$$

27.
$$\begin{array}{rl} 97 & \\ \times\ 16 & \\ \hline 582 & \leftarrow 6(97) \\ 970 & \leftarrow 10(97) \\ \hline 1{,}552 & \text{Add} \end{array}$$

29.
$$\begin{array}{rl} 168 & \\ \times\ 25 & \\ \hline 840 & \leftarrow 5(168) \\ 3{,}360 & \leftarrow 20(168) \\ \hline 4{,}200 & \text{Add} \end{array}$$

31.
$$\begin{array}{rl} 728 & \\ \times\ 91 & \\ \hline 728 & \leftarrow 1(728) \\ 65{,}520 & \leftarrow 90(728) \\ \hline 66{,}248 & \text{Add} \end{array}$$

33.
$$\begin{array}{rl} 698 & \\ \times\ 400 & \\ \hline 279{,}200 & \leftarrow 400(698) \end{array}$$

35.
$$\begin{array}{rl} 111 & \\ \times\ 111 & \\ \hline 111 & \leftarrow 1(111) \\ 1{,}110 & \leftarrow 10(111) \\ 11{,}100 & \leftarrow 100(111) \\ \hline 12{,}321 & \text{Add} \end{array}$$

37.
$$\begin{array}{rl} 532 & \\ \times\ 200 & \\ \hline 106{,}400 & \leftarrow 200(532) \end{array}$$

39.
$$\begin{array}{rl} 856 & \\ \times\ 232 & \\ \hline 1{,}712 & \leftarrow 2(856) \\ 25{,}680 & \leftarrow 30(856) \\ 171{,}200 & \leftarrow 200(856) \\ \hline 198{,}592 & \text{Add} \end{array}$$

41.
$$\begin{array}{rl} 976 & \\ \times\ 628 & \\ \hline 7{,}808 & \leftarrow 8(976) \\ 19{,}520 & \leftarrow 20(976) \\ 585{,}600 & \leftarrow 600(976) \\ \hline 612{,}928 & \text{Add} \end{array}$$

43.
$$\begin{array}{rl} 2{,}468 & \\ \times\ 135 & \\ \hline 12{,}340 & \leftarrow 5(2{,}468) \\ 74{,}040 & \leftarrow 30(2{,}468) \\ 246{,}800 & \leftarrow 100(2{,}468) \\ \hline 333{,}180 & \text{Add} \end{array}$$

45.
$$\begin{array}{rl} 24{,}563 & \\ \times\ 735 & \\ \hline 122{,}815 & \leftarrow 5(24{,}563) \\ 736{,}890 & \leftarrow 30(24{,}563) \\ 17{,}194{,}100 & \leftarrow 700(24{,}563) \\ \hline 18{,}053{,}805 & \text{Add} \end{array}$$

47.
$$\begin{array}{rl} 44{,}777 & \\ \times\ 5{,}888 & \\ \hline 358{,}216 & \leftarrow 8(44{,}777) \\ 3{,}582{,}160 & \leftarrow 80(44{,}777) \\ 35{,}821{,}600 & \leftarrow 800(44{,}777) \\ 223{,}885{,}000 & \leftarrow 5{,}000(44{,}777) \\ \hline 263{,}646{,}976 & \text{Add} \end{array}$$

49.
$$\begin{array}{r} 11 \\ \times 11 \\ \hline 11 \\ 110 \\ \hline 121 \end{array} \quad \begin{array}{r} 11 \\ \times 22 \\ \hline 22 \\ 220 \\ \hline 242 \end{array} \quad \begin{array}{r} 22 \\ \times 22 \\ \hline 44 \\ 440 \\ \hline 484 \end{array} \quad \begin{array}{r} 22 \\ \times 44 \\ \hline 88 \\ 880 \\ \hline 968 \end{array}$$

Note: In each case, one of the factors is multiplied by 2, and the resulting product is twice the previous product.

51.
$$\begin{array}{r} 25 \\ \times 15 \\ \hline 125 \\ 250 \\ \hline 375 \end{array} \quad \begin{array}{r} 25 \\ \times 30 \\ \hline 750 \end{array} \quad \begin{array}{r} 50 \\ \times 15 \\ \hline 250 \\ 500 \\ \hline 750 \end{array} \quad \begin{array}{r} 50 \\ \times 30 \\ \hline 1{,}500 \end{array}$$

Note: In each case, one of the factors is multiplied by 2, and the resulting product is twice the previous product.

53.
$$\begin{array}{r} 25 \\ \times 10 \\ \hline 250 \end{array} \quad \begin{array}{r} 25 \\ \times 100 \\ \hline 2{,}500 \end{array} \quad \begin{array}{r} 25 \\ \times 1{,}000 \\ \hline 25{,}000 \end{array} \quad \begin{array}{r} 25 \\ \times 10{,}000 \\ \hline 250{,}000 \end{array}$$

Note: When multiplying by a power of 10, add the same number of zeros in the product as are in the power of 10.

55. $6 \cdot 7$
→ The product of 6 and 7

57. $2 \cdot n$
→ The product of 2 and n

59. $9 \cdot 7 = 63$
→ The product of 9 and 7 is 63

61. The product of
7 | and | n
$7 \cdot n$

63. The product of
6 | and | 7 | is | 42
$6 \cdot 7 = 42$

65. The product of
0 | and | 6 | is | 0
$0 \cdot 6 = 0$

67. Products: $9 \cdot 7, 63$

69. Products: $4(4), 16$

71. Factors: 2,3,4

73. Factors: 2,2,3

75. $5(9) = 9(5)$

77. $6 \cdot 7 = 7 \cdot 6$

79. $2 \cdot (7 \cdot 6) = (2 \cdot 7) \cdot 6$

81. $3 \times (9 \times 1) = (3 \times 9) \times 1$

83. $7(2+3) = 7 \cdot 2 + 7 \cdot 3$ *Distributive Prop.*
$= 14 + 21$ *Multiply*
$= 35$ *Add*

85. $9(4+7) = 9 \cdot 4 + 9 \cdot 7$ *Distributive Prop.*
$= 36 + 63$ *Multiply*
$= 99$ *Add*

87. $3(x+1) = 3 \cdot x + 3 \cdot 1$ *Distributive Prop.*
$= 3x + 3$ *Multiply*

89. $2(x+5) = 2 \cdot x + 2 \cdot 5$ *Distributive Prop.*
$= 2x + 10$ *Multiply*

91. The solution to $4 \cdot n = 12$ is $n = 3$, because $4 \cdot 3 = 12$

93. The solution to $9 \cdot n = 81$ is $n = 9$, because $9 \cdot 9 = 81$

95. The solution to $0 = n \cdot 5$ is $n = 0$, because $0 = 0 \cdot 5$

97. The solution to $27 = 9 \cdot n$ is $n = 3$, because $27 = 9 \cdot 3$

Applying the Concepts

99.

Number of adults:	300
+ Number of children:	+ 80
Number of people attended:	380

101. $300 \text{ adults} \times \dfrac{\$6}{\text{adult ticket}} = \$1{,}800$

103. Gallons of gas 130
× miles per gallon × 22
Total miles 260
2,600
2,860

2,860 miles.

105. 1st 8 hours: $8 \text{ hours} \times \frac{\$12}{\text{hour}} = \$96$

Next 4 hours: $4 \text{ hours} \times \frac{\$18}{\text{hour}} = \$72$

Total wages: $96
+ 72
$168

107. Calories burned playing handball:
Calories per hour: 816
× hours : ×2
Total calories burned: 1,632

Calories burned bicycling:
Calories per hour: 449
× hours : ×1
Total calories burned: 449

Handball calories: 1,632
+ Bicycling calories : + 449
Total calories: 2,081

109. Calories per serving: 140
× servings per bag: × 2
Total calories per bag: 280

111. Calories per serving: 140
× 3 servings: ×3
Calories per 3 servings: 420

Since you burn 449 calories per hour riding the bike, you would burn off the 420 calories from the chips.

Estimating

113. 750 rounds to 800
× 12 rounds to × 10
8,000

115. 3,472 rounds to 3,000
× 511 rounds to × 500
1,500,000

117. 2,399 rounds to 2,000
× 698 rounds to × 700
1,400,000

Extending the Concepts

119. 5, 10, 20, **40**,...
The geometric sequence starts with 5. Note that $5 \cdot 2 = 10$, $10 \cdot 2 = 20$. Each number thereafter comes from multiplying the previous number by 2. The next number in the sequence is $20 \cdot 2 = 40$.

121. 2, 6, 18, **54**,...
The geometric sequence starts with 2. Note that $2 \cdot 3 = 6$, $6 \cdot 3 = 18$., i.e., each number after 2 in the sequence comes from multiplying the previous number by 3. The next number in the sequence is $18 \cdot 3 = 54$.

Improving Your Quantitative Literacy

123. Each phone pictured in the graph is worth 31 million phones.

a. 31 million phones: 31
× phones in graph: × 3
Total phones: 93 million

b. There are between 4 and 5 phones in the graph. Using 5 phones:
31 million phones: 31
× phones in graph: × 5
Total phones: 155 million
Since there are less than 5 phones in the graph, the statement is false. There will be a less than 155 million.

c. There is between 9 and 10 phones in the graph. Using 10 phones:
31 million phones: 31
× phones in graph: × 10
Total phones: 310 million
Since there are less than 10 phones in the graph, the statement is true. There will be less than 310 million.

Section 1.6

1.

The quotient of		
6	and	3
6	÷	3

3.

The quotient of		
45	and	9
45	÷	9

5.

The quotient of		
12	and	4
12	÷	4

7.

The quotient of				
20	and	4	is	5
20	÷	4	=	5

9. $6 \div 2 = 3 \Leftrightarrow 2 \cdot 3 = 6$

11. $\frac{36}{9} = 4 \Leftrightarrow 9 \cdot 4 = 36$

13. $\frac{48}{6} = 8 \Leftrightarrow 6 \cdot 8 = 48$

15. $28 \div 7 = 4 \Leftrightarrow 7 \cdot 4 = 28$

17. $25 \div 5 = 5,$ because $5 \cdot 5 = 25$

19. $40 \div 5 = 8,$ because $5 \cdot 8 = 40$

21. $9 \div 0$ is undefined. *There is no number that when multiplied by 0 will give you 9.*

23.

```
     45
 8)360
   32↓
   40
  -40
    0
```

25.

```
    23
 6)138
   12↓
    18
   -18
     0
```

27.

```
   1,530
 5)7,650
   5↓
   26
   25↓
    15
    15↓
    00
```

29.

```
   1,350
 5)6,750
   5↓
   17
   15↓
   25
   25↓
   00
```

31.
$$\begin{array}{r} 18{,}000 \\ 3\overline{)54{,}000} \\ \underline{3\downarrow}\phantom{0{,}000} \\ 24\phantom{{,}000} \\ \underline{24\downarrow}\phantom{{,}000} \\ 00 \end{array}$$

Note: Fill in the quotient with as many zeros as there are in the dividend.

33.
$$\begin{array}{r} 16{,}680 \\ 3\overline{)50{,}040} \\ \underline{3\downarrow}\phantom{0{,}040} \\ 20\phantom{{,}040} \\ \underline{18\downarrow}\phantom{{,}040} \\ 20 \\ \underline{18\downarrow} \\ 24 \\ \underline{24\downarrow} \\ 00 \end{array}$$

Estimating

35. $845 \div 93$ is approximately $900 \div 90 = 10$.
Round the divisor to 90, then round the dividend such that it is easily divisible by 90, such as 900.

37. $15{,}208 \div 771$ is approximately $16{,}000 \div 800 = 20$.
Round the divisor to 800, then round the dividend such that it is easily divisible by 800, such as 16,000.

39. $316 \div 289$ is approximately $300 \div 300 = 1$
Round the divisor to 300, then round the dividend to 300 because it is easily divisible by 300.

41. $728 \div 355$ is approximately $800 \div 400 = 2$
Round the divisor to 400, then round the dividend to 800 because it is easily divisible by 400.

43. $921 \div 243$ is approximately $800 \div 200 = 4$
Round the divisor to 200, then round the dividend to 800 because it is easily divisible by 200.

45. $673 \div 109$ is approximately $600 \div 100 = 6$
Round the divisor to 100, then round the dividend to 600 because it is easily divisible by 100.

47.
$$\begin{array}{r} 45 \\ 32\overline{)1{,}440} \\ \underline{128\downarrow} \\ 160 \\ \underline{160} \\ 0 \end{array}$$

49.
$$\begin{array}{r} 49 \\ 49\overline{)2{,}401} \\ \underline{196\downarrow} \\ 441 \\ \underline{441} \\ 0 \end{array}$$

51.
$$\begin{array}{r} 432 \\ 28\overline{)12{,}096} \\ \underline{112\downarrow} \\ 89 \\ \underline{84\downarrow} \\ 56 \\ \underline{56} \\ 0 \end{array}$$

53.
$$\begin{array}{r} 1{,}438 \\ 63\overline{)90{,}594} \\ \underline{63\downarrow} \\ 275 \\ \underline{252\downarrow} \\ 239 \\ \underline{189\downarrow} \\ 504 \\ \underline{504} \\ 0 \end{array}$$

55.
$$\begin{array}{r} 4 \\ 25\overline{)100} \\ \underline{100} \\ 0 \end{array} \qquad \begin{array}{r} 3 \text{ R22} \\ 26\overline{)100} \\ \underline{78} \\ 22 \end{array}$$

$$\begin{array}{r} 3 \text{ R19} \\ 27\overline{)100} \\ \underline{81} \\ 19 \end{array} \qquad \begin{array}{r} 3 \text{ R16} \\ 28\overline{)100} \\ \underline{84} \\ 16 \end{array}$$

57.
$$\begin{array}{r} 61 \text{ R4} \\ 6\overline{)370} \\ \underline{36\downarrow} \\ 10 \\ \underline{6} \\ 4 \end{array}$$

59.
$$\begin{array}{r} 90 \text{ R1} \\ 3\overline{)271} \\ \underline{27\downarrow} \\ 01 \\ \underline{0} \\ 1 \end{array}$$

61.
$$\begin{array}{r} 13 \text{ R7} \\ 26\overline{)345} \\ \underline{26\downarrow} \\ 85 \\ \underline{78} \\ 7 \end{array}$$

63.
$$\begin{array}{r} 234 \text{ R6} \\ 71\overline{)16{,}620} \\ \underline{142\downarrow} \\ 242 \\ \underline{213\downarrow} \\ 290 \\ \underline{284} \\ 6 \end{array}$$

65.
$$\begin{array}{r} 452 \text{ R4} \\ 23\overline{)10{,}400} \\ \underline{92\downarrow} \\ 120 \\ \underline{115\downarrow} \\ 50 \\ \underline{46} \\ 4 \end{array}$$

67.
$$\begin{array}{r} 35 \text{ R35} \\ 169\overline{)5{,}950} \\ \underline{507\downarrow} \\ 880 \\ \underline{845} \\ 35 \end{array}$$

Applying the Concepts

69. $\dfrac{\$22{,}200 \text{ yearly income}}{12 \text{ months per year}} = \text{Monthly income}$

$$\begin{array}{r} 1{,}850 \\ 12\overline{)22{,}200} \\ \underline{12\downarrow} \\ 102 \\ \underline{96\downarrow} \\ 60 \\ \underline{60\downarrow} \\ 00 \end{array}$$

The monthly income is \$1,850.

71. $\dfrac{96¢ \text{ (for 6 pounds)}}{6 \text{ pounds}} = \text{cost per pound}$

$$\begin{array}{r} 16 \\ 6\overline{)96} \\ \underline{1\downarrow} \\ 36 \\ \underline{36} \\ 0 \end{array}$$

One pound costs 16¢.

73. $\dfrac{10{,}000 \text{ steps}}{2{,}000 \text{ steps per mile}} = \text{Number of miles}$

$$\begin{array}{r} 5 \\ 2000\overline{)10000} \\ \underline{10000} \\ 0 \end{array}$$

You need to walk 5 miles.

75. $\dfrac{32 \text{ total ounces}}{5 \text{ ounces per glass}} = \text{Number of glasses}$

$$\begin{array}{r} 6 \\ 5\overline{)32} \\ \underline{30} \\ 2 \end{array}$$

6 glasses with 2 ounces left over.

77.

16 glasses	16
× 6 ounces per glass:	× 6
Total ounces needed:	96

$\dfrac{96 \text{ total ounces needed}}{32 \text{ ounces per bottle}} = \text{Number of bottles}$

$$\begin{array}{r} 3 \\ 32\overline{)96} \\ \underline{96} \\ 0 \end{array}$$

You will need 3 bottles.

79. $\dfrac{\$192 \text{ (for 16 bottles)}}{16 \text{ bottles}} = \text{Cost per bottles}$

$$\begin{array}{r} 12 \\ 16\overline{)192} \\ \underline{16\downarrow} \\ 32 \\ \underline{32} \\ 0 \end{array}$$

\$12 per bottles.

81. Eggs: 25 mg per egg · 3 eggs = 75 mg
Milk: 215 mg per glass · 2 glasses = 430 mg
Toast: 40 mg per slice · 4 slices = 160 mg

Calcium in eggs:	75 mg
Calcium in milk:	430 mg
+ Calcium in toast:	+ 160 mg
Calcium in breakfast:	665 mg

Calculator Problems

83. $305{,}026 \div 698 = 437$

85. $18{,}436{,}466 \div 5{,}678 = 3{,}247$

87. $695\overline{)603{,}955} = 603{,}955 \div 695 = 869$

89. $4{,}903\overline{)27{,}868{,}652} = 27{,}868{,}652 \div 4{,}903$
$= 5{,}684$

91. $\dfrac{79{,}768 \text{ total gallons}}{472 \text{ minutes}} = \text{Gallons per minute}$

$79{,}768 \div 472 = 169$
169 gallons each minute.

Section 1.7

1. 4^5: base = 4, exponent = 5

3. 3^6: base = 3, exponent = 6

5. 8^2: base = 8, exponent = 2

7. 9^1: base = 9, exponent = 1

9. 4^0: base = 4, exponent = 0

11. $6^2 = 6 \cdot 6 = 36$

13. $2^3 = 2 \cdot 2 \cdot 2 = 8$

15. $1^4 = 1 \cdot 1 \cdot 1 \cdot 1 = 1$

17. $9^0 = 1$ *by definition:* $a^0 = 1$

19. $9^2 = 9 \cdot 9 = 81$

21. $10^1 = 10$

23. $12^1 = 12$

25. $45^0 = 1$ *by definition:* $a^0 = 1$

27.
$3 + 5 \cdot 8$ *multiply*
$= 3 + 40$ *add*
$= 43$

29.
$3 \cdot 6 - 2$ *multiply*
$= 18 - 2$ *subtract*
$= 16$

31.
$6 \cdot 2 + 9 \cdot 8$ *multiply 1st*
$= 12 + 72$ *add*
$= 84$

33.
$8 \cdot 10^2 - 6 \cdot 4^3$ *exponentiate*
$= 8 \cdot 100 - 6 \cdot 64$ *multiply*
$= 800 - 384$ *subtract*
$= 416$

35.
$2(3 + 6 \cdot 5)$ *inside parenthesis*
$= 2(3 + 30)$ *inside parenthesis*
$= 2(33)$ *multiply*
$= 66$

37.
$19 + 50 \div 5^2$ *exponentiate*
$= 19 + 50 \div 25$ *divide*
$= 19 + 2$ *add*
$= 21$

39.
$9 - 2(4 - 3)$ *parenthesis*
$= 9 - 2(1)$ *multiply*
$= 9 - 2$ *subtract*
$= 7$

41.
$8 \cdot 2^4 + 25 \div 5 - 3^2$ *exponentiate*
$= 8 \cdot 16 + 25 \div 5 - 9$ *multiply & divide*
$= 128 + 5 - 9$ *add & subtract*
$= 124$

43.
$5 + 2[9 - 2(4 - 1)]$ *inside parenthesis*
$= 5 + 2[9 - 2 \cdot 3]$ *inside brackets*
$= 5 + 2[9 - 6]$ *inside brackets*
$= 5 + 2 \cdot 3$ *multiply*
$= 5 + 6$ *add*
$= 11$

45. $3+4\left[6+8(2-0)\right]$ *inside parenthesis*

$=3+4\left[6+8\cdot 2\right]$ *inside brackets*

$=3+4\left[6+16\right]$ *inside brackets*

$=3+4\cdot 22$ *multiply*

$=3+88$ *add*

$=91$

47.

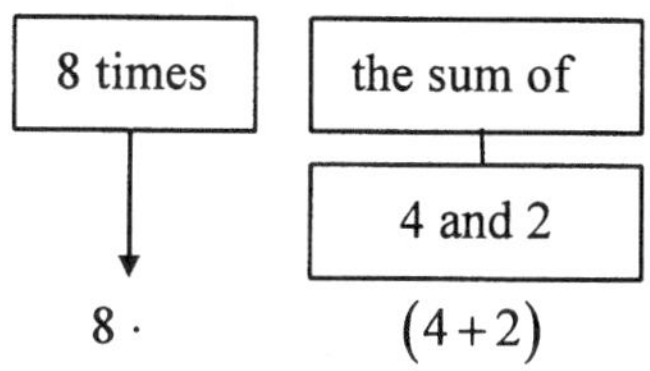

$8(4+2)=8\cdot 6$

$=48$

49.

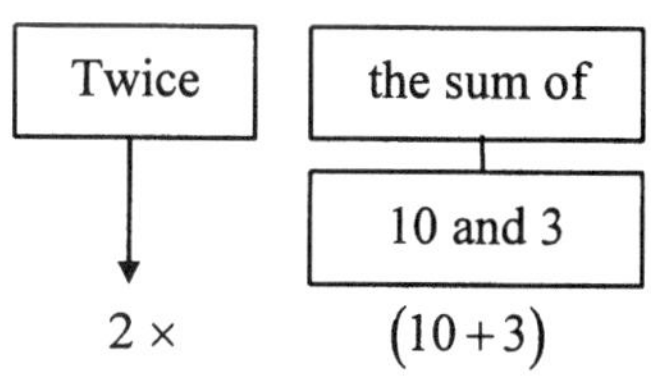

$2(10+3)=2\cdot 13$

$=26$

51.

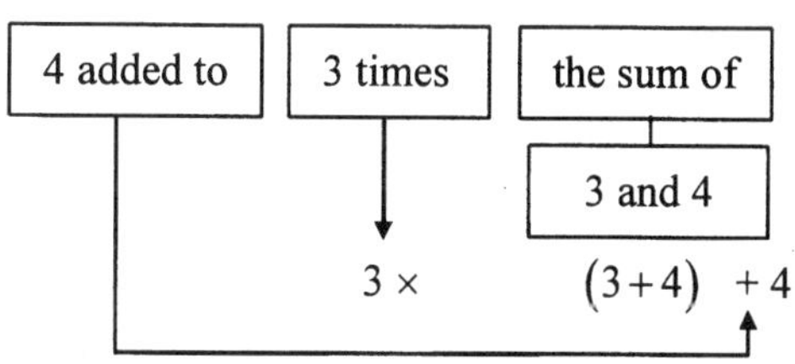

$3(3+4)+4=3\cdot 7+4$

$=21+4$

$=25$

53.

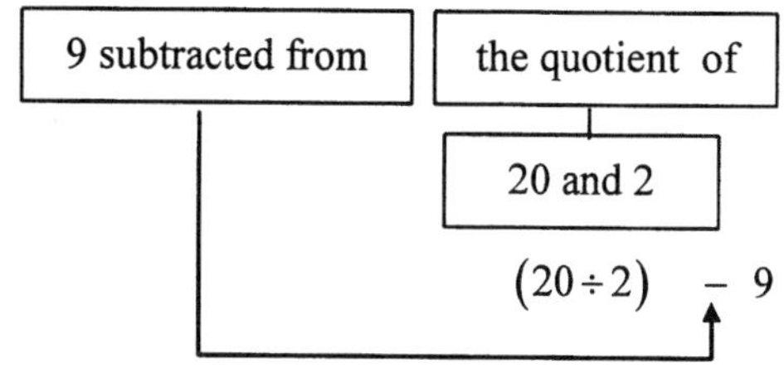

$(20\div 2)-9=10-9$

$=1$

55.

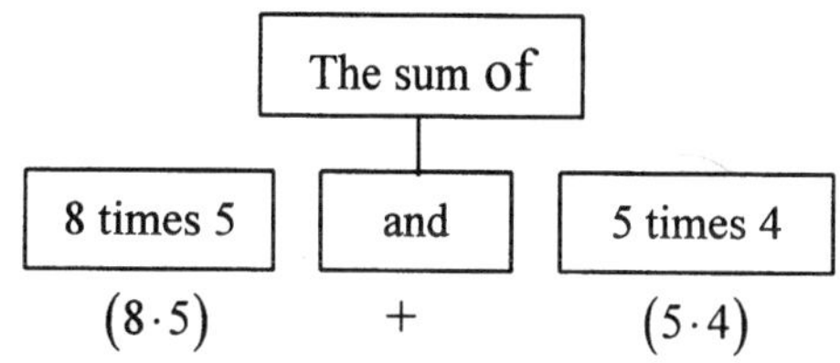

$(8\cdot 5)+(5\cdot 4)=40+20$

$=60$

57. $\text{Mean}=\dfrac{\text{sum of values}}{\text{number of values}}$

$=\dfrac{1+2+3+4+5}{5}$

$=\dfrac{15}{5}$

$=3$

Range = high value − low value

$=5-1$

$=4$

59. $\text{Mean}=\dfrac{\text{sum of values}}{\text{number of values}}$

$=\dfrac{1+3+9+11}{4}$

$=\dfrac{24}{4}$

$=6$

Range = high value − low value

$=11-1$

$=10$

61. $5, 9, 11, 13, 15$
Because there is an odd number of values, the median is the middle value: Median = 11
Range = high − low = $15 - 5 = 10$

63. $10, 20, 50, 90, 100$
Because there is an odd number of values, the median is the middle value: Median = 50
Range = high − low = $100 - 10 = 90$

65. $14, 18, 27, 36, 18, 73$
The mode is the value that occurs most often. Mode = 18.
Range = high − low = $73 - 14 = 59$

Applying the Concepts

67. Calories in:

1 serving of spaghetti:	210
1 serving of tomatoes:	25
+ 1 serving of cheese:	+ 20
Total calories:	255

69. Calories in:

2 servings of spaghetti:	420
1 serving of tomatoes:	25
+ 1 serving of cheese:	+ 20
Total calories:	465

71. Fat calories in:

2 serving of spaghetti:	20
1 serving of tomatoes:	0
+ 1 serving of cheese:	+ 10
Total calories from fat:	30

73.

Big Mac calories:	510
− Meal calories:	− 255
Difference in calories:	255

The Big Mac has 255 more calories, or twice as many calories, as the meal.

75. a. $\text{Mean} = \dfrac{\text{sum of points}}{\text{number of games}}$

$= \dfrac{61 + 76 + 98 + 55 + 76 + 102}{6}$

$= \dfrac{468}{6}$

$= 78$ points

b. Put in order from smallest to largest:
55, 61 ,76, 76, 98, 102
Because there is an even number of values, the median is the mean of the middle two values:

$\text{Median} = \dfrac{76 + 76}{2} = \dfrac{152}{2} = 76$

c. The mode is the value that occurs most often: mode = 76

d. Range = high − low = $102 - 55 = 47$

77. $\text{Mean} = \dfrac{\text{sum of stuents}}{\text{number of years}}$

$= \dfrac{6{,}789 + 6{,}970 + 7{,}242 + 6{,}981 + 6{,}423}{5}$

$= \dfrac{34{,}405}{5}$

$= 6{,}881$ students

Range = high − low
$= 7{,}242 - 6{,}423$
$= 819$ students

Extending the Concepts

79. $1, 3, 5, 7...$
Add the 1st 2 numbers: $1 + 3 = 4$

81. $1, 3, 5, 7...$
Add the 1st 4 numbers: $1 + 3 + 5 + 7 = 16$

83. Answers will vary.

85. 1, 1, 2, 3, 5, 8, …
The first term in the sequence is 1. The second term in the sequence is 1. Thereafter, each term in the sequence is obtained by taking the sum of the *2* previous terms, i.e.,

$1+1=2$

$1+2=3$

$2+3=5$

$3+5=8$

$5+8=13$

$8+13=21$

$13+21=34$

The next three numbers in the sequence are 13, 21, 34.

Section 1.8

1. Square:

$A = s^2$

$A = (5 \text{ cm})^2$

$A = 25 \text{ cm}^2$

3. Rectangle:

$A = l \cdot w$

$A = (24 \text{ m})(14 \text{ m})$

$A = 336 \text{ m}^2$

5. Parallelogram:

$A = b \cdot h$

$A = (10 \text{ ft})(6 \text{ ft})$

$A = 60 \text{ ft}^2$

7.

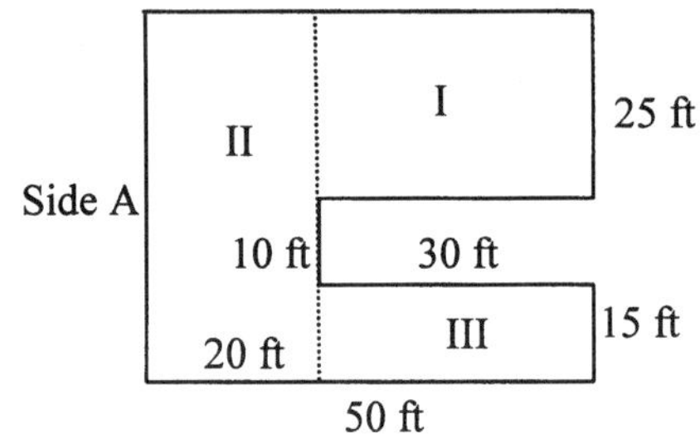

Separate into 3 areas.
Find missing dimension:

Side A = 15 ft. + 10 ft. + 25 ft. = 50 ft.

Area I = lw

$= (25 \text{ ft})(30 \text{ ft})$

$= 750 \text{ ft}^2$

Area II = lw

$= (50 \text{ ft})(20 \text{ ft})$

$= 1,000 \text{ ft}^2$

Area III = lw

$= (30 \text{ ft})(15 \text{ ft})$

$= 450 \text{ ft}^2$

Area = Area I + Area II + Area III

$= 750 \text{ ft}^2 + 1,000 \text{ ft}^2 + 450 \text{ ft}^2$

$= 2,200 \text{ ft}^2$

9.

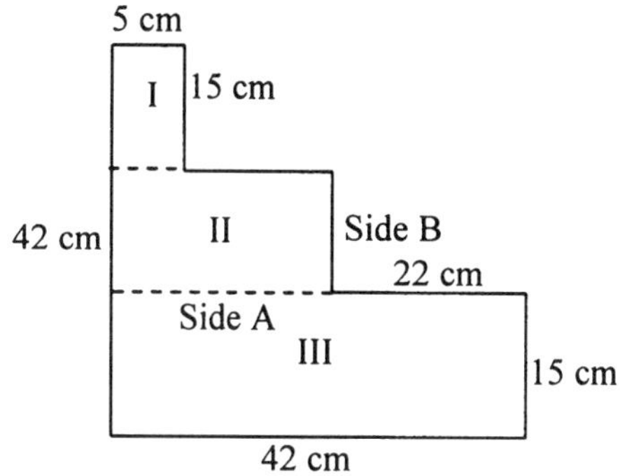

Separate into 3 areas.
Find missing dimension:

Side A = 42 cm − 22 cm = 20 cm

Side B =

42 cm − 15 cm − 15 cm = 12 cm

Area I = lw

$= (5 \text{ cm})(15 \text{ cm})$

$= 75 \text{ cm}^2$

Area II = lw

$= (20 \text{ cm})(12 \text{ cm})$

$= 240 \text{ cm}^2$

Area III = lw

$= (42 \text{ cm})(15 \text{ cm})$

$= 630 \text{ cm}^2$

Area = Area I + Area II + Area III

$= 75 \text{ cm}^2 + 240 \text{ cm}^2 + 630 \text{ cm}^2$

$= 945 \text{ cm}^2$

11. Square:

$A = s^2$

$A = (10 \text{ in})^2$

$A = 100 \text{ in}^2$

13. Cube:

$V = s^3$

$V = (4 \text{ cm})^3$

$V = 64 \text{ cm}^3$

15.

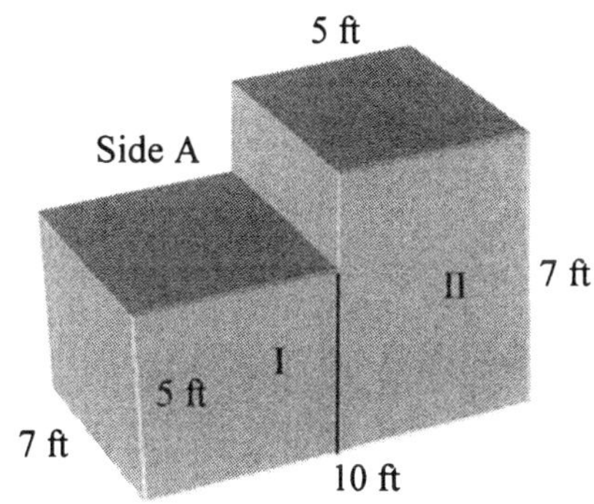

Separate into 2 rectangular solids.
Find missing dimension:

Side A = 10 ft − 5 ft = 5 ft

Volume I = $l \cdot w \cdot h$

$= (5 \text{ ft})(7 \text{ ft})(5 \text{ ft})$

$= 175 \text{ ft}^3$

Volume II = $l \cdot w \cdot h$

$= (5 \text{ ft})(7 \text{ ft})(7 \text{ ft})$

$= 245 \text{ ft}^3$

Volume = Volume I + Volume II

$= 175 \text{ ft}^3 + 420 \text{ ft}^3$

$= 420 \text{ ft}^3$

17.

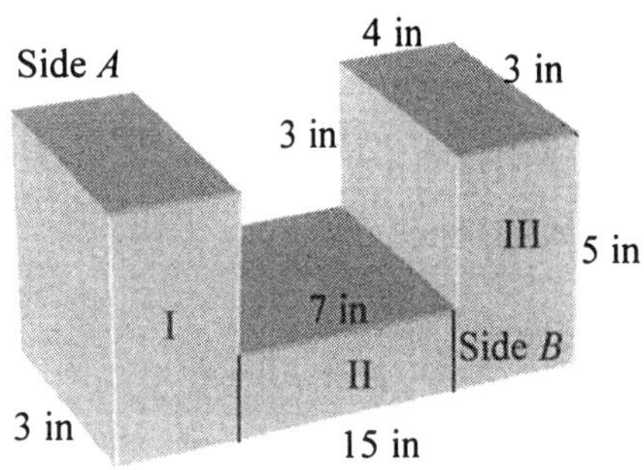

Separate into 3 rectangular solids.
Find in missing dimensions:
Side A = 15 in. – 7 in. – 4 in. = 4 in.
Side B = 5 in. – 3 in. = 2 in.

$$\text{Volume I} = l \cdot w \cdot h = (3 \text{ in})(4 \text{ in})(5 \text{ in}) = 60 \text{ in}^3$$

$$\text{Volume II} = l \cdot w \cdot h = (3 \text{ in})(7 \text{ in})(2 \text{ in}) = 42 \text{ in}^3$$

$$\text{Volume III} = l \cdot w \cdot h = (4 \text{ in})(5 \text{ in})(3 \text{ in}) = 60 \text{ in}^3$$

$$\text{Volume} = \text{Vol I} + \text{Vol II} + \text{Vol III} = 60 \text{ in}^3 + 42 \text{ in}^3 + 60 \text{ in}^3 = 162 \text{ in}^3$$

19. $S = 2lw + 2lh + 2hw$
$= 2(4)(4) + 2(4)(4) + 2(4)(4)$
$= 32 \text{ cm}^2 + 32 \text{ cm}^2 + 32 \text{ cm}^2$
$= 96 \text{ cm}^2$

21. $S = 2lw + 2lh + 2hw$
$= 2\cdot(5\text{ft})(4\text{ft}) + 2(5\text{ft})(6\text{ft}) + 2(6\text{ft})(4\text{ft})$
$= 40 \text{ ft}^2 + 60 \text{ ft}^2 + 48 \text{ ft}^2$
$= 148 \text{ ft}^2$

23. $S = 2lw + 2lh + 2hw$
$= 2(10)(8) + 2(10)(8) + 2(8)(8)$
$= 160 \text{ ft}^2 + 160 \text{ ft}^2 + 128 \text{ ft}^2$
$= 448 \text{ ft}^2$

Applying the Concepts

25.

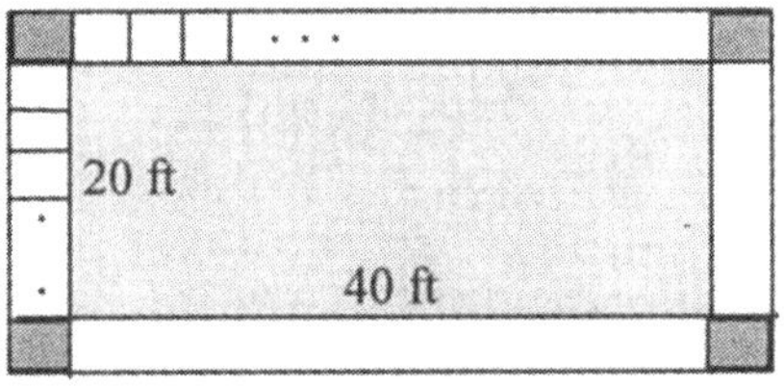

Make a sketch. Find the perimeter of the pool:
$P = 2l + 2w$
$P = 2 \cdot 40 \text{ ft} + 2 \cdot 20 \text{ ft}$
$P = 80 \text{ ft} + 40 \text{ ft}$
$P = 120 \text{ ft}$
Since each tile is 1 foot square, 120 tiles will cover the *inside* perimeter. However, from the diagram you can see you need *4 additional tiles* for the corners. Total tiles: 124

27. 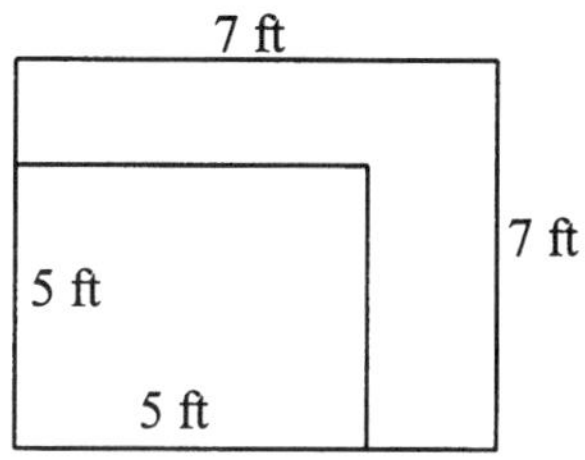

Make a sketch. Find the area of the large and small squares:

Small square:	Large square:
$A = s^2$	$A = s^2$
$A = (5\text{ ft})^2$	$A = (7\text{ ft})$
$A = 25\text{ ft}^2$	$A = 49\text{ ft}^2$

Increase in area $= 49\text{ ft}^2 - 25\text{ ft}^2 = 24\text{ ft}^2$

29. Rectangle:

$A = l \cdot w$

$A = (127\text{ mm})(67\text{ mm})$

$A = 8{,}509\text{ mm}^2$

31. Rectangle:

$A = l \cdot w$

$A = (36\text{ mm})(20\text{ mm})$

$A = 720\text{ mm}^2$

33. $V = l \cdot w \cdot h$

$= (13\text{ ft})(13\text{ ft})(8\text{ ft})$

$= 1{,}352\text{ ft}^3$

35. a. The squares are as follows:

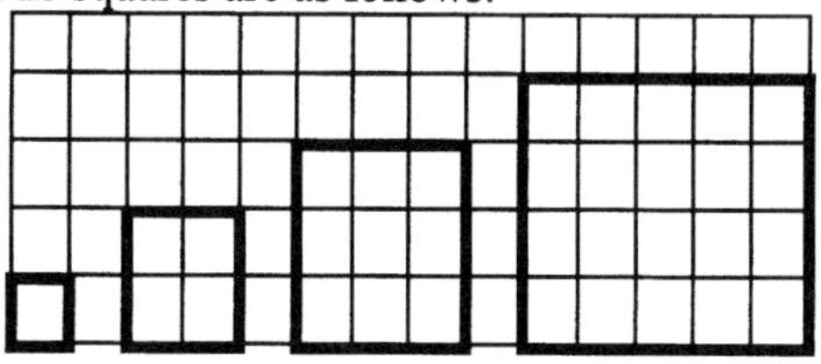

b. The tables are as follows:

Length of side, s (in cm)	$P = 4s$ (in cm)
1	$P = 4 \cdot 1 = 4$
2	$P = 4 \cdot 2 = 8$
3	$P = 4 \cdot 3 = 12$
4	$P = 4 \cdot 4 = 16$

Length of side, s (in cm)	$A = s^2$ (in cm^2)
1	$P = 1^2 = 1$
2	$P = 2^2 = 4$
3	$P = 3^2 = 9$
4	$P = 4^2 = 16$

Extending the Concepts

37. $A = s^2$

$49\text{ ft}^2 = s^2$

$49\text{ ft}^2 = (7\text{ ft})^2$ *by inspection,* $s = 7$ ft

Each side is 7 ft.

39. $A = l \cdot w$

$36\text{ ft}^2 = l \cdot 4\text{ ft}$

$36\text{ ft}^2 = 9\text{ ft} \cdot 4\text{ ft}$ *by inspection,* $l = 9$ ft

The length is 9 ft.

Improving Your Quantitative Literacy

41. a.

2008 passengers:	653 million
− 1998 passengers:	− 577 million
Increase:	76 million

b. Since there were 615 million passengers in 2000, that number will next occur in 2007 (since 2006 has 610 million, and 2008 has 653 million).

c. Answers will vary.

Chapter 1 Review

1. 1,376 = one thousand, three hundred seventy-six

3. Five million, two hundred forty-five thousand, six hundred fifty-two = 5,245,652

5. $1{,}025{,}639 = 1{,}000{,}000 + 20{,}000 + 5{,}000 + 600 + 30 + 9$

7. $5+7=7+5$
d, Commutative Property of Addition

9. $6 \cdot 1 = 6$
c, Multiplicative Property of 1

11. $5 \cdot 0 = 0$
b, Multiplicative Property of 0

13. $5\cdot(3\cdot 2)=(5\cdot 3)\cdot 2$
g, Associative Property of Multiplication

15.
$$\begin{array}{r} \scriptstyle 1 \\ 498 \\ +\ 251 \\ \hline 749 \end{array}$$

17.
$$\begin{array}{r} \scriptstyle 1\ \ 11 \\ 7{,}384 \\ 251 \\ +\ 637 \\ \hline 8{,}272 \end{array}$$

19.
$$\begin{array}{r} 789 \\ -\ 475 \\ \hline 314 \end{array}$$

21.
$$\begin{array}{r} \scriptstyle 8\ \ 9\ \ 18 \\ 5{,}\not{9}\,\not{0}\,\not{8} \\ -2{,}7\ 5\ 9 \\ \hline 3{,}1\ 4\ 9 \end{array}$$

23.
$$\begin{array}{r} \scriptstyle 2 \\ 73 \\ \times\ 8 \\ \hline 584 \end{array}$$

25.
$$\begin{array}{r} 63 \\ \times 59 \\ \hline 567 \\ 3{,}150 \\ \hline 3{,}717 \end{array}$$

27.
$$\begin{array}{r} 173 \\ 4\overline{)692} \\ 4\downarrow \\ \hline 29 \\ 28\downarrow \\ \hline 12 \\ 12 \\ \hline 0 \end{array}$$

29.
$$\begin{array}{r} 428 \\ 36\overline{)15{,}408} \\ 144\downarrow \\ \hline 100 \\ -72\downarrow \\ \hline 288 \\ 288 \\ \hline 0 \end{array}$$

31. Round 3,781,092 to the nearest ten:
locate the digit to the right of the tens:
3,781,09**<u>2</u>**: *2 < 5 so replace it with 0 :*
3,781,090

33. Round 3,781,092 to the nearest hundred thousand:
locate the digit to the right of the hundred thousands:
3,7**8**1,092: *8 ≥ 5 so replace it and all digits to the right with 0 and add 1 to the hundred thousands place:*
3,800,000

35.
$$\begin{aligned} &4+3\cdot 5^2 && \textit{exponentiate}\\ &=4+3\cdot 25 && \textit{multiply}\\ &=4+75 && \textit{add}\\ &=79 \end{aligned}$$

37.
$$\begin{aligned} &3(2+8\cdot 9) && \textit{parenthesis (multiply)}\\ &=3(2+72) && \textit{parenthesis (add)}\\ &=3(74) && \textit{multiply}\\ &=222 \end{aligned}$$

39.
$$\begin{aligned} &24\div 6\cdot 2 && \textit{divide}\\ &=4\cdot 2 && \textit{multiply}\\ &=8 \end{aligned}$$

41.
$$\begin{aligned} &4(3-1)^3 && \textit{parenthesis}\\ &=4(2)^3 && \textit{exponentiate}\\ &=4(8) && \textit{multiply}\\ &=32 \end{aligned}$$

43.
$$\begin{aligned} \text{Mean} &= \frac{\text{sum of values}}{\text{number of values}}\\ &=\frac{80+67+78+91}{4}\\ &=\frac{316}{4}\\ &=79 \end{aligned}$$

Put the grades in order: 67, 78, 80, 91
Since there are an even number of values, the median is the mean of the middle two:

$$\text{Median} = \frac{78+80}{2} = \frac{158}{2} = 79$$

45.

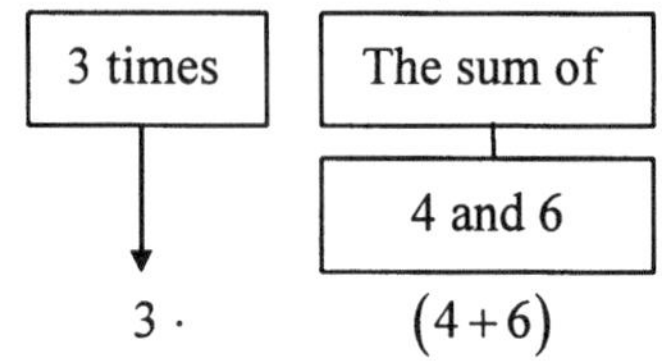

$$\begin{aligned} 3(4+6) &= 3\cdot 10\\ &=30 \end{aligned}$$

47.

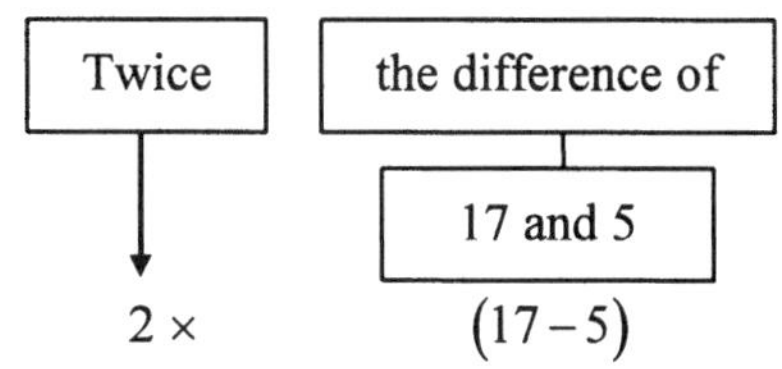

$$\begin{aligned} 2(17-5) &= 2\cdot 12\\ &=24 \end{aligned}$$

Applying the Concepts

49.

Income :	$1,783
− Expenses :	− 1,295
Difference:	$488

51. Smallest soccer field:
$$\begin{aligned} P &= 2l+2w\\ &=2(100\text{ yd})+2(55\text{ yd})\\ &=200\text{ yd}+110\text{ yd}\\ &=310\text{ yd} \end{aligned}$$
Rose Bowl:
$$\begin{aligned} P &= 2l+2w\\ &=2(116\text{ yd})+2(72\text{ yd})\\ &=232\text{ yd}+144\text{ yd}\\ &=376\text{ yd} \end{aligned}$$
Largest soccer field:
$$\begin{aligned} P &= 2l+2w\\ &=2(120\text{ yd})+2(75\text{ yd})\\ &=240\text{ yd}+150\text{ yd}\\ &=390\text{ yd} \end{aligned}$$

53.

Rent:	$1,150
Food:	625
+ Entertainment:	+ 257
Total:	$2,032

55. $\dfrac{\$23{,}256 \text{ yearly income}}{12 \text{ months per year}} = \text{Monthly income}$

$$\begin{array}{r} 1{,}938 \\ 12\overline{)23{,}256} \\ \underline{12\downarrow} \\ 112 \\ \underline{108\downarrow} \\ 45 \\ \underline{36\downarrow} \\ 96 \\ \underline{96} \\ 0 \end{array}$$

The monthly income is $1,938.

57. 1st 40 hours: $40 \text{ hours} \times \dfrac{\$16}{\text{hour}} = \$640$

Overtime: $5 \text{ hours} \times \dfrac{\$24}{\text{hour}} = \$120$

Gross pay: $640
+ 120
$760

Take-home pay:

Gross pay:	$760
– deductions:	– 228
Take-home pay:	$532

59.

Calories in Big Mac:	510
× 2:	×2
Calories in 2 Big Macs:	1,020

Calories in:

2 Big Macs:	1,020
+ 1 order fries:	+ 450
Total calories:	1,470

61.

Calories in burrito:	345
× 2:	× 2
Calories in 2 burritos:	690

Calories in:

1 Colossus burger:	940
– 2 burritos:	– 690
Difference:	250

There are 250 more calories.

63. Calories burned in 30 minutes: 235
Calories burned in 60 minutes:
$235 \cdot 2 = 470$
Since a Whopper contains 630 calories, you would not burn enough calories skating.

65. Calories in:

1 Big Mac:	510
+ 1 large fries:	+ 450
Total calories:	960

You could choose many combinations of activities to total at least 960 calories.

67.
$$\begin{aligned} V &= l \cdot w \cdot h \\ &= (8 \text{ cm})(4 \text{ cm})(5 \text{ cm}) \\ &= 160 \text{ cm}^3 \end{aligned}$$

$$\begin{aligned} S &= 2lw + 2lh + 2hw \\ &= 2(8)(4) + 2(8)(5) + 2(5)(4) \\ &= 64 \text{ cm}^2 + 80 \text{ cm}^2 + 40 \text{ cm}^2 \\ &= 184 \text{ cm}^2 \end{aligned}$$

Chapter 1 Test

1. 20,347 = Twenty thousand, three hundred forty-seven.

3. $123,407 = 100,000 + 20,000 + 3,000 + 400 + 7$

5. $7 \cdot 1 = 7$
c, Multiplicative Property of 1

7. $5 \cdot 6 = 6 \cdot 5$
Commutative Property of Multiplication

9.
$$\begin{array}{r} \overset{1}{5},4\overset{1}{0}1 \\ 329 \\ +\ 10,653 \\ \hline 16,383 \end{array}$$

11.
$$\begin{array}{r} \overset{6}{\not 7},\overset{9}{\not 0}\overset{14}{\not 5}\overset{12}{\not 2} \\ -3,967 \\ \hline 3,085 \end{array}$$

13.
$$\begin{array}{r} 359 \\ \times 62 \\ \hline 718 \\ 21,540 \\ \hline 22,258 \end{array}$$

15.
$$\begin{array}{r} 21 \\ 583\overline{)12243} \\ 1166\downarrow \\ \hline 583 \\ 583 \\ \hline 0 \end{array}$$

17.
$$\begin{aligned} &8(5)^2 - 7(3)^3 && \textit{exponentiate} \\ &= 8 \cdot 25 - 7 \cdot 27 && \textit{multiply} \\ &= 200 - 189 && \textit{subtract} \\ &= 11 \end{aligned}$$

19.
$$\begin{aligned} &7 + 2(53 - 3) && \textit{parenthesis} \\ &= 7 + 2(50) && \textit{multiply} \\ &= 7 + 100 && \textit{add} \\ &= 107 \end{aligned}$$

21. Sum of prices = \$210,000 + \$139,000 + \$122,000 + \$145,000 + \$120,000 + \$540,000 + \$167,000 + \$125,000 + \$125,000 + \$950,000 = \$2,643,000

$$\text{Mean} = \frac{\text{sum of values}}{\text{number of values}} = \frac{\$2,643,000}{10} = \$264,300$$

Median: Order the prices from low to high: \$120,000, \$122,000, \$125,000, \$125000, \$139,000, \$145,000, \$167,000, \$210,000, \$540,000, \$950,000

Since there are an even number of values, the median is the mean of the middle two:

$$\text{Median} = \frac{\$139,000 + \$145,000}{2} = \frac{\$284,000}{2} = \$142,000$$

Mode: occurs most often: \$125,000

23.

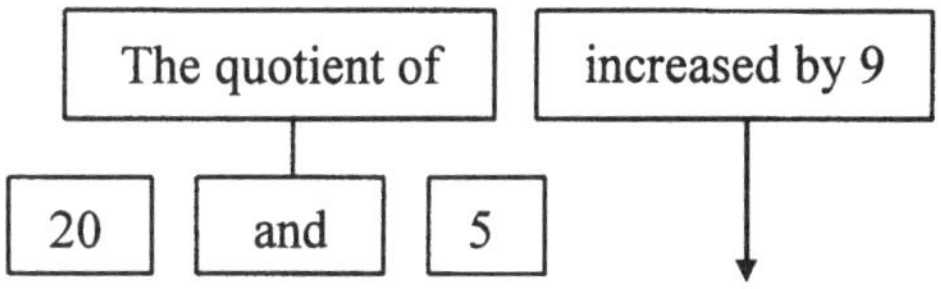

$20 \quad \div \quad 5 \quad +9$

$20 \div 5 + 9$ *divide*

$= 4 + 9$ *add*

$= 13$

25.

$$\begin{aligned} P &= 2l + 2w \\ &= 2(4\text{ ft}) + 2(3\text{ ft}) \\ &= 8\text{ ft} + 6\text{ ft} \\ &= 14\text{ ft} \end{aligned}$$

$$\begin{aligned} A &= lw \\ &= (4\text{ ft})(3\text{ ft}) \\ &= 12\text{ ft}^2 \end{aligned}$$

Chapter 2 Preview

1. Addition facts:

+	0	1	2	3	4	5	6	7	8	9
0	0	1	2	3	4	5	6	7	8	9
1	1	2	3	4	5	6	7	8	9	10
2	2	3	4	5	6	7	8	9	10	11
3	3	4	5	6	7	8	9	10	11	12
4	4	5	6	7	8	9	10	11	12	13
5	5	6	7	8	9	10	11	12	13	14
6	6	7	8	9	10	11	12	13	14	15
7	7	8	9	10	11	12	13	14	15	16
8	8	9	10	11	12	13	14	15	16	17
9	9	10	11	12	13	14	15	16	17	18

3. $5-3=2$

5. $6-1=5$

7. $8-7=1$

9. $14-8=6$

11. $6\div 2=3$

13. $9\div 3=3$

15. $12\div 4=3$

17. $5(3\cdot 2)=(5\cdot 3)\cdot 2$

19. $$\begin{aligned} P &= 4s \\ &= 4(6\text{ in}) \\ &= 24\text{ in} \end{aligned}$$

Chapter 2 Pretest

1. The opposite of 10 is −10.

3. $-3 > -5$, because −3 is right of −5 on the number line.

5. $-|-8| = -8$

7. $9 - 19 = 9 + (-19)$
$= -10$

9. $-6 - 8 = -6 + (-8)$
$= -14$

11. $-40 - (-7) = -40 + 7$
$= -33$

13. $(-2)(-8) = 16$

15. $\dfrac{-40}{-5} = 8$

17. $(-2)^3$ *definition of exponent*
$= (-2)(-2)(-2)$ *multiply left to right*
$= 4(-2)$ *multiply*
$= -8$

19. Simplify above and below fraction bar first:
$\dfrac{7 + 4(-1)}{2 - 3}$ *multiply first*
$= \dfrac{7 + (-4)}{2 + (-3)}$ *add*
$= \dfrac{3}{-1}$ *divide*
$= -3$

21. $5(3y - 2) = 5 \cdot 3y - 5 \cdot 2$
$= 15y - 10$

23. High temp – low temp = difference
$30° - (-5°)$
$= 30° + 5°$
$= 35°$

Section 2.1

1. $4 < 7$
4 is less than 7.

3. $5 > -2$
5 is greater than -2.

5. $-10 < -3$
-10 is less than -3.

7. $0 > -4$
0 is greater than -4.

9. 30 is greater than -30.
$30 > -30$

11. -10 is less than 0.
$-10 < 0$

13. -3 is greater than -15.
$-3 > -15$

15. $3 < 7$, because 3 is left of 7 on the number line.

17. $7 > -5$, because 7 is right of -5 on the number line.

19. $-6 < 0$, because -6 is left of 0 on the number line.

21. $-12 < -2$, because -12 is left of -2 on the number line.

23. $-10 > -100$, because -10 is right of -100 on the number line.

25. $-4 < 4$, because -4 is left of 4 on the number line.

27. $-3 < -2$, because -3 is left of -2 on the number line.

29. $-3 \underline{\ ?\ } |-6|$ *replace* $|-6| = 6$
$-3 \underline{\ ?\ } 6$
$-3 < 6$

31. $15 \underline{\ ?\ } |-4|$ *replace* $|-4| = 4$
$15 \underline{\ ?\ } 4$
$15 > 4$

33. $|-2| \underline{\ ?\ } |-7|$ *replace* $|-7| = 7,\ |-2| = 2$
$2 \underline{\ ?\ } 7$
$2 < 7$

35. $|2| = 2$, because 2 is 2 units from 0.

37. $|100| = 100$, because 100 is 100 units from 0.

39. $|-8| = 8$, because -8 is 8 units from 0.

41. $|-231| = 231$, because -231 is 231 units from 0.

43. $|-6| = 6$, because -6 is 6 units from 0.

45. $|-200| = 200$, because -200 is 200 units from 0.

47. $|8| = 8$, because 8 is 8 units from 0.

49. $|231| = 231$, because 231 is 231 units from 0.

51. The opposite of 3 is -3

53. The opposite of -2 is $-(-2) = 2$.

55. The opposite of 75 is -75.

57. The opposite of 0 is 0.

59. The opposite of -121 is $-(-121) = 121$.

61. The opposite of 555 is –555.

63. $-(-2)=2$
The opposite of –2 is 2.

65. $-(-8)=8$
The opposite of –8 is 8.

67. $-|-2|=-2$

69. $-|-8|=-8$

71. The only number that is its own opposite is 0.

73. If n is negative:
$-n$ is $-$(a negative number)
$-n$ is a positive number

Estimating

75. –60 is closer to –100 on the number line.

77. –10 is closer to –20 on the number line.

79. –362 is closer to –360 on the number line.

Applying the Concepts

81. 61 degrees *below zero* Fahrenheit: –61 °F
10 degrees warmer: –51 °F
Note: below zero indicates a negative number.

83. 5 degrees *below zero* Fahrenheit: –5 °F
10 degrees colder: –15 °F
Note: below zero indicates a negative number.

85. From the 25 mph row, 25°F column: –7 °F

87. 10°F and 25 mph wind gives wind chill temperature of –29°F.
–5 °F and 10 mph wind gives wind chill temperature of –27°F.
Because $-29<-27$, 10°F and 25 mph wind feels colder.

89. Above each month (numbered 1-12, beginning with January), draw a dot corresponding with the low temperature as measured by the vertical axis. To sketch the line graph, connect the dots.

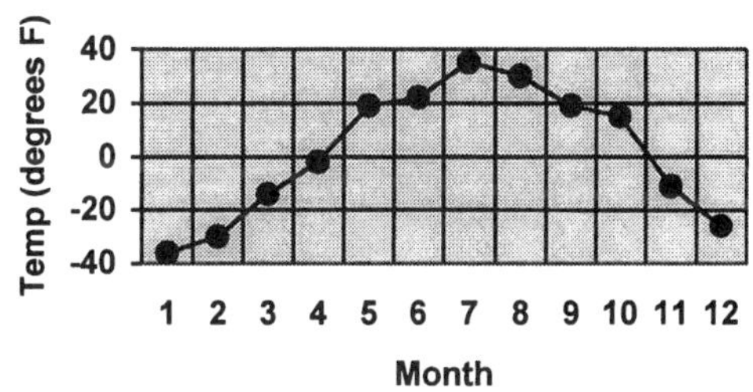

Review Problems

91. By the Commutative Property of Addition,
$3+5=5+3$

93. By the Associative Property of Addition,
$7+(2+6)=(7+2)+6$

95.

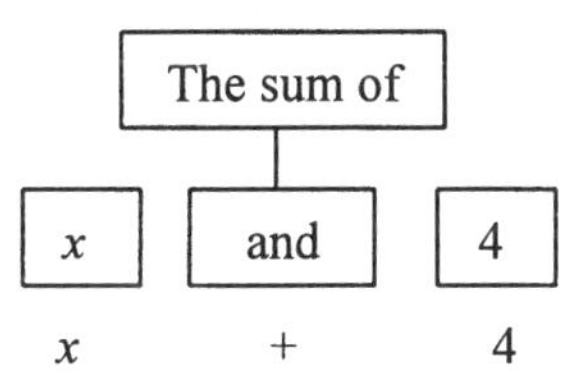

97.

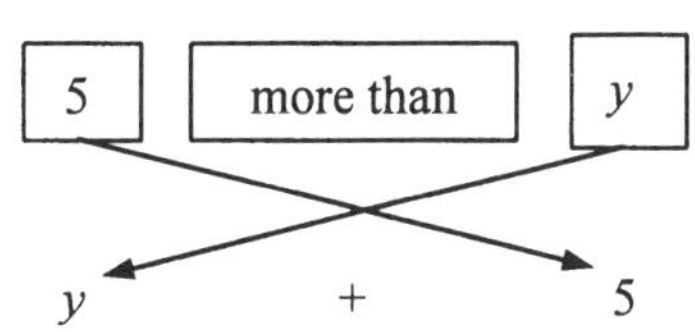

99.
$$\begin{array}{r} \overset{1}{2}\overset{1}{7}6 \\ 32 \\ +\ 4,005 \\ \hline 4,313 \end{array}$$

101. $(43,567+76,543)+11,111$
$=120,110+11,111$
$=131,221$

Extending the Concepts

103.

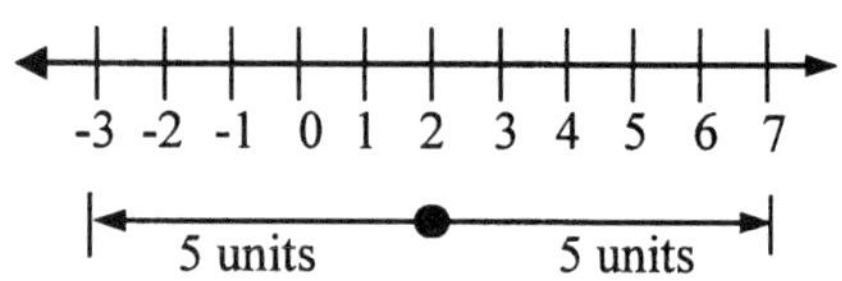

-3 is also 5 units from 2.

105. Answers will vary.

107. Answers will vary.

Improving Your Quantitative Literacy

109. a. The highest wind speed came on the day with the lowest wind chill:
$-59°F$ wind chill: 1/10/82 Cincinnati

b. The lowest wind speed cam on the day with the highest wind chill:
No wind chill: 1/4/81 Cleveland

c. The highest wind speed has the lowest wind chill, which is at Green Bay ($-48°F$ wind chill).

Section 2.2

1.

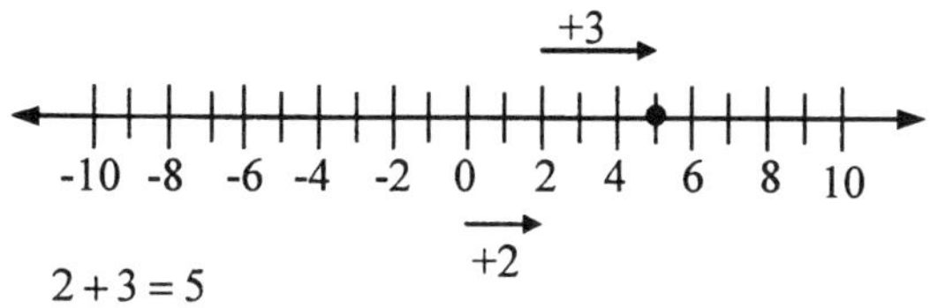

$2+3=5$

3.

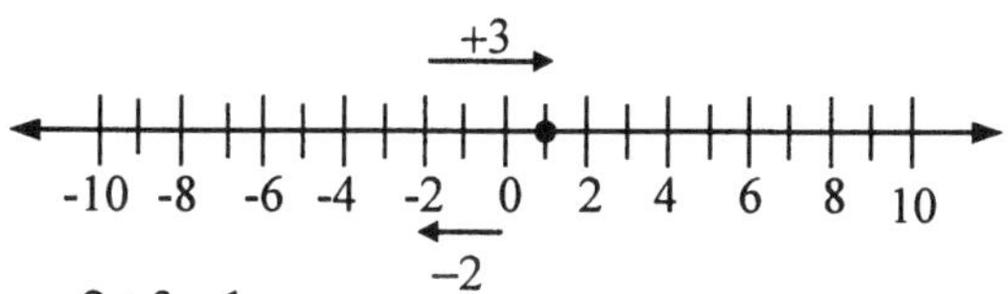

$-2+3=1$

5.

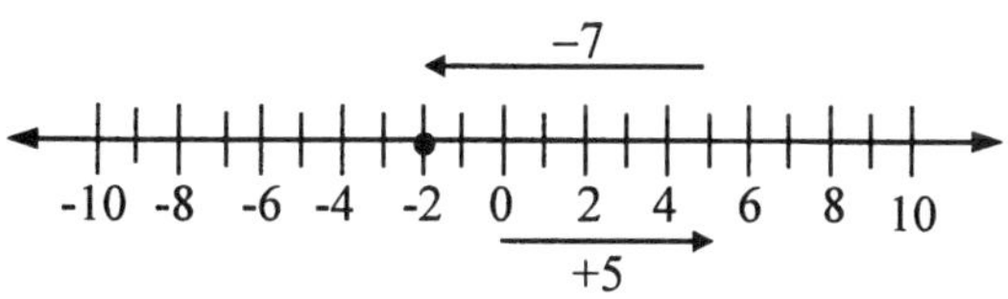

$5+(-7)=-2$

7.

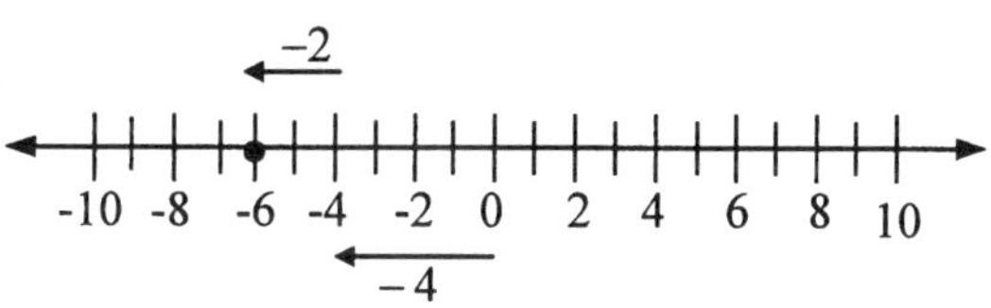

$-4+(-2)=-6$

9.

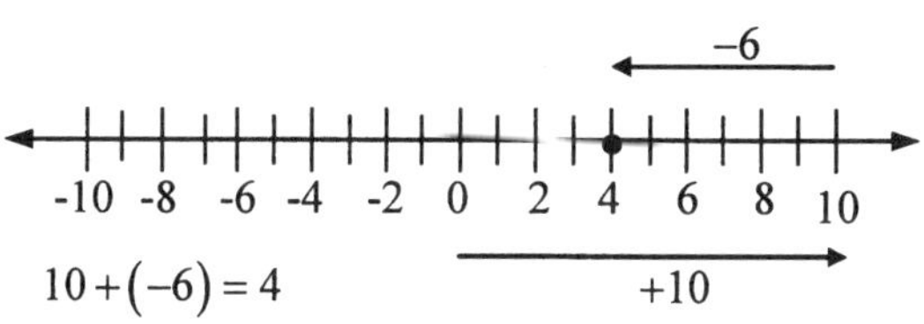

$10+(-6)=4$

11.

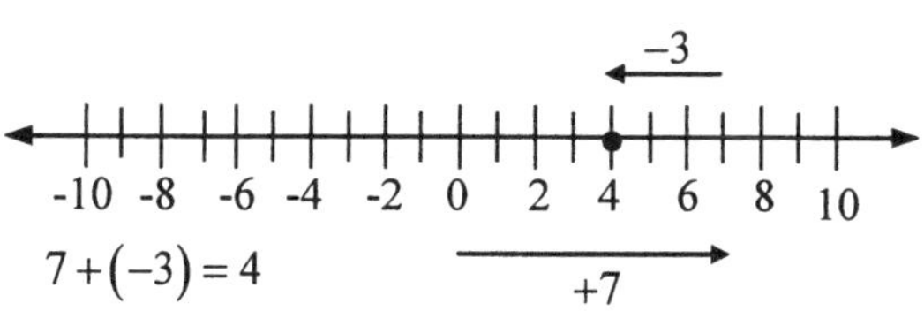

$7+(-3)=4$

13.

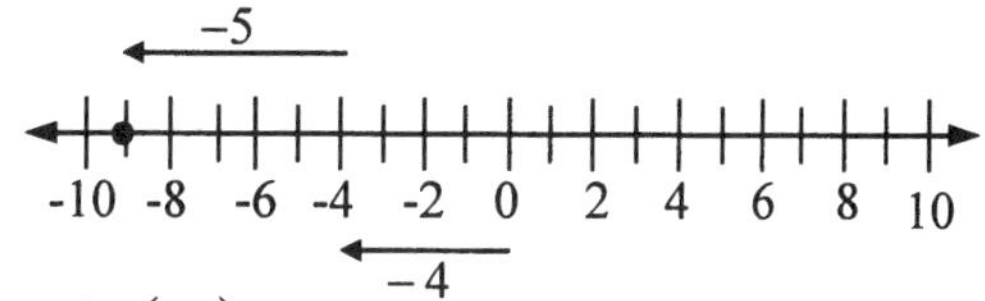

$-4+(-5)=-9$

15. $7+8$: *add absolute values*

$|7|+|8|=7+8=15$

The common sign is positive, so the answer is $+15$.

17. $5+(-8)$: *subtract absolute values*

$|-8|-|5|=8-5=3$

The number with the larger absolute value is –8, so the answer is –3.

19. $-6+(-5)$: *add absolute values*

$|-6|+|-5|=6+5=11$

The common sign is negative, so the answer is -11.

21. $-10+3$: *subtract absolute values*

$|-10|-|3|=10-3=7$

The number with the larger absolute value is –10, so the answer is –7.

23. $-1+(-2)$: *add absolute values*

$|-1|+|-2|=1+2=3$

The common sign is negative, so the answer is –3.

25. $-11+(-5)$: *add absolute values*

$|-11|+|-5|=11+5=16$

The common sign is negative, so the answer is –16.

27. $4+(-12)$: *subtract absolute values*

$|-12|-|4|=12-4=8$

The number with the larger absolute value is –12, so the answer is –8.

29. $-85+(-42)$: *add absolute values*

$|-85|+|-42|=85+42=127$

The common sign is negative, so the answer is -127.

31. $-121+170$: *subtract absolute values*

$|170|-|-121|=170-121=49$

The number with the larger absolute value is $+170$, so the answer is $+49$.

33. $-375+409$: *subtract absolute values*

$|409|-|-375|=409-375=34$

The number with the larger absolute value is $+409$, so the answer is $+34$.

35. The table is as follows:

a	b	$a+b$
5	−3	$5+(-3)=2$
5	−4	$5+(-4)=1$
5	−5	$5+(-5)=0$
5	−6	$5+(-6)=-1$
5	−7	$5+(-7)=-2$

37. The table is as follows:

x	y	$x+y$
−5	−3	$-5+(-3)=-8$
−5	−4	$-5+(-4)=-9$
−5	−5	$-5+(-5)=-10$
−5	−6	$-5+(-6)=-11$
−5	−7	$-5+(-7)=-12$

39. $10+(-18)+4$ *add left to right*

$=-8+4$ *add*

$=-4$

41. $24+(-6)+(-8)$ *add left to right*

$=18+(-8)$ *add*

$=10$

43. $-201+(-143)+(-101)$ *add left to right*

$=-344+(-101)$ *add*

$=-445$

45. $-321+752+(-324)$ *add left to right*

$=431+(-324)$ *add*

$=107$

47. $-8+3+(-5)+9$ *add left to right*

$=-5+(-5)+9$ *add left to right*

$=-10+9$ *add*

$=-1$

49. $-2+(-5)+(-6)+(-7)$ *add left to right*

$=-7+(-6)+(-7)$ *add left to right*

$=-13+(-7)$ *add*

$=-20$

51. $15+(-30)+18+(-20)$ *add left to right*

$=-15+18+(-20)$ *add left to right*

$=3+(-20)$ *add*

$=-17$

53. $-78+(-42)+57+13$ *add left to right*

$=-120+57+13$ *add left to right*

$=-63+13$ *add*

$=-50$

55. $(-8+5)+(-6+2)$ *parenthesis first*

$=-3+(-4)$ *add*

$=-7$

57. $(-10+4)+(-3+12)$ *parenthesis first*

$=-6+9$ *add*

$=3$

59. $20+(-30+50)+10$ *parenthesis first*
$=20+20+10$ *add left to right*
$=40+10$ *add*
$=50$

61. $108+(-456+275)$ *parenthesis*
$=108+(-181)$ *add left to right*
$=-73$

63. $[5+(-8)]+[3+(-11)]$ *brackets first*
$=-3+(-8)$ *add*
$=-11$

65. $[57+(-35)]+[19+(-24)]$ *brackets first*
$=22+(-5)$ *add*
$=17$

67. $-8+(-10)+(-3)$ *add left to right*
$=-18+(-3)$ *add*
$=-21$

69.

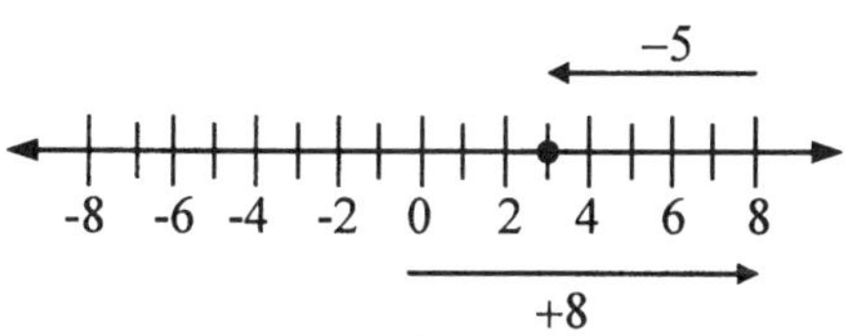

–5, because $8+(-5)=3$

71.

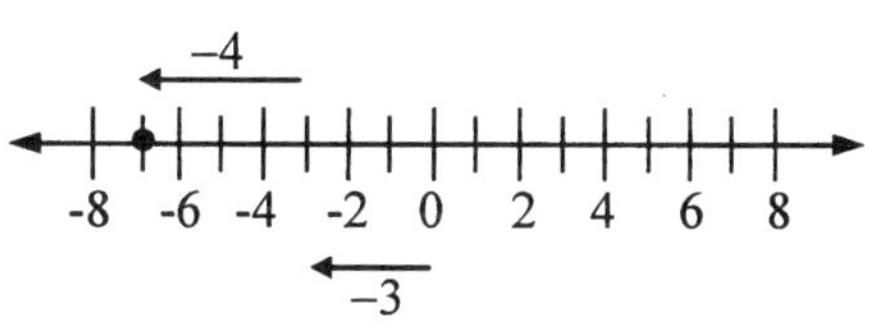

–4, because $-3+(-4)=-7$

73.

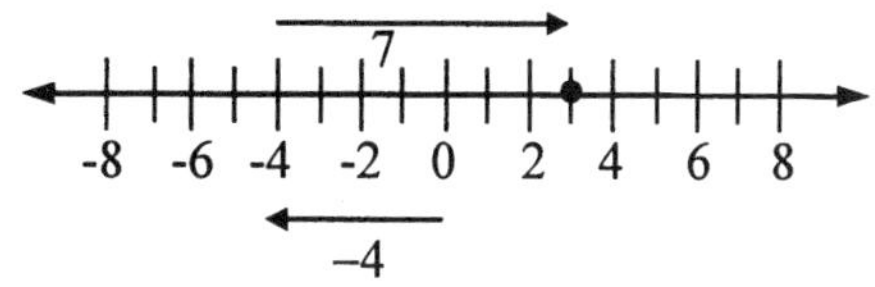

7, because $-4+7=3$

75.

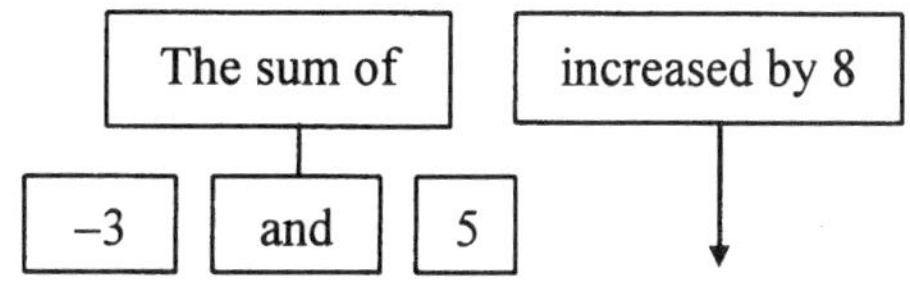

$-3 \quad + \quad 5 \quad +8$

$-3+5+8$ *add left to right*
$=2+8$ *add*
$=10$

Estimating

77. a. $251+249$ rounds to $250+250=500$

79. b. $-251+249$ rounds to $-250+250=0$

81. d. $77+22$ rounds to $80+20=100$

83. c. $77+(-22)$ rounds to $80+(-20)=60$

Applying the Concepts

85. Balance + deposit + check = new balance
$-\$40+\$100+(-\$50)$
$=\$60+(-\$50)$
$=\$10$

87. Wins + losses = net loss/gain
$\$74+(-\$141)=-\$67$

89. Gains + losses = net loss/gain
3 pts + (–5 pts) = – 2 pts

91. 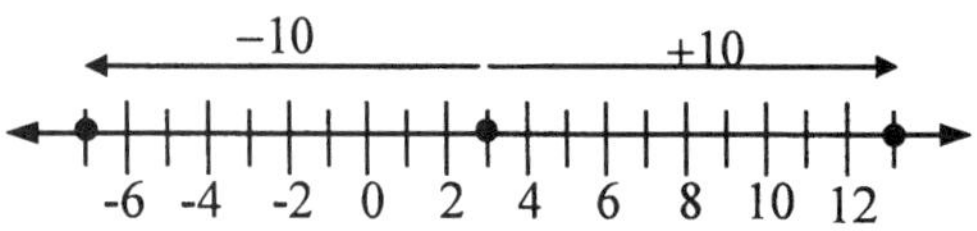

The answers are –7 and 13, because
$3+(-10)=-7$ and $3+10=13$

Review Problems

93. 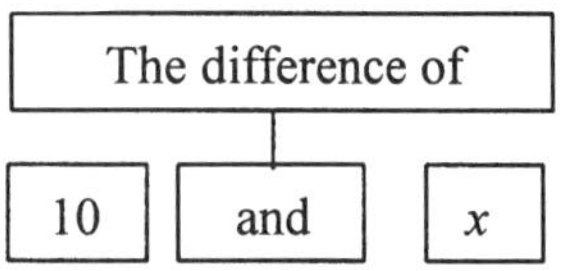

$10 \quad - \quad x$

95. 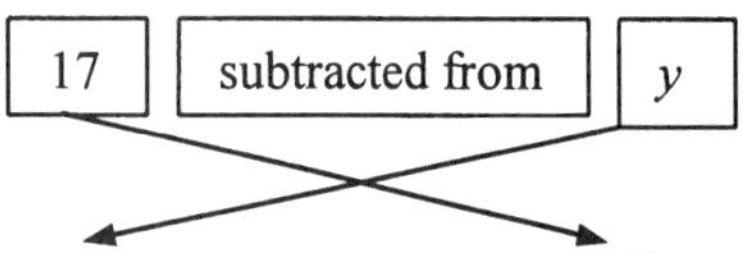

$y \quad - \quad 17$

97.

$$\begin{array}{r} \overset{6}{\not7}\,\overset{15}{\not6}\,\overset{13}{\not3} \\ -\,2\,8\,9 \\ \hline 4\,7\,4 \end{array}$$

99. $20-9-3$ *subtract left to right*
$=11-3$ *subtract*
$=8$

101. $20-(9-3)$ *parenthesis first*
$=20-6$ *subtract*
$=14$

103.

$$\begin{array}{r} 9\,\overset{5}{\not6}\,\overset{13}{\not3} \\ -2\,3\,8 \\ \hline 7\,2\,5 \end{array}$$

To nearest ten, 725 rounds to 730.

Section 2.3

1. $7-5=7+(-5)=2$

3. $8-6=8+(-6)=2$

5. $-3-5=-3+(-5)=-8$

7. $-4-1=-4+(-1)=-5$

9. $5-(-2)=5+2=7$

11. $3-(-9)=3+9=12$

13. $-4-(-7)=-4+7=3$

15. $-10-(-3)=-10+3=-7$

17. $15-18=15+(-18)=-3$

19. $100-113=100+(-113)=-13$

21. $-30-20=-30+(-20)=-50$

23. $-79-21=-79+(-21)=-100$

25. $156-(-243)=156+243=399$

27. $-35-(-14)=-35+14=-21$

29. The table is as follows:

x	y	$x-y$
8	6	$8-6=8+(-6)=2$
8	7	$8-7=8+(-7)=1$
8	8	$8-8=8+(-8)=0$
8	9	$8-9=8+(-9)=-1$
8	10	$8-10=8+(-10)=-2$

31. The table is as follows:

x	y	$x-y$
8	−6	$8-(-6)=8+6=14$
8	−7	$8-(-7)=8+7=15$
8	−8	$8-(-8)=8+8=16$
8	−9	$8-(-9)=8+9=17$
8	−10	$8-(-10)=8+10=18$

33. $4-5-6$ *change subtraction to addition of the opposite*

$=4+(-5)+(-6)$ *add from left to right*

$=-1+(-6)$ *add*

$=-7$

35. $-8+3-4$ *change subtraction to addition of the opposite*

$=-8+3+(-4)$ *add from left to right*

$=-5+(-4)$ *add*

$=-9$

37. $-8-4-2$ *change subtraction to addition of the opposite*

$=-8+(-4)+(-2)$ *add from left to right*

$=-12+(-2)$ *add*

$=-14$

39. $-12-30-47$ *change subtraction to addition of the opposite*

$=-12+(-30)+(-47)$ *add left to right*

$=-42+(-47)$ *add*

$=-89$

41. $33-(-22)-66$ *change subtraction to addition of the opposite*

$=33+22+(-66)$ *add from left to right*

$=55+(-66)$ *add*

$=-11$

43. $-101-(-95)+6$ *change subtraction to addition of the opposite*

$=-101+95+6$ *add from left to right*

$=-6+6$ *add*

$=0$

45. $-900+400-(-100)$ *change subtraction to addition of the opposite*

$=-900+400+100$ *add from left to right*

$=-500+100$ *add*

$=-400$

47. $5-(-6)=5+6=11$

49. $-5-(-1)=-5+1=-4$

51. $(-8+12)-(-4)$ *change subtration to addition of the opposite*

$=(-8+12)+4$ *add from left to right*

$=4+4$ *add*

$=8$

53.

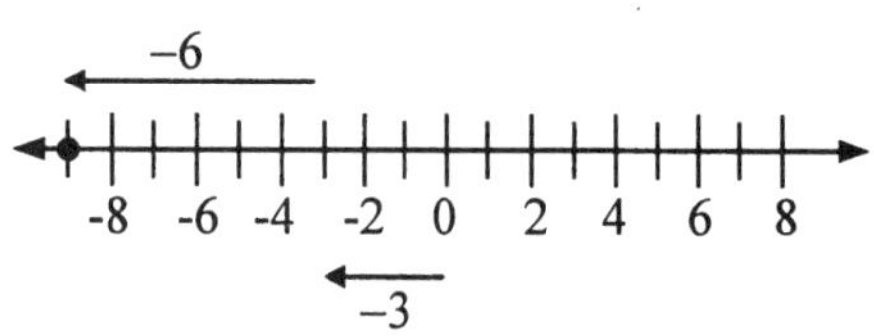

From the number line: $-3+(-6)=-9$
However, adding −6 is the same as subtracting 6.

Estimating

55. $52-49$ rounds to $50-50$

$=50+(-50)$

$=0$

57. $52-(-49)$ rounds to $50-(-50)$

$=50+50$

$=100$

59. $-161-(-62)$ rounds to $-150-(-50)$

$=-150+50$

$=-100$

61. $37-61$ rounds to $40-60$

$=40+(-60)$

$=-20$

63. $-37-61$ rounds to $-40-60$

$=-40+(-60)$

$=-100$

Applying the Concepts

65. High temp – low temp = difference

$28°-(-16°)$

$=28°+16°$

$=44°$

67. 2008 price – 2000 price = difference

$\$1,903-\$9,809=\$1,903+(-\$9,809)$

$=-\$7,906$

69. To find the difference, subtract the 25 mph wind chill temp from the 15 mph wind chill temp:

$-11°F-(-22°F)$

$=-11°F+22°F$

$=11°F$

71. To find the difference, subtract the 20 mph wind chill temp from the 10 mph wind chill temp:

$3°F-(-24°F)$

$=3°F+24°F$

$=27°F$

73. To find the difference, subtract the low temperature from the high temperature:

$60°F-(-26°F)$

$=60°F+26°F$

$=86°F$

75. To find the difference:
March low temp – December low temp

$-14°F-(-26°F)$

$=-14°F+26°F$

$=12°F$

Review Problems

77.

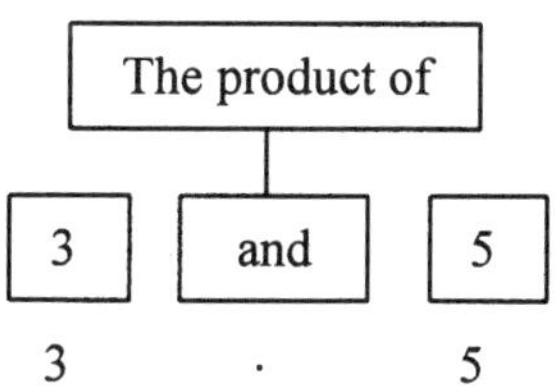

79.

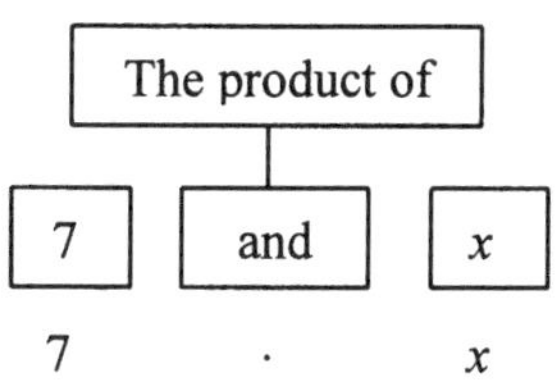

81. By Commutative Property of Multiplication:
$3(5)=5(3)$

83. By Associative Property of Multiplication:
$5(7\cdot 8)=(5\cdot 7)8$

85.

$2(3+4)$	*distribute*
$=2\cdot 3+2\cdot 4$	*multiply*
$=6+8$	*add*
$=14$	

87.

$$\begin{array}{r} 42 \\ \times\ 56 \\ \hline 252 \\ 2,100 \\ \hline 2,352 \end{array}$$

Extending the Concepts

89. Choose any two different integers and find the values of $a-b$ and $b-a$. For example,

$2-6=2+(-6)$ $\quad$ $6-2=6+(-2)$

$=-4$ $\quad$ $=4$

Because $-4\neq 4$, $2-6\neq 6-2$. Therefore, subtraction is not commutative.

91. $\$10-\$12=-\$2$
A checkbook balance is \$10, and you write a check for \$12. The new balance is –\$2.

93. 10, 5, 0,...is an arithmetic sequence starting with 10. Each number thereafter comes from adding –5 to the previous number. The next numbers in the sequence are –5, –10.

95. –10, –6, –2,...is an arithmetic sequence starting with –10. Each number thereafter comes from adding 4 to the previous number. The next numbers in the sequence are 2, 6.

Improving Your Quantitative Literacy

97. a. 2003 grants – 1998 grants = difference
$\$2,484-\$1,940=\$544$

b. 2003 costs – 1998 costs = difference
$\$1,115-\$1,636=\$1,115+(-\$1,636)$
$=-\$521$

c. Students are receiving more grants and tax breaks.

Section 2.4

1. $7(-8)=-56$
unlike signs; negative answer

3. $-6(10)=-60$
unlike signs; negative answer

5. $-7(-8)=56$
like signs; positive answer

7. $-9(-9)=81$
like signs; positive answer

9. $4(-6)=-24$
unlike signs; negative answer

11. $-6(-5)=30$
like signs; positive answer

13. $-5(0)(10)$ *multiply left to right*
$=0(10)$ *multiply*
$=0$

15. $3(-2)(4)$ *multiply left to right*
$=-6(4)$ *multiply*
$=-24$

17. $-4(3)(-2)$ *multiply left to right*
$=-12(-2)$ *multiply*
$=24$

19. $-1(-2)(-3)$ *multiply left to right*
$=2(-3)$ *multiply*
$=-6$

21. a. $(-4)^2=(-4)(-4)$
$=16$
b. $-4^2=-(4\cdot 4)=-16$

23. a. $(-5)^3$ *definition of exponent*
$=(-5)(-5)(-5)$ *multiply left to right*
$=25(-5)$ *multiply*
$=-125$
b. $-5^3=-(5\cdot 5\cdot 5)=-125$

25. a. $(-2)^4$ *exponent*
$=(-2)(-2)(-2)(-2)$ *multiply left to right*
$=4(-2)(-2)$ *multiply left to right*
$=-8(-2)$ *multiply*
$=16$
b. $-2^4=-(2\cdot 2\cdot 2\cdot 2)=-16$

27. The table is as follows:

x	x^2
-3	$(-3)^2=(-3)(-3)=9$
-2	$(-2)^2=(-2)(-2)=4$
-1	$(-1)^2=(-1)(-1)=1$
0	$0^2=(0)(0)=0$
1	$1^2=(1)(1)=1$
2	$2^2=(2)(2)=4$
3	$3^2=(3)(3)=9$

29. The table is as follows:

x	y	xy
6	2	$6(2)=12$
6	1	$6(1)=6$
6	0	$6(0)=0$
6	-1	$6(-1)=-6$
6	-2	$6(-2)=-12$

31. The table is as follows:

a	b	ab
−5	3	$(-5)(3)=-15$
−5	2	$(-5)(2)=-10$
−5	1	$(-5)(1)=-5$
−5	0	$(-5)(0)=0$
−5	−1	$(-5)(-1)=5$
−5	−2	$(-5)(-2)=10$
−5	−3	$(-5)(-3)=15$

33. $4(-3+2)$ *parenthesis first*
$=4(-1)$ *multiply*
$=-4$

35. $-10(-2-3)$ *parenthesis first*
$=-10(-5)$ *multiply*
$=50$

37. $-3+2(5-3)$ *parenthesis first*
$=-3+2(2)$ *multiply*
$=-3+4$ *add*
$=1$

39. $-7+2[-5-9]$ *brackets first*
$=-7+2[-14]$ *multiply*
$=-7+(-28)$ *add*
$=-35$

41. $2(-5)+3(-4)$ *multiply first*
$=-10+(-12)$ *add*
$=-22$

43. $3(-2)4+3(-2)$ *multiply first*
$=-6(4)+(-6)$ *multiply*
$=-24+(-6)$ *add*
$=-30$

45. $(8-3)(2-7)$ *parenthesis first*
$=5(-5)$ *multiply*
$=-25$

47. $(2-5)(3-6)$ *parenthesis first*
$=-3(-3)$ *multiply*
$=9$

49. $3(5-8)+4(6-7)$ *parenthesis first*
$=3(-3)+4(-1)$ *multiply*
$=-9+(-4)$ *add*
$=-13$

51. $-3(4-7)-2(-3-2)$ *parenthesis first*
$=-3(-3)-2(-5)$ *multiply left to right*
$=9-(-10)$ *change to addition*
$=9+10$ *add*
$=19$

53. $3(-2)(6-7)$ *parenthesis first*
$=3(-2)(-1)$ *multiply left to right*
$=-6(-1)$ *multiply*
$=6$

55. $-3(-2)(-1)$ *multiply left to right*
$=6(-1)$ *multiply*
$=-6$

57. $-3(?)=12$
The answer is –4, because $-3(-4)=12$

59. $-5(4)-(-3)$ *multiply first*
$=-20-(-3)$ *change to addition*
$=-20+3$ *add*
$=-17$

Applying the Concepts

61. Number of shares × \$gain/loss per share = total \$gain/loss

$$(100 \text{ shares})(-\$2)+(50 \text{ shares})(\$8)$$
$$=-\$200+\$400$$
$$=\$200$$

63. Set up a table showing a –4°F temperature change for each 1000 ft altitude change:

Altitude Change	Temperature Change
2000 ft to 3000 ft	–4°F
3000 ft to 4000 ft	–4°F
4000 ft to 5000 ft	–4°F
5000 ft to 6000 ft	–4°F
Net Temperature Change	–16°F

65. $\text{Pts} = W(2)+L(-2)+S(3)+TS(4)+BS(-2)$

Gagne:

$$\text{Pts} = 2(2)+3(-2)+53(3)+2(4)+0(-2)$$
$$=4+(-6)+159+8+0$$
$$=165$$

Smoltz:

$$\text{Pts} = 0(2)+2(-2)+42(3)+3(4)+4(-2)$$
$$=0+(-4)+126+12+(-8)$$
$$=126$$

Wagner:

$$\text{Pts} = 1(2)+4(-2)+40(3)+4(4)+3(-2)$$
$$=2+(-8)+120+16+(-6)$$
$$=124$$

Worrell:

$$\text{Pts} = 4(2)+4(-2)+36(3)+2(4)+7(-2)$$
$$=8+(-8)+108+8+(-14)$$
$$=102$$

Borowksi:

$$\text{Pts} = 2(2)+2(-2)+32(3)+2(4)+4(-2)$$
$$=4+(-4)+96+8+(-8)$$
$$=96$$

67. The table is as follows:

	Value	Number	Product
Eagle	–2	0	$-2(0)=0$
Birdie	–1	7	$-1(7)=-7$
Par	0	7	$0(7)=0$
Bogie	+1	3	$1(3)=3$
Double Bogie	+2	1	$2(1)=2$
			Total: –2

The total of –2 means Garcia was 2 under par. If par was 72, his score was $72+(-2)=70$.

Estimating

69. $-32(-522)$ rounds to $-30(-500)=15{,}000$

71. $-47(470)$ rounds to $-50(500)=-25{,}000$

73. $-222(-987)$ rounds to $-200(-1{,}000)=200{,}000$

75. $-222-(-987)$ rounds to

$$-200-(-1{,}000)=-200+1{,}000$$
$$=800$$

Review Problems

77.

The quotient of

12 | and | 6

12 ÷ 6

79. $2(3)=6 \Leftrightarrow 6\div 3=2$

81. $10\div 5=2 \Leftrightarrow 5\cdot 2=10$

83.

$$\begin{array}{r} 89 \\ 56\overline{)4{,}984} \\ \underline{448\downarrow} \\ 504 \\ \underline{504} \\ 0 \end{array}$$

Extending the Concepts

85. a. True. The sum of two negative numbers is always a negative number.

b. False. The statement does not indicate if it is about addition or multiplication. From 85a, we know the *sum* of two negatives is negative.

c. True. The product of two negative numbers is always a positive number.

87. You own 3 shares of a stock which loses \$4 per share on a trading day. Your net loss is 3(–\$4) = –\$12.

89. 2, –6, 18,…
The geometric sequence starts with 2. Note that $2(-3)=-6$ and $-6(-3)=18$. Each number after 2 comes from multiplying the previous number by –3. The next numbers in the sequence are –54 and 162.

91. $-2, 6, -18, \ldots$
The geometric sequence starts with –2. Note that $-2(-3)=6$ and $6(-3)=-18$.
Each number after –2 comes from multiplying the previous number by –3. The next numbers in the sequence are 54 and –162.

93.

$5(-2)^2-3(-2)^3$	*exponents first*
$=5(4)-3(-8)$	*multiply*
$=20-(-24)$	*change to addition*
$=20+24$	*add*
$=44$	

95.

$7-3(4-8)$	*inside parenthesis first*
$=7-3(-4)$	*multiply*
$=7-(-12)$	*change to addition*
$=7+12$	*add*
$=19$	

97.

$5-2[3-4(6-8)]$	*grouping symbols first*
$=5-2[3-4(-2)]$	*multiply inside brackets*
$=5-2[3-(-8)]$	*brackets, change to add*
$=5-2[3+8]$	*inside brackets*
$=5-2[11]$	*multiply*
$=5-22$	*subtract*
$=-17$	

Section 2.5

1. $-15 \div 5 = -3$ *unlike signs, negative answer*

3. $20 \div (-4) = -5$ *unlike signs, negative answer*

5. $-30 \div (-10) = 3$ *like signs, positive answer*

7. $\frac{-14}{-7} = 2$ *like signs, positive answer*

9. $\frac{12}{-3} = -4$ *unlike signs, negative answer*

11. $\frac{-22}{11} = -2$ *unlike signs, negative answer*

13. $\frac{0}{-3} = 0$, because $-3(0) = 0$

15. $\frac{125}{-25} = -5$ *unlike signs, negative answer*

17. The table is as follows:

a	b	$\frac{a}{b}$
100	−5	$\frac{100}{-5} = -20$
100	−10	$\frac{100}{-10} = -10$
100	−25	$\frac{100}{-25} = -4$
100	−50	$\frac{100}{-50} = -2$

19. The table is as follows:

a	b	$\frac{a}{b}$
−100	−5	$\frac{-100}{-5} = 20$
−100	5	$\frac{-100}{5} = -20$
100	−5	$\frac{100}{-5} = -20$
100	5	$\frac{100}{5} = 20$

21. $\frac{4(-7)}{-28}$ *simplify above fraction bar first*

$= \frac{-28}{-28}$ *divide*

$= 1$

23. $\frac{-3(-10)}{-5}$ *simplify above fraction bar first*

$= \frac{30}{-5}$ *divide*

$= -6$

25. Simplify above and below fraction bar first:

$\frac{2(-3)}{6-3}$

$= \frac{-6}{3}$ *divide*

$= -2$

27. Simplify above and below fraction bar first:

$\frac{4-8}{8-4}$ *subtract*

$= \frac{-4}{4}$ *divide*

$= -1$

29. Simplify above and below fraction bar first:

$\frac{2(-3)+10}{-4}$ *multiply before adding*

$=\frac{-6+10}{-4}$ *add*

$=\frac{4}{-4}$ *divide*

$=-1$

31. Simplify above and below fraction bar first:

$\frac{2+3(-6)}{4-12}$ *multiply before adding*

$=\frac{2+(-18)}{4+(-12)}$ *change subtraction to addition*

$=\frac{-16}{-8}$ *divide*

$=2$

33. Simplify above and below fraction bar first:

$\frac{6(-7)+3(-2)}{20-4}$ *multiply before adding*

$=\frac{-42+(-6)}{20+(-4)}$ *change subtraction to addition*

$=\frac{-48}{16}$ *divide*

$=-3$

35. Simplify above and below fraction bar before dividing:

$\frac{3(-7)(-4)}{6(-2)}$ *multiply from left to right*

$=\frac{84}{-12}$ *divide*

$=-7$

37. $(-5)^2+20\div 4$ *exponents first*

$=25+20\div 4$ *divide before adding*

$=25+5$ *add*

$=30$

39. $100\div(-5)^2$ *exponents before division*

$=100\div 25$ *divide*

$=4$

41. $-100\div 10\div 2$ *divide from left to right*

$=-10\div 2$ *divide*

$=-5$

43. $-100\div(10\div 2)$ *inside parenthesis first*

$=-100\div 5$ *divide*

$=-20$

45. $(-100\div 10)\div 2$ *inside parenthesis first*

$=-10\div 2$ *divide*

$=-5$

47. The quotient of –25 and 5:

$\frac{-25}{5}=-5$

49. If n represents the number,

$\frac{n}{-5}=-7$

$n=35$, because $\frac{35}{-5}=-7$

51. Subtract –3 from the quotient of 27 and 9:

$\frac{27}{9}-(-3)$ *divide before subtracting*

$=3-(-3)$ *change subtraction to addition*

$=3+3$ *add*

$=6$

Estimating

53. $397\div(-401)$ rounds to $400\div(-400)=-1$

55. $-121\div 27$ rounds to $-120\div 30=-4$
Note: round to numbers that are easily divisible

57. $-151+(-40)$ rounds to
$-150+(-50)=-200$

59. $-151(-49)$ rounds to $-150(-50)=7,500$

Applying the Concepts

61. $$\text{Mean}=\frac{\text{sum of values}}{\text{number of values}}=\frac{-5+0+5}{3}=\frac{0}{3}=0$$

63. a. $$\text{Mean}=\frac{\text{sum of values}}{\text{number of values}}=\frac{-5+(-4)+0+2+2+2+3}{7}=\frac{0}{7}=0$$

b. $-5,-4,0,2,2,2,3$
The values are already ordered. Because there is an odd number of values, the median is the middle value
Median = 2

c. $-5,-4,0,2,2,2,3$
The mode is the value that occurs most often. Mode = 2.

65. Above each day, draw a dot corresponding with the low temperature as measured by the vertical axis. To sketch the line graph, connect the dots.

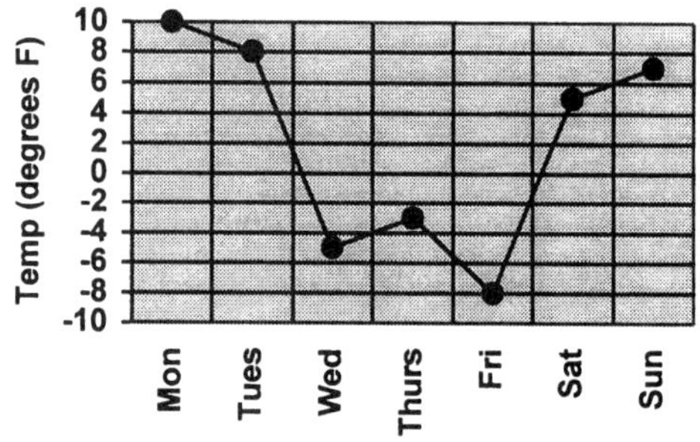

67. a. $$\text{Mean}=\frac{\text{sum of values}}{\text{number of values}}=\frac{10+8+(-5)+(-3)+(-8)+5+7}{7}=\frac{14}{7}=2°\text{F}$$
The mean low temperature is 2°F.

b. Order the values from smallest to largest:
–8°F, –5°F, –3°F, 5°F, 7°F, 8°F, 10°F
With an odd number of values, the median is the middle value:
Median = 5°F

Review Problems

69. By the Commutative Property of Addition,
$3+x=x+3$

71. By the Associative Property of Addition,
$5+(7+a)=(5+7)+a$

73. By the Associative Property of Multiplication, $3(4y)=(3\cdot4)y$

75. By the Distributive Property,
$5(3+7)=5(3)+5(7)$

77. $a+0=a$
Additive Property of 0

79. $1\cdot x=x$
Multiplicative Property of 1

Extending the Concepts

81. Choose 2 different numbers to show $a\div b\neq b\div a$

For example, $3\div 1=3$, but $1\div 3=\frac{1}{3}$

Therefore, $3\div 1\neq 1\div 3$.

83. Answers will vary.

85. $32,-16,8,\ldots$
The first number in the sequence is 32. Note that $32\div(-2)=-16$ and $-16\div(-2)=8$, i.e., each number after 32 in the sequence comes from *dividing* the previous number by –2. The next number in the sequence is $8\div(-2)=-4$.

87. $-32,16,-8,\ldots$
The first number in the sequence is –32. Note that $-32\div(-2)=16$ and $16\div(-2)=-8$, i.e., each number after –32 in the sequence comes from *dividing* the previous number by –2. The next number in the sequence is $-8\div(-2)=4$.

89. Simplify above and below fraction bar before dividing:

$\frac{-5-3}{-5+3}$ *change subtraction to addition*

$=\frac{-5+(-3)}{-5+3}$ *add*

$=\frac{-8}{-2}$ *divide*

$=4$

91. Simplify above and below fraction bar before dividing:

$\frac{6-3(2-11)}{6-3(2+11)}$ *inside parenthesis first*

$=\frac{6-3(-9)}{6-3(13)}$ *multiply before adding*

$=\frac{6-(-27)}{6-39}$ *change to addition*

$=\frac{6+27}{6+(-39)}$ *add*

$=\frac{33}{-33}$ *divide*

$=-1$

93. Simplify above and below fraction bar before dividing:

$\frac{6-(3-4)-3}{1-2-3}$ *inside parenthesis first*

$=\frac{6-(-1)-3}{1-2-3}$ *subtract from left to right*

$=\frac{4}{-4}$ *divide*

$=-1$

Section 2.6

1. $5(4a) = (5 \cdot 4)a$
$= 20a$

3. $6(8a) = (6 \cdot 8)a$
$= 48a$

5. $-6(3x) = (-6 \cdot 3)x$
$= -18x$

7. $-3(9x) = (-3 \cdot 9)x$
$= -27x$

9. $5(-2y) = (5 \cdot -2)y$
$= -10y$

11. $6(-10y) = (6 \cdot -10)y$
$= -60y$

13. $2 + (3 + x) = (2 + 3) + x$
$= 5 + x$

15. $5 + (8 + x) = (5 + 8) + x$
$= 13 + x$

17. $4 + (6 + y) = (4 + 6) + y$
$= 10 + y$

19. $7 + (1 + y) = (7 + 1) + y$
$= 8 + y$

21. $(5x + 2) + 4 = 5x + (2 + 4)$
$= 5x + 6$

23. $(6y + 4) + 3 = 6y + (4 + 3)$
$= 6y + 7$

25. $(12a + 2) + 19 = 12a + (2 + 19)$
$= 12a + 21$

27. $(7x + 8) + 20 = 7x + (8 + 20)$
$= 7x + 28$

29. $7(x + 5) = 7 \cdot x + 7 \cdot 5$
$= 7x + 35$

31. $6(a - 7) = 6 \cdot a - 6 \cdot 7$
$= 6a - 42$

33. $2(x - y) = 2 \cdot x - 2 \cdot y$
$= 2x - 2y$

35. $4(5 + x) = 4 \cdot 5 + 4 \cdot x$
$= 20 + 4x$

37. $3(2x + 5) = 3 \cdot 2x + 3 \cdot 5$
$= 6x + 15$

39. $6(3a + 1) = 6 \cdot 3a + 6 \cdot 1$
$= 18a + 6$

41. $2(6x - 3y) = 2 \cdot 6x - 2 \cdot 3y$
$= 12x - 6y$

43. $5(7 - 4y) = 5 \cdot 7 - 5 \cdot 4y$
$= 35 - 20y$

45. $3x + 5x = (3 + 5)x$
$= 8x$

47. $3a + a = 3a + 1a$
$= (3 + 1)a$
$= 4a$

49. $-2x + 6x = (-2 + 6)x$
$= 4x$

51. $6y - y = 6y - 1y$
$= (6-1)y$
$= 5y$

53. $-8a - 2a = (-8-2)a$
$= -10a$

55. $4x - 9x = (4-9)x$
$= -5x$

Applying the Concepts

57. $A = s^2$
$= (6 \text{ ft})^2$
$= 36 \text{ ft}^2$

$P = 4s$
$= 4(6 \text{ ft})$
$= 24 \text{ ft}$

59. $A = s^2$
$= (9 \text{ in})^2$
$= 81 \text{ in}^2$

$P = 4s$
$= 4(9 \text{ in})$
$= 36 \text{ in}$

61. $A = lw$
$= (20 \text{ in})(10 \text{ in})$
$= 200 \text{ in}^2$

$P = 2l + 2w$
$= 2(20 \text{ in}) + 2(10 \text{ in})$
$= 40 \text{ in} + 20 \text{ in}$
$= 60 \text{ in}$

63. $A = lw$
$= (25 \text{ ft})(12 \text{ ft})$
$= 300 \text{ ft}^2$

$P = 2l + 2w$
$= 2(25 \text{ ft}) + 2(12 \text{ ft})$
$= 50 \text{ ft} + 24 \text{ ft}$
$= 74 \text{ ft}$

65. $\frac{5(F-32)}{9}$ *substitute value for F*
$= \frac{5(68-32)}{9}$ *parenthesis first*
$= \frac{5 \cdot 36}{9}$ *divide before multiplying*
$= 5 \cdot \frac{36}{9}$ *divide*
$= 5 \cdot 4$ *multiply*
$= 20$
68°F converts to 20°C.

67. $\frac{5(F-32)}{9}$ *substitute value for F*
$= \frac{5(41-32)}{9}$ *inside parenthesis first*
$= \frac{5 \cdot 9}{9}$ *divide before multiplying*
$= 5 \cdot \frac{9}{9}$ *divide*
$= 5 \cdot 1$ *multiply*
$= 5$
41°F converts to 5°C.

69. $\frac{5(F-32)}{9}$ *substitute value for F*
$= \frac{5(14-32)}{9}$ *inside parenthesis first*
$= \frac{5 \cdot (-18)}{9}$ *divide before multiplying*
$= 5 \cdot \frac{-18}{9}$ *divide*
$= 5 \cdot -2$ *multiply*
$= -10$
14°F converts to −10°C.

Review Problems

71. $8 + (-4) = 4$

73. $-8+(-4)=-12$

75. $8-(-4)=8+4$
$=12$

77. $8(-4)=-32$

79. $-8(-4)=32$

81. $-8\div 4=-2$

Chapter 2 Review

1. The opposite of 17 is -17.

3. $6 > -6$, because 6 is right of -6 on the number line.

5. $|-3| \underline{\ ?\ } 2$ *replace* $|-3| = 3$

$3 \underline{\ ?\ } 2$

$3 > 2$

7. $-(-4) = 4$

The opposite of -4 is 4.

9. $5 + (-7)$: *subtract absolute values*

$|-7| - |5| = 7 - 5 = 2$

The number with the larger absolute value is negative, so the answer is -2.

11. $-345 + (-626)$: *add absolute values*

$|-345| + |-626| = 345 + 626 = 971$

The common sign is negative, so the answer is -971.

13. $-7 - 5 - 2 - 3$ *change to addition*

$= -7 + (-5) + (-2) + (-3)$ *add left to right*

$= -12 + (-2) + (-3)$ *add left to right*

$= -14 + (-3)$ *add*

$= -17$

15. $5(-4) = -20$

unlike signs; negative answer

17. $\dfrac{48}{-16} = -3$

unlike signs; negative answer

19. $\dfrac{-14}{-7} = 2$

like signs; positive answer

21. $(-2)^3 = (-2)(-2)(-2)$

$= -8$

23. $(-3)(-4) + 2(-5)$ *multiply before add*

$= 12 + (-10)$ *add*

$= 2$

25. $3(-6) + 8(2-5)$ *parenthesis first*

$= 3(-6) + 8(-3)$ *multiply before add*

$= -18 + (-24)$ *add*

$= -42$

27. $\dfrac{-4 + 2(-5)}{6-4}$ *simplify above and below fraction bar first*

$= \dfrac{-4 + (-10)}{6-4}$ *add and subtract*

$= \dfrac{-14}{2}$ *divide*

$= -7$

29. $\dfrac{-2(5) + 4(-3)}{10-8}$ *simplify above and below fraction bar first*

$= \dfrac{-10 + (-12)}{2}$ *add*

$= \dfrac{-22}{2}$ *divide*

$= -11$

31. $-78 + (-51) = -129$

33. $-10 - (-8) = -10 + 8$

$= -2$

35. $-3[-9 + (-4)] = -3[-13]$

$= 39$

37. $\frac{-38}{2} = -19$

39. True, because $10-(-5)=10+5=15$.

41. $-6-(-2)=-8$ is false, because
$-6-(-2)=-6+2=-4$.

43. On the horizontal axis, mark the speeds at 10 mph intervals from 0 mph to 80 mph. On the vertical axis, mark the scale from 0 ft to 100 ft at even intervals. Above each speed on the horizontal axis, draw a dot corresponding to the distance as measured on the vertical axis. To sketch the line graph, connect the dots.

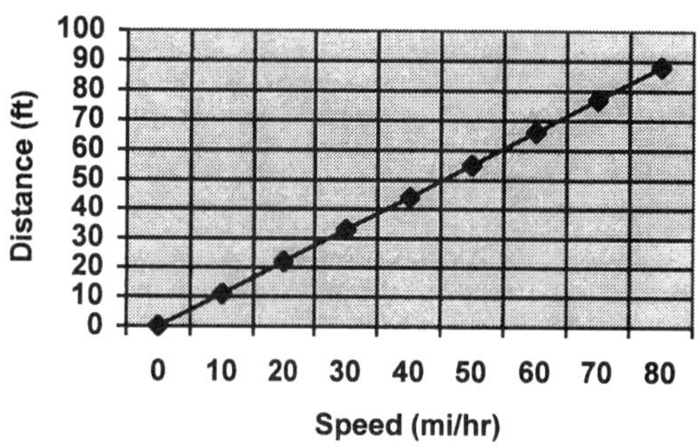

45.

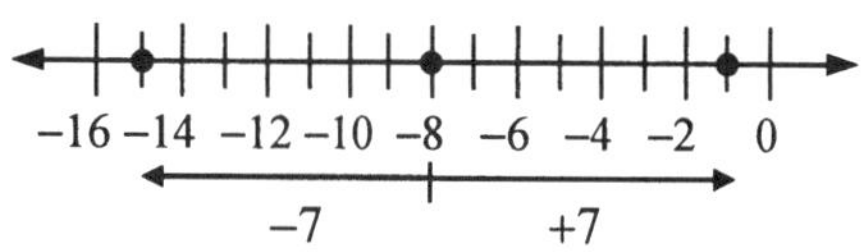

The two numbers are –15 and –1.

47. Let n be the number. There are 2 cases:

$5-n=-3$	or	$n-5=-3$
$5+(-n)=-3$		$n+(-5)=-3$
$5+(-8)=-3$		$2+(-5)=-3$
$n=8$		$n=2$

The other number is 8 or 2.

49. $8(3x)=(8\cdot3)x$
$=24x$

51. $6(-5y)=(6\cdot-5)y$
$=-30y$

53. $2(x-5)=2\cdot x-2\cdot 5$
$=2x-10$

55. $3(2a+5b)=3(2a)+3(5b)$
$=(3\cdot2)a+(3\cdot5)b$
$=6a+15b$

57. $-8a+10a=(-8+10)a=2a$

59. $12x+4x=(12+4)x=16x$

Chapters 1-2 Cumulative Review

1.
$$\begin{array}{r} \scriptstyle 1\ 1\ 1 \\ 6{,}801 \\ 539 \\ +\ 374 \\ \hline 7{,}714 \end{array}$$

3.
$$\begin{array}{r} \scriptstyle 4\ \ 9\ 12\ 18 \\ \not{5},\not{0}\not{3}\not{8} \\ -\,2,\,7\,6\,9 \\ \hline 2,\,2\,6\,9 \end{array}$$

5.
$$\begin{array}{r} 867 \\ \times\ 52 \\ \hline 1{,}734 \\ \underline{43{,}350} \\ 45{,}084 \end{array}$$

7. $5(2x-8)=5\cdot 2x-5\cdot 8$
$=10x-40$

9.
$$\begin{array}{r} 68\ \text{R3} \\ 15\overline{)1023} \\ \underline{90\downarrow} \\ 123 \\ \underline{120} \\ 3 \end{array}$$

11. $\dfrac{-24}{-6}=4$

13. $9-3(8-5)=9-3(3)$
$=9-9$
$=0$

15. $-|-11|$ *replace* $|-11|$ *with 11*
$=-11$
Note: The *outer* negative sign is carried down.

17. $(-3)^3$ *definition of exponent*
$=(-3)(-3)(-3)$ *multiply left to right*
$=9(-3)$ *multiply*
$=-27$

19. $(6-3)(8-13)$ *inside parenthesis first*
$=(3)(-5)$ *multiply*
$=-15$

21. $\dfrac{-8(6)}{-4}=\dfrac{-48}{-4}$
$=12$

23. $|3|\ \underline{\ ?\ }\ |-5|$ *replace* $|3|=3,\ |-5|=5$
$3\ \underline{\ ?\ }\ 5$
$3<5$

25.

Twice	the difference of		
	19	and	7
2 ×	(19	−	7)

$2(19-7)=2(12)$
$=24$

27. For each speed, draw a horizontal line from the top of the bar over to the vertical axis to find the corresponding stopping distance.

Speed (mi/hr)	Distance (ft)
20	22
30	49
40	88
50	137
60	198
70	269
80	352

29. $\dfrac{\$56 \text{ income}}{8 \text{ hours worked}} = \text{wage per hour}$

$\dfrac{\$56}{8} = \$7 \text{ per hour}$

31. To find the difference, subtract the low temperature from the high temperature:

$$\begin{aligned} 18° - (-5°) &= 18° + 5° \\ &= 23° \end{aligned}$$

33.

Number of tickets:	250
× cost per ticket:	× 7
Total ticket sales:	1,750

Total sales are $1,750.

Chapter 2 Test

1. The opposite of 14 is –14.

3. –1 > –4, because –1 is right of –4 on the number line.

5. $-(-7)=7$
The opposite of –7 is 7.

7. $8+(-17)=-9$

9. $-9+(-12)=-21$

11. $(-6)(-7)=42$

13. $\frac{-80}{16}=-5$

15. $(-3)^2=(-3)(-3)$
$=9$

17. $(-7)(3)+(-2)(-5)$ *multiply before add*
$=-21+10$ *add*
$=-11$

19. Simplify above and below fraction bar first:
$\frac{-5+3(-3)}{5-7}$
$=\frac{-5+(-9)}{5-7}$ *add and subtract*
$=\frac{-14}{-2}$ *divide*
$=7$

21. $-15+(-46)=-61$

23. $(-8)(-3)=24$

25. Above each year, draw a dot corresponding to the amount of garbage for that year as measured on the vertical axis. To draw the line graph, connect the dots.

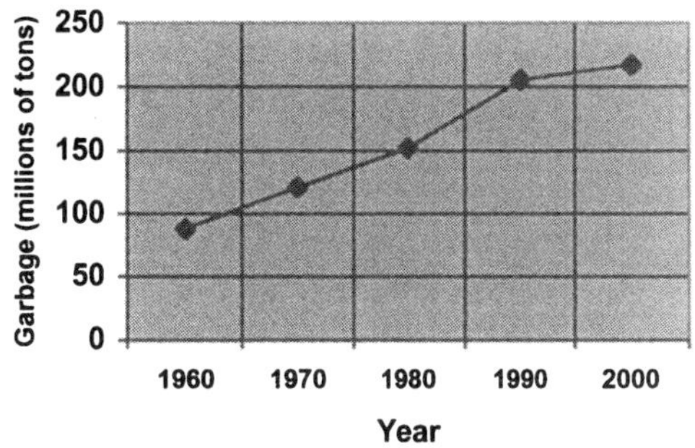

27. To find the difference, subtract the low temperature from the high temperature:
$21°-(-4°)=21°+4°$
$=25°$

29. $4(5x-1)=4\cdot 5x-4\cdot 1$
$=(4\cdot 5)x-4$
$=20x-4$

31. $12x+20x=(12+20)x=32x$

Chapter 3 Preview

1. The table is as follows:

$16+5=21$	$9-3=6$	$2(3)=6$	$\frac{24}{1}=24$
$9+2=11$	$12-8=4$	$2(-3)=-6$	$\frac{24}{24}=1$
$20+3=23$	$14-9=5$	$(-3)(-2)=6$	$\frac{120}{10}=12$
$5+9=14$	$21-10=11$	$11\cdot 29=319$	$\frac{105}{5}=21$
$15+14=29$	$7-6=1$	$2\cdot 20=40$	$\frac{54}{2}=27$
$30+55=85$	$18-5=13$	$59\cdot 1=59$	$\frac{342}{3}=114$

3. $2\cdot 5\cdot x=(2\cdot 5)\cdot x=10x$

5. $3+2-9=5-9=5+(-9)=-4$

7. $9\cdot 6+5=54+5=59$

9. $(3+5)(2+1)=(8)(3)=24$

11. $$\begin{aligned}32\div 4^2+75\div 5^2 &= (32\div 16)+(75\div 25)\\ &= 2+3\\ &= 5\end{aligned}$$

13. $\frac{208}{24}$: $$\begin{array}{r} 8 \text{ R16} \\ 24\overline{)208} \\ \underline{192} \\ 16 \end{array}$$

15. $\frac{111}{8}$: $$\begin{array}{r} 13 \text{ R7} \\ 8\overline{)111} \\ \underline{8\downarrow} \\ 31 \\ \underline{24} \\ 7 \end{array}$$

17. $2\cdot 2\cdot 3\cdot 3\cdot 3=2^2 3^3$

19. 4 six-packs – 4 cans = cans left
$4\cdot 6-4=24-4=20$ cans

Chapter 3 Pretest

1. $\dfrac{2\text{ girls}}{4\text{ children}}=\dfrac{1}{2}$

3. $\dfrac{16xy^2}{20xy}=\dfrac{\cancel{2}\cdot\cancel{2}\cdot 2\cdot 2\cdot\cancel{x}\cdot\cancel{y}\cdot y}{\cancel{2}\cdot\cancel{2}\cdot 5\cdot\cancel{x}\cdot\cancel{y}}=\dfrac{4y}{5}$

5. $-\dfrac{3}{4}\cdot\dfrac{2}{3}=-\dfrac{\cancel{3}\cdot\cancel{2}}{\cancel{2}\cdot 2\cdot\cancel{3}}=-\dfrac{1}{2}$

7. $\dfrac{3}{8}\cdot 16=\dfrac{3\cdot 16}{8}=\dfrac{3\cdot\cancel{8}\cdot 2}{\cancel{8}}=6$

9. $\dfrac{32}{45}\div\dfrac{40}{63}=\dfrac{32}{45}\cdot\dfrac{63}{40}=\dfrac{\cancel{8}\cdot 4\cdot\cancel{9}\cdot 7}{\cancel{9}\cdot 5\cdot\cancel{8}\cdot 5}=\dfrac{28}{25}=1\dfrac{3}{25}$

11. $\dfrac{5}{8}-\dfrac{1}{8}=\dfrac{5-1}{8}=\dfrac{4}{8}=\dfrac{1}{2}$

13. $\dfrac{5}{8}-\left(-\dfrac{1}{6}\right)=\dfrac{5}{8}+\dfrac{1}{6}$

$=\dfrac{5\cdot 3}{8\cdot 3}+\dfrac{1\cdot 4}{6\cdot 4}$

$=\dfrac{15}{24}+\dfrac{4}{24}$

$=\dfrac{19}{24}$

15. $\dfrac{21}{8}$: $8\overline{)21}$ quotient 2, 16, remainder 5 so $\dfrac{21}{8}=2+\dfrac{5}{8}=2\dfrac{5}{8}$

17. $12\div 3\dfrac{1}{6}=\dfrac{12}{1}\div\dfrac{19}{6}=\dfrac{12}{1}\cdot\dfrac{6}{19}=\dfrac{72}{19}=3\dfrac{15}{19}$

19.

$$\begin{array}{rrrr} 6\frac{1}{6}= & 6\frac{1}{6}= & 6\frac{1}{6}= & 5\frac{7}{6} \\ -\;3\frac{1}{3}= & -\;3\frac{1\cdot 2}{3\cdot 2}= & -\;3\frac{2}{6}= & -\;3\frac{2}{6} \\ \hline & & & 2\frac{5}{6} \end{array}$$

21. $\left(\dfrac{1}{3}\right)^2\cdot 27+\left(\dfrac{3}{2}\right)^2\cdot 4$

$=\left(\dfrac{1}{3}\right)\left(\dfrac{1}{3}\right)\cdot 27+\left(\dfrac{3}{2}\right)\left(\dfrac{3}{2}\right)\cdot 4$

$=\dfrac{27}{9}+\dfrac{36}{4}$

$=3+9$

$=12$

23. $\left(1-\dfrac{1}{4}\right)\left(1\dfrac{2}{3}-\dfrac{1}{3}\right)=\left(\dfrac{4}{4}-\dfrac{1}{4}\right)\left(\dfrac{5}{3}-\dfrac{1}{3}\right)$

$=\left(\dfrac{3}{4}\right)\left(\dfrac{4}{3}\right)$

$=1$

25. $\dfrac{2+\frac{1}{4}}{2-\frac{1}{4}}=\dfrac{4\left(2+\frac{1}{4}\right)}{4\left(2-\frac{1}{4}\right)}=\dfrac{8+1}{8-1}=\dfrac{9}{7}=1\dfrac{2}{7}$

27. $20\div\dfrac{4}{5}+1=\dfrac{20}{1}\cdot\dfrac{5}{4}+1$

$=25+1$

$=26$

29. $A=\dfrac{1}{2}bh=\dfrac{1}{2}(3\text{ ft})(4\text{ ft})=\dfrac{12}{2}\text{ ft}^2=6\text{ ft}^2$

Section 3.1

1. $\frac{1}{3}$: numerator = 1

3. $\frac{2}{3}$: numerator = 2

5. $\frac{x}{8}$: numerator = x

7. $\frac{a}{b}$: numerator = a

9. $\frac{2}{5}$: denominator = 5

11. $6 = \frac{6}{1}$: denominator = 1

13. $\frac{a}{12}$: denominator = 12

15. The table is as follows:

Numerator a	Denominator b	Fraction $\frac{a}{b}$
3	5	$\frac{3}{5}$
1	7	$\frac{1}{7}$
x	y	$\frac{x}{y}$
$x+1$	x	$\frac{x+1}{x}$

17. $\left\{\frac{3}{4}, \frac{6}{5}, \frac{12}{3}, \frac{1}{2}, \frac{9}{10}, \frac{20}{10}\right\}$

Proper fractions: $\frac{3}{4}, \frac{1}{2}, \frac{9}{10}$

19. True. If a is any whole number greater than 1, a can be written as $a = \frac{a}{1}$, which is improper.

21. False. For example, add 1 to the numerator and denominator of the fraction $\frac{1}{2}$:

$\frac{1+1}{2+1} = \frac{2}{3}$, and $\frac{1}{2} \neq \frac{2}{3}$.

23. $\frac{6}{8} = \frac{6 \div 2}{8 \div 2} = \frac{3}{4}$

25. $\frac{86}{94} = \frac{86 \div 2}{94 \div 2} = \frac{43}{47}$

27. $\frac{12}{9} = \frac{12 \div 3}{9 \div 3} = \frac{4}{3}$

29. $\frac{39}{51} = \frac{39 \div 3}{51 \div 3} = \frac{13}{17}$

31. $\frac{2}{3} = \frac{2 \cdot 2}{3 \cdot 2} = \frac{4}{6}$

33. $\frac{55}{66} = \frac{55 \div 11}{66 \div 11} = \frac{5}{6}$

35. $\frac{2}{3} = \frac{2 \cdot 4}{3 \cdot 4} = \frac{8}{12}$

37. $\frac{56}{84} = \frac{56 \div 7}{84 \div 7} = \frac{8}{12}$

39. $\frac{1}{6} = \frac{1 \cdot 2x}{6 \cdot 2x} = \frac{2x}{12x}$

41. $\frac{3}{8} = \frac{3 \cdot 3a}{8 \cdot 3a} = \frac{9a}{24a}$

43. $\frac{5}{6}=\frac{5\cdot 4a}{6\cdot 4a}=\frac{20a}{24a}$

45. Answers will vary. Because $\frac{1}{4}=\frac{1\cdot 2}{4\cdot 2}=\frac{2}{8}$, $\frac{1}{4}$ is equivalent to 2 of 8 parts. Shade any 2 of the 8 parts:

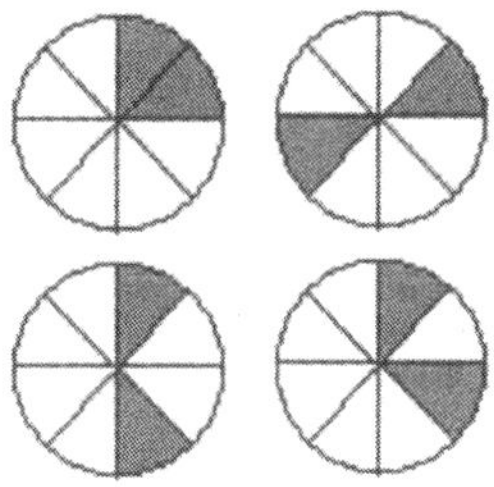

47. $\frac{3}{1}=3$

49. $\frac{6}{3}=2$

51. $\frac{37}{1}=37$

53.

a.  $\frac{\text{Number parts shaded}}{\text{Number equal parts}}=\frac{1}{2}$

b.  $\frac{\text{Number parts shaded}}{\text{Number equal parts}}=\frac{1}{2}$

c. 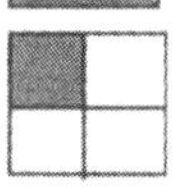 $\frac{\text{Number parts shaded}}{\text{Number equal parts}}=\frac{1}{4}$

d.  $\frac{\text{Number parts shaded}}{\text{Number equal parts}}=\frac{1}{4}$

55. Divide the distance from 0 to 1 into 4 equal parts. Then start at 0 and count over 1 part.

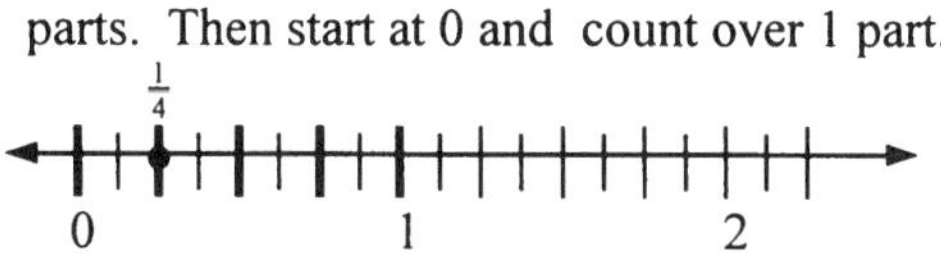

57. Divide the distance from 0 to 1 into 16 equal parts. Then start at 0 and count over 1 part.

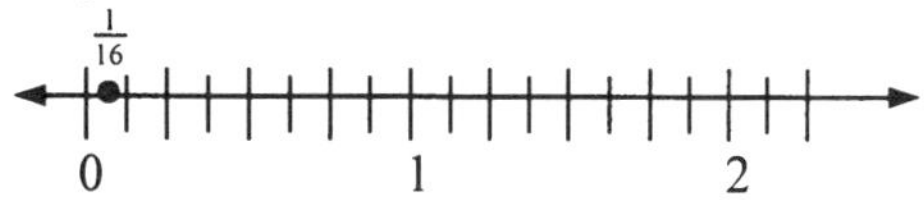

59. Divide the distance from 0 to 1 into 4 equal parts. Then start at 0 and count over 3 parts.

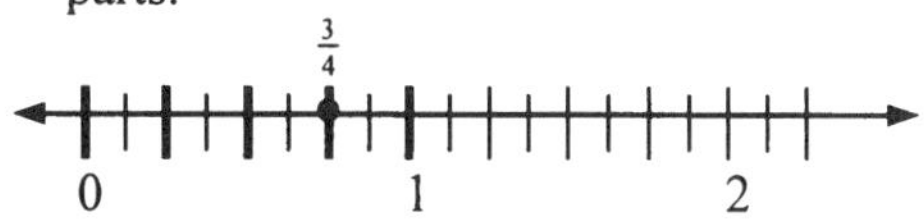

61. Divide the distance from 0 to 1 into 2 equal parts, and divide the distance from 1 to 2 into 2 equal parts. Then start at 0 and count over 3 parts.

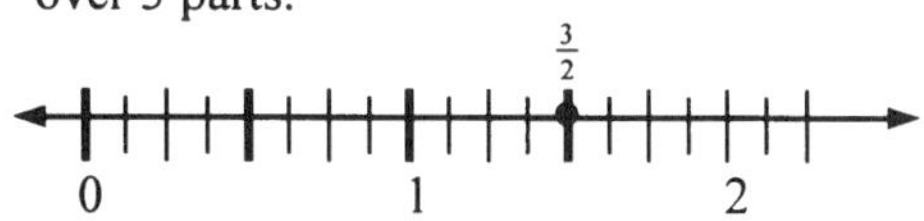

63. Divide the distance from 0 to 1 into 16 equal parts, and divide the distance from 1 to 2 into 16 equal parts. Then start at 0 and count over 31 parts.

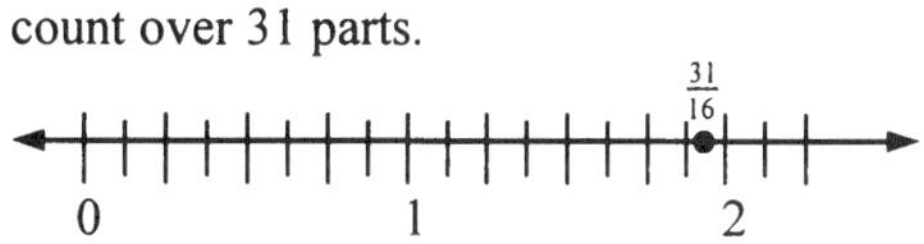

Applying the Concepts

65. The table is as follows:

non-work-related e-mail sent	Fraction of respondents
Never	$\frac{4}{25}$
1 to 5 times a day	$\frac{47}{100}$
5 to 10 times a day	$\frac{8}{25}$
more than 10 times a day	$\frac{1}{20}$

67. $\frac{\text{Number of girls}}{\text{Total children}}=\frac{4}{5}$

69. $\frac{\text{Amount spent on clothes}}{\text{Total amount earned}}=\frac{\$2}{\$15}=\frac{2}{15}$

71. $\dfrac{\text{Students finished}}{\text{Students started}} = \dfrac{29}{43}$

73. $\dfrac{\text{House payment}}{\text{Total income}} = \dfrac{\$1{,}121}{\$1{,}791} = \dfrac{1{,}121}{1{,}791}$

75. Find the day on the horizontal axis, then go up to the graph and find the matching fraction value on the vertical axis.

Days since discontinuing	Fraction remaining
0	1
5	$\frac{1}{2}$
10	$\frac{1}{4}$
15	$\frac{1}{8}$
20	$\frac{1}{16}$

77. $\dfrac{\text{Runners under 3 hours}}{\text{Total runners}} = \dfrac{238}{527}$

Estimating

79. 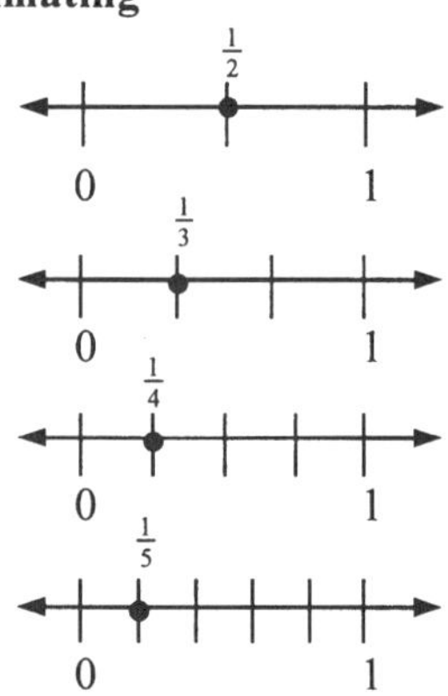

$\dfrac{1}{5}$ is closest to 0.

81. 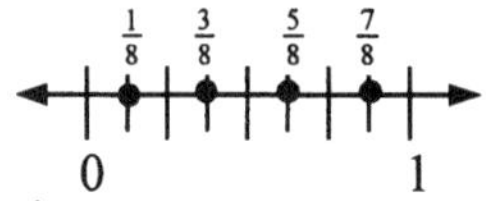

$\dfrac{1}{8}$ is closest to 0.

Note: In comparing positive fractions, if the denominators are the same, the largest fraction has the largest numerator.

Calculator Problems

83. $\dfrac{596}{1{,}788} \stackrel{?}{=} \dfrac{1}{3}$. Yes. Because $1{,}788 \div 3 = 596$,

$\dfrac{596}{1{,}788} = \dfrac{596 \div 3}{1{,}788 \div 3} = \dfrac{1}{3}$

85. $\dfrac{3}{9} = \dfrac{?}{2{,}205}$. Because $2{,}205 \div 9 = 245$,

$\dfrac{3}{9} = \dfrac{3 \cdot 245}{9 \cdot 245} = \dfrac{735}{2{,}205}$

87. $\dfrac{1}{4} = \dfrac{?}{384}$. Because $384 \div 4 = 96$,

$\dfrac{1}{4} = \dfrac{1 \cdot 96}{4 \cdot 96} = \dfrac{96}{384}$

The missing term is 96.

89. $\dfrac{5}{6} = \dfrac{1{,}580}{?}$. Because $1{,}580 \div 5 = 316$,

$\dfrac{5}{6} = \dfrac{5 \cdot 316}{6 \cdot 316} = \dfrac{1{,}580}{1{,}896}$

The missing term is 1,896.

Review Problems

91. $3 + 4 \cdot 5 = 3 + 20$
$= 23$

93. $5 \cdot 2^4 - 3 \cdot 4^2 = 5 \cdot 16 - 3 \cdot 16$
$= 80 - 48$
$= 32$

95. $4\cdot 3+2(5-3)=4\cdot 3+2\cdot 5-2\cdot 3$

$$=12+10-6$$
$$=22-6$$
$$=16$$

97. $18+12\div 4-3$ *divide before adding and subtracting*

$$=18+3-3$$
$$=21-3$$
$$=18$$

Improving Your Quantitative Literacy

99. True. For example, $1<2<3$, and $\frac{1}{4}<\frac{2}{4}<\frac{3}{4}$.

101. a.

Twins	119,648
Triplets	6,742
Quadruplets	506
+ Quintuplets and up	+ 77
	126,973

b. $\frac{9}{10}$ of total multiple births

$$=\frac{9}{10}\cdot 126{,}973$$
$$=\frac{9\cdot 126{,}973}{10}$$
$$=\frac{1{,}142{,}757}{10}\approx 114{,}276$$

Yes. There were 119,648 sets of twins, which is greater than 114,276.

Section 3.2

1. 11 is prime, because the only factors of 11 are 1 and 11.

3. 105 is composite. Some factors of 105 are: 3,35,5,21,7,15.

5. 81 is composite. Some factors of 81 are: 3,9,27.

7. 13 is prime, because the only factors of 13 are 1 and 13.

9. $$\begin{array}{l} 12 = 2 \quad \cdot \quad 6 \\ \qquad\ \ \downarrow \qquad \downarrow \\ \quad = 2 \quad \cdot \quad 2 \cdot 3 = 2^2 \cdot 3 \end{array}$$

11. $$\begin{array}{l} 81 = \quad 9 \quad \cdot \quad 9 \\ \qquad\quad \downarrow \qquad \downarrow \\ \quad = 3 \cdot 3 \quad \cdot \quad 3 \cdot 3 = 3^4 \end{array}$$

13. $215 = 5 \cdot 43$

15. $15 = 3 \cdot 5$

17. $\frac{5}{10} = \frac{1 \cdot \cancel{5}}{2 \cdot \cancel{5}} = \frac{1}{2}$

19. $\frac{4}{6} = \frac{\cancel{2} \cdot 2}{\cancel{2} \cdot 3} = \frac{2}{3}$

21. $\frac{8x}{10x} = \frac{\cancel{2} \cdot 4 \cdot \cancel{x}}{\cancel{2} \cdot 5 \cdot \cancel{x}} = \frac{4}{5}$

23. $\frac{36}{20} = \frac{\cancel{2} \cdot \cancel{2} \cdot 3 \cdot 3}{\cancel{2} \cdot \cancel{2} \cdot 5} = \frac{9}{5}$

25. $\frac{42}{66} = \frac{\cancel{2} \cdot \cancel{3} \cdot 7}{\cancel{2} \cdot \cancel{3} \cdot 11} = \frac{7}{11}$

27. $\frac{24xy}{40y} = \frac{\cancel{2} \cdot \cancel{2} \cdot \cancel{2} \cdot 3 \cdot x \cdot \cancel{y}}{\cancel{2} \cdot \cancel{2} \cdot \cancel{2} \cdot 5 \cdot \cancel{y}} = \frac{3x}{5}$

29. $\frac{14}{98} = \frac{\cancel{2} \cdot \cancel{7}}{\cancel{2} \cdot \cancel{7} \cdot 7} = \frac{1}{7}$

31. $\frac{70}{90} = \frac{\cancel{2} \cdot \cancel{5} \cdot 7}{\cancel{2} \cdot 3 \cdot 3 \cdot \cancel{5}} = \frac{7}{9}$

33. $\frac{42x^2}{30x} = \frac{\cancel{2} \cdot \cancel{3} \cdot 7 \cdot \cancel{x} \cdot x}{\cancel{2} \cdot \cancel{3} \cdot 5 \cdot \cancel{x}} = \frac{7x}{5}$

35. $\frac{18a^2b}{90ab} = \frac{\cancel{2} \cdot \cancel{3} \cdot \cancel{3} \cdot \cancel{a} \cdot a \cdot \cancel{b}}{\cancel{2} \cdot \cancel{3} \cdot \cancel{3} \cdot 5 \cdot \cancel{a} \cdot \cancel{b}} = \frac{a}{5}$

37. $\frac{110}{70} = \frac{\cancel{2} \cdot \cancel{5} \cdot 11}{\cancel{2} \cdot \cancel{5} \cdot 7} = \frac{11}{7}$

39. $\frac{180xyz}{108xy} = \frac{\cancel{2} \cdot \cancel{2} \cdot \cancel{3} \cdot \cancel{3} \cdot 5 \cdot \cancel{x} \cdot \cancel{y} \cdot z}{\cancel{2} \cdot \cancel{2} \cdot \cancel{3} \cdot \cancel{3} \cdot 3 \cdot \cancel{x} \cdot \cancel{y}} = \frac{5z}{3}$

41. $\frac{96x^2y}{108xy^2} = \frac{\cancel{2} \cdot \cancel{2} \cdot 2 \cdot 2 \cdot 2 \cdot \cancel{3} \cdot \cancel{x} \cdot x \cdot \cancel{y}}{\cancel{2} \cdot \cancel{2} \cdot \cancel{3} \cdot 3 \cdot 3 \cdot \cancel{x} \cdot \cancel{y} \cdot y} = \frac{8x}{9y}$

43. $\frac{126}{165} = \frac{2 \cdot \cancel{3} \cdot 3 \cdot 7}{\cancel{3} \cdot 5 \cdot 11} = \frac{42}{55}$

45. $\frac{102a^2bc^3}{114ab^2c^2}$

$= \frac{\cancel{2} \cdot \cancel{3} \cdot 17 \cdot \cancel{a} \cdot a \cdot \cancel{b} \cdot \cancel{c} \cdot \cancel{c} \cdot c}{\cancel{2} \cdot \cancel{3} \cdot 19 \cdot \cancel{a} \cdot \cancel{b} \cdot b \cdot \cancel{c} \cdot \cancel{c}} = \frac{17ac}{19b}$

47. $\frac{294}{693} = \frac{2 \cdot \cancel{3} \cdot \cancel{7} \cdot 7}{\cancel{3} \cdot 3 \cdot \cancel{7} \cdot 11} = \frac{14}{33}$

49. a. $\frac{6}{51}=\frac{2\cdot\cancel{3}}{\cancel{3}\cdot17}=\frac{2}{17}$

b. $\frac{6}{52}=\frac{\cancel{2}\cdot3}{\cancel{2}\cdot2\cdot13}=\frac{3}{26}$

c. $\frac{6}{54}=\frac{\cancel{2}\cdot\cancel{3}}{\cancel{2}\cdot\cancel{3}\cdot3\cdot3}=\frac{1}{9}$

d. $\frac{6}{56}=\frac{\cancel{2}\cdot3}{\cancel{2}\cdot2\cdot2\cdot7}=\frac{3}{28}$

e. $\frac{6}{57}=\frac{2\cdot\cancel{3}}{\cancel{3}\cdot19}=\frac{2}{19}$

51. a. $\frac{2}{90}=\frac{2\div2}{90\div2}=\frac{1}{45}$

b. $\frac{3}{90}=\frac{3\div3}{90\div3}=\frac{1}{30}$

c. $\frac{5}{90}=\frac{5\div5}{90\div5}=\frac{1}{18}$

d. $\frac{6}{90}=\frac{6\div6}{90\div6}=\frac{1}{15}$

e. $\frac{9}{90}=\frac{9\div9}{90\div9}=\frac{1}{10}$

53. a. $\frac{5}{15}=\frac{\cancel{5}}{3\cdot\cancel{5}}=\frac{1}{3}$

Note: The slashes indicate *division* by 5. $5\div5=1$, not 0.

b. $\frac{5}{6}=\frac{5}{2\cdot3}=\frac{5}{6}$

c. $\frac{6}{30}=\frac{\cancel{2}\cdot\cancel{3}}{\cancel{2}\cdot\cancel{3}\cdot5}=\frac{1}{5}$

Note: The slashes indicate *division* by 2 and 3. $2\div2=1, 3\div3=1$, not 0.

55. $\frac{6}{8}=\frac{6\div2}{8\div2}=\frac{3}{4}$

$\frac{15}{20}=\frac{15\div5}{20\div5}=\frac{3}{4}$

$\frac{9}{16}=\frac{3\cdot3}{4\cdot4}=\frac{9}{16}$

$\frac{21}{28}=\frac{21\div7}{28\div7}=\frac{3}{4}$

$\frac{9}{16}\neq\frac{3}{4}$

57. First reduce all fractions to lowest terms:

$\frac{1}{2}$ is reduced $\qquad \frac{2}{4}=\frac{2\div2}{4\div2}=\frac{1}{2}$

$\frac{4}{8}=\frac{4\div4}{8\div4}=\frac{1}{2} \qquad \frac{8}{16}=\frac{8\div8}{16\div8}=\frac{1}{2}$

$\frac{1}{2}=\frac{2}{4}=\frac{4}{8}=\frac{8}{16}$

59. First reduce all fractions to lowest terms:

$\frac{5}{4}$ is reduced $\qquad \frac{10}{8}=\frac{10\div2}{8\div2}=\frac{5}{4}$

$\frac{20}{16}=\frac{20\div4}{16\div4}=\frac{5}{4}$

$\frac{5}{4}=\frac{10}{8}=\frac{20}{16}$

Applying the Concepts

61.
$$\frac{\text{Amount spent on food}}{\text{Total income}}=\frac{\$600}{\$2,400}=\frac{600}{2,400}=\frac{600\div600}{2,400\div600}=\frac{1}{4}$$

63. a. Total babies born: 4,021,726 ;

$$\frac{1}{2} \text{ of total babies born} = \frac{1}{2} \cdot 4{,}021{,}726 = 2{,}010{,}863$$

The number of babies born to 20-29 year olds is more than half of all babies born:
$2{,}082{,}487 > 2{,}010{,}863$

b. Total babies born: 4,021,726 ;

$$\frac{1}{3} \text{ of total babies born} = \frac{1}{3} \cdot 4{,}021{,}726 = \frac{4{,}021{,}726}{3} \approx 1{,}340{,}575$$

The number of babies born to 30-39 year olds is more than one-third of all babies born:
$1{,}405{,}146 > 1{,}340{,}575$

65. $\frac{\text{Number who didn't get an A}}{\text{Total number taking the test}} = \frac{33-11}{33}$;

$$= \frac{22}{33} = \frac{22 \div 11}{33 \div 11} = \frac{2}{3}$$

67. $\frac{\text{Fat calories}}{\text{Total calories}} = \frac{70}{210} = \frac{70 \div 70}{210 \div 70} = \frac{1}{3}$

69. $\frac{\text{Saturated fat}}{\text{Total fat}} = \frac{1\text{ g}}{8\text{ g}} = \frac{1}{8}$

71. $\frac{\text{Sugar carbohydrates}}{\text{Total carbohydrates}} = \frac{12\text{g}}{32\text{g}} = \frac{12}{32}$;

$$\frac{12}{32} = \frac{12 \div 4}{32 \div 4} = \frac{3}{8}$$

Estimating

73.

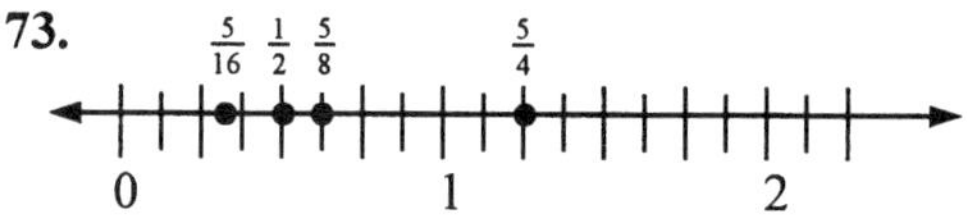

$\frac{5}{8}$ is closest to $\frac{1}{2}$.

75.

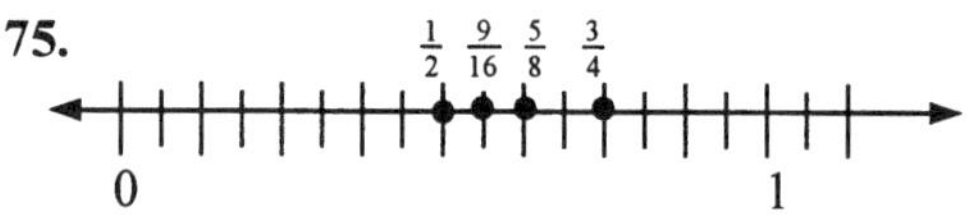

$\frac{9}{16}$ is closest to $\frac{1}{2}$.

Calculator Problems

77. Because $574 \div 7 = 82$,

$$\frac{5}{7} = \frac{5 \cdot 82}{7 \cdot 82} = \frac{410}{574}.$$

The missing term is 410.

79. Because $728 \div 8 = 91$,

$$\frac{8}{9} = \frac{8 \cdot 91}{9 \cdot 91} = \frac{728}{819}$$

The missing term is 819.

81. Because $4{,}869 \div 9 = 541$,

$$\frac{3{,}787}{4{,}869} = \frac{3{,}787 \div 541}{4{,}869 \div 541} = \frac{7}{9}.$$

The missing term is 7.

83. Because $1{,}965 \div 15 = 131$,

$$\frac{262}{1{,}965} = \frac{262 \div 131}{1{,}965 \div 131} = \frac{2}{15}.$$

The missing term is 2.

Review Problems

85.

$$6(4+8) = 6 \cdot 4 + 6 \cdot 8 = 24 + 48 = 72$$

87. $9(3+2)=9\cdot3+9\cdot2$
$=27+18$
$=45$

89. $5(x+4)=5\cdot x+5\cdot4$
$=5x+20$

91. $3(4x+5)=3\cdot4x+3\cdot5$
$=12x+15$

Improving Your Quantitative Literacy

93. $\frac{\text{U.S population}}{\text{World population}}=\frac{293{,}131{,}160}{6{,}363{,}296{,}6399}$;

$\frac{293{,}131{,}160}{6{,}363{,}296{,}6399}$ rounds to

$\frac{300{,}000{,}000}{6{,}000{,}000{,}000}=\frac{3}{60}=\frac{1}{20}$

Section 3.3

1. $\frac{2}{3}\cdot\frac{4}{5}=\frac{2\cdot 4}{3\cdot 5}=\frac{8}{15}$

3. $\frac{1}{2}\cdot\frac{7}{4}=\frac{1\cdot 7}{2\cdot 4}=\frac{7}{8}$

5. $-\frac{5}{3}\cdot\frac{3}{5}=-\frac{\cancel{5}\cdot\cancel{3}}{\cancel{3}\cdot\cancel{5}}=-\frac{1}{1}=-1$

7. $\frac{3}{4}\cdot 9=\frac{3}{4}\cdot\frac{9}{1}=\frac{3\cdot 9}{4\cdot 1}=\frac{27}{4}$

9. $-\frac{6}{7}\left(-\frac{7}{6}\right)=\frac{\cancel{6}\cdot\cancel{7}}{\cancel{7}\cdot\cancel{6}}=\frac{1}{1}=1$

11. $\frac{1}{2}\cdot\frac{1}{3}\cdot\frac{1}{4}=\frac{1\cdot 1\cdot 1}{2\cdot 3\cdot 4}=\frac{1}{24}$

13. $\frac{2}{5}\cdot\frac{3}{5}\cdot\frac{4}{5}=\frac{2\cdot 3\cdot 4}{5\cdot 5\cdot 5}=\frac{24}{125}$

15. $\frac{x}{y}\cdot\frac{y}{z}\cdot\frac{z}{x}=\frac{\cancel{x}\cdot\cancel{y}\cdot\cancel{z}}{\cancel{y}\cdot\cancel{z}\cdot\cancel{x}}=\frac{1}{1}=1$

17. The table is as follows:

x	y	xy
$\frac{1}{2}$	$\frac{2}{3}$	$\frac{1}{2}\cdot\frac{2}{3}=\frac{1\cdot 2}{2\cdot 3}=\frac{1}{3}$
$\frac{2}{3}$	$\frac{3}{4}$	$\frac{2}{3}\cdot\frac{3}{4}=\frac{2\cdot 3}{3\cdot 4}=\frac{2\cdot 3}{3\cdot(2\cdot 2)}=\frac{1}{2}$
$\frac{3}{4}$	$\frac{4}{5}$	$\frac{3}{4}\cdot\frac{4}{5}=\frac{3\cdot 4}{4\cdot 5}=\frac{3}{5}$
$\frac{5}{a}$	$\frac{a}{6}$	$\frac{5}{a}\cdot\frac{a}{6}=\frac{5\cdot a}{a\cdot 6}=\frac{5}{6}$

19. The table is as follows:

x	y	xy
$\frac{1}{2}$	30	$\frac{1}{2}\cdot\frac{30}{1}=\frac{1\cdot 30}{2\cdot 1}=\frac{2\cdot 3\cdot 5}{2}=\frac{3\cdot 5}{1}$ $=15$
$\frac{1}{5}$	30	$\frac{1}{5}\cdot\frac{30}{1}=\frac{1\cdot 30}{5\cdot 1}=\frac{2\cdot 3\cdot 5}{5}=\frac{2\cdot 3}{1}$ $=6$
$\frac{1}{6}$	30	$\frac{1}{6}\cdot\frac{30}{1}=\frac{1\cdot 30}{6\cdot 1}=\frac{2\cdot 3\cdot 5}{2\cdot 3}=\frac{5}{1}$ $=5$
$\frac{1}{15}$	30	$\frac{1}{15}\cdot\frac{30}{1}=\frac{1\cdot 30}{15\cdot 1}=\frac{2\cdot 3\cdot 5}{3\cdot 5}=\frac{2}{1}$ $=2$

21.
$$\begin{aligned}\frac{9}{20}\cdot\frac{4}{3}&=\frac{9\cdot 4}{20\cdot 3}\\&=\frac{(\cancel{3}\cdot 3)(\cancel{2}\cdot\cancel{2})}{(\cancel{2}\cdot\cancel{2}\cdot 5)(\cancel{3})}\\&=\frac{3}{5}\end{aligned}$$

23.
$$\begin{aligned}\frac{3}{4}\cdot 12&=\frac{3}{4}\cdot\frac{12}{1}\\&=\frac{3\cdot 12}{4\cdot 1}\\&=\frac{3(\cancel{2}\cdot\cancel{2}\cdot 3)}{\cancel{2}\cdot\cancel{2}}\\&=9\end{aligned}$$

25.
$$\begin{aligned}-\frac{1}{3}(-3)&=\frac{1}{3}\cdot\frac{3}{1}\\&=\frac{1\cdot\cancel{3}}{\cancel{3}\cdot 1}\\&=1\end{aligned}$$

27. $$\frac{2}{5}\cdot 20=\frac{2}{5}\cdot\frac{20}{1}=\frac{2\cdot(2\cdot 2\cdot \cancel{5})}{\cancel{5}\cdot 1}=\frac{8}{1}=8$$

29. $$\frac{72}{35}\cdot\frac{55}{108}\cdot\frac{7}{110}=\frac{(\cancel{2}\cdot\cancel{2}\cdot\cancel{2}\cdot\cancel{3}\cdot\cancel{3})\cdot(\cancel{5}\cdot\cancel{11})\cdot\cancel{7}}{(\cancel{5}\cdot\cancel{7})(\cancel{2}\cdot\cancel{2}\cdot\cancel{3}\cdot\cancel{3}\cdot 3)(\cancel{2}\cdot 5\cdot\cancel{11})}=\frac{1}{15}$$

31. $$\frac{a^2b}{c}\cdot\frac{c^3}{ab^2}=\frac{a\cdot a\cdot b\cdot c\cdot c\cdot c}{c\cdot a\cdot b\cdot b}=\frac{\cancel{a}\cdot a\cdot\cancel{b}\cdot\cancel{c}\cdot c\cdot c}{\cancel{c}\cdot\cancel{a}\cdot\cancel{b}\cdot b}=\frac{ac^2}{b}$$

33. $$\frac{1}{2}(2x)=\left(\frac{1}{2}\cdot 2\right)x=1\cdot x=x$$

35. $$\frac{1}{5}(5y)=\left(\frac{1}{5}\cdot 5\right)y=1\cdot y=y$$

37. $$\frac{1}{8}(8a)=\left(\frac{1}{8}\cdot 8\right)a=1\cdot a=a$$

39. $$\left(\frac{2}{3}\right)^2=\frac{2}{3}\cdot\frac{2}{3}=\frac{2\cdot 2}{3\cdot 3}=\frac{4}{9}$$

41. $$\left(\frac{3}{4}\right)^2=\frac{3}{4}\cdot\frac{3}{4}=\frac{3\cdot 3}{4\cdot 4}=\frac{9}{16}$$

43. $$\left(-\frac{1}{2}\right)^2=\left(-\frac{1}{2}\right)\left(-\frac{1}{2}\right)=\frac{1\cdot 1}{2\cdot 2}=\frac{1}{4}$$

45. $$\left(-\frac{2}{3}\right)^3=\left(-\frac{2}{3}\right)\left(-\frac{2}{3}\right)\left(-\frac{2}{3}\right)=-\frac{2\cdot 2\cdot 2}{3\cdot 3\cdot 3}=-\frac{8}{27}$$

47. $$\left(\frac{3}{4}\right)^2\cdot\left(\frac{8}{9}\right)=\frac{3}{4}\cdot\frac{3}{4}\cdot\frac{8}{9}=\frac{3\cdot 3\cdot 8}{4\cdot 4\cdot 9}=\frac{\cancel{3}\cdot\cancel{3}\cdot\cancel{2}\cdot\cancel{2}\cdot\cancel{2}}{(\cancel{2}\cdot\cancel{2})\cdot(\cancel{2}\cdot 2)\cdot(\cancel{3}\cdot\cancel{3})}=\frac{1}{2}$$

49. $\left(\frac{1}{2}\right)^2\left(\frac{3}{5}\right)^2 = \frac{1}{2}\cdot\frac{1}{2}\cdot\frac{3}{5}\cdot\frac{3}{5}$

$$= \frac{1\cdot 1\cdot 3\cdot 3}{2\cdot 2\cdot 5\cdot 5}$$

$$= \frac{9}{100}$$

51. $\left(\frac{1}{2}\right)^2\cdot 8 + \left(\frac{1}{3}\right)^2\cdot 9$

$$= \frac{1}{2}\cdot\frac{1}{2}\cdot\frac{8}{1} + \frac{1}{3}\cdot\frac{1}{3}\cdot\frac{9}{1}$$

$$= \frac{\cancel{2}\cdot\cancel{2}\cdot 2}{\cancel{2}\cdot\cancel{2}} + \frac{\cancel{3}\cdot\cancel{3}}{\cancel{3}\cdot\cancel{3}}$$

$$= 2 + 1$$

$$= 3$$

53. $\frac{1}{2}(12x+6) = \frac{1}{2}(12x) + \frac{1}{2}(6)$

$$= \frac{1}{2}\cdot\frac{12}{1}x + \frac{1}{2}\cdot\frac{6}{1}$$

$$= \frac{12}{2}x + \frac{6}{2}$$

$$= 6x + 3$$

55. $\frac{2}{3}(-3x-6) = \frac{2}{3}(-3x) - \frac{2}{3}(6)$

$$= \left(-\frac{2}{\cancel{3}}\cdot\frac{\cancel{3}}{1}\right)x - \frac{2\cdot 2\cdot\cancel{3}}{\cancel{3}}$$

$$= -2x - 4$$

57. $3\left(\frac{5}{6}a + \frac{4}{9}\right) = 3\left(\frac{5}{6}a\right) + 3\left(\frac{4}{9}\right)$

$$= \left(\frac{\cancel{3}\cdot 5}{\cancel{3}\cdot 2}\right)a + \frac{\cancel{3}\cdot 2\cdot 2}{\cancel{3}\cdot 3}$$

$$= \frac{5}{2}a + \frac{4}{3}$$

59. $-3\left(\frac{2}{3}y + \frac{5}{6}\right) = -3\left(\frac{2}{3}y\right) - 3\left(\frac{5}{6}\right)$

$$= \left(-\frac{\cancel{3}\cdot 2}{1\cdot\cancel{3}}\right)y - \frac{\cancel{3}\cdot 5}{\cancel{3}\cdot 2}$$

$$= -2y - \frac{5}{2}$$

61. $\frac{4}{5}\left(\frac{5}{6}x - 10\right) = \frac{4}{5}\left(\frac{5}{6}x\right) - \frac{4}{5}\left(\frac{10}{1}\right)$

$$= \left(\frac{\cancel{2}\cdot 2\cdot\cancel{5}}{\cancel{5}\cdot\cancel{2}\cdot 3}\right)x - \frac{2\cdot 2\cdot 2\cdot\cancel{5}}{\cancel{5}\cdot 1}$$

$$= \frac{2}{3}x - 8$$

63. $\frac{3}{8}$ of $64 = \frac{3}{8}\cdot 64$

$$= \frac{3}{8}\cdot\frac{64}{1}$$

$$= \frac{3\cdot 64}{8\cdot 1}$$

$$= \frac{3\cdot\left(\cancel{2}\cdot\cancel{2}\cdot\cancel{2}\cdot 2\cdot 2\cdot 2\right)}{\left(\cancel{2}\cdot\cancel{2}\cdot\cancel{2}\right)\cdot 1}$$

$$= \frac{24}{1}$$

$$= 24$$

65. $\frac{1}{3}(8+4) = \frac{1}{3}(12)$

$$= \frac{1}{3}\cdot\frac{12}{1}$$

$$= \frac{1\cdot 12}{3\cdot 1}$$

$$= \frac{1\cdot\left(2\cdot 2\cdot\cancel{3}\right)}{\cancel{3}\cdot 1}$$

$$= \frac{4}{1}$$

$$= 4$$

67. $\frac{1}{2}$ of $\frac{3}{4}$ of $24 = \frac{1}{2}\cdot\frac{3}{4}\cdot\frac{24}{1}$

$$= \frac{1\cdot 3\cdot 24}{2\cdot 4\cdot 1}$$

$$= \frac{1\cdot 3\cdot(\not{2}\cdot\not{2}\cdot\not{2}\cdot 3)}{\not{2}\cdot(\not{2}\cdot\not{2})\cdot 1}$$

$$= \frac{9}{1}$$

$$= 9$$

69. $\frac{1}{2}\cdot\frac{3}{5} = \frac{1\cdot 3}{2\cdot 5} = \frac{3}{10}$

Note: The mistake in the original problem was that the numerators were added, not multiplied.

71. a. The table is as follows:

x	x^2
1	$1^2 = 1\cdot 1 = 1$
2	$2^2 = 2\cdot 2 = 4$
3	$3^2 = 3\cdot 3 = 9$
4	$4^2 = 4\cdot 4 = 16$
5	$5^2 = 5\cdot 5 = 25$
6	$6^2 = 6\cdot 6 = 36$
7	$7^2 = 7\cdot 7 = 49$
8	$8^2 = 8\cdot 8 = 64$

b. For numbers larger than 1, the square of a number is *larger* than the number.

73. $A = \frac{1}{2}bh$

$$= \frac{1}{2}\cdot 19\text{ in}\cdot 14\text{ in}$$

$$= \frac{1\cdot 19\cdot 14}{2}\text{in}^2$$

$$= \frac{1\cdot 19\cdot \not{2}\cdot 7}{\not{2}}\text{in}^2$$

$$= 133\text{ in}^2$$

75. $A = \frac{1}{2}bh$

$$= \frac{1}{2}\cdot\frac{4}{3}\text{ft}\cdot\frac{2}{3}\text{ft}$$

$$= \frac{1\cdot 4\cdot\not{2}}{\not{2}\cdot 3\cdot 3}\text{ft}^2$$

$$= \frac{4}{9}\text{ft}^2$$

77. $A = \frac{1}{2}bh$

$$= \frac{1}{2}\cdot 2\text{yd}\cdot 3\text{yd}$$

$$= \frac{1\cdot\not{2}\cdot 3}{\not{2}}\text{yd}^2$$

$$= 3\text{ yd}^2$$

79.

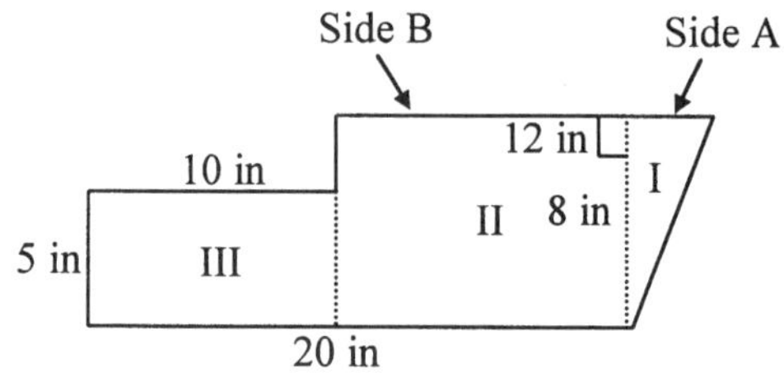

Separate the figure into 3 parts.
Find the length of the missing sides:

Side A = 10 in + 12 in – 20 in = 2 in.

Side B = 12 in – 2 in = 10 in

$$\text{Area I} = \frac{1}{2}bh = \frac{1}{2}\cdot 2\text{ in}\cdot 8\text{ in} = 1\text{ in}\cdot 8\text{ in} = 8\text{ in}^2$$

$$\text{Area II} = lw = 10\text{ in}\cdot 8\text{ in} = 80\text{ in}^2$$

$$\text{Area III} = lw = 10\text{ in}\cdot 5\text{ in} = 50\text{ in}^2$$

$$\text{Area} = \text{Area I} + \text{Area II} + \text{Area III} = 8\text{ in}^2 + 80\text{ in}^2 + 50\text{ in}^2 = 138\text{ in}^2$$

Applying the Concepts

81. $$\frac{11}{50}\text{ of }15{,}800 = \frac{11}{50}\cdot 15{,}800 = \frac{11}{50}\cdot\frac{15{,}800}{1} = \frac{11\cdot(\cancel{2}\cdot 2\cdot 2\cdot\cancel{5}\cdot\cancel{5}\cdot 79)}{\cancel{2}\cdot\cancel{5}\cdot\cancel{5}} = 3{,}476$$

3,476 students.

83. $\frac{6}{5}$ of volume of Rally 105

$$= \frac{6}{5}\text{ of }105{,}400 = \frac{6}{5}\cdot\frac{105{,}400}{1} = \frac{6\cdot\cancel{5}\cdot 21{,}080}{\cancel{5}\cdot 1}$$

$= 126{,}480 \approx 126{,}500$ (to the nearest hundred)

The volume of the Rally 126 is 126,500 ft^3.

85. First, find the number of children who were wearing helmets:

$\frac{2}{5}$ of children

$$= \frac{2}{5}\text{ of }8{,}159 = \frac{2\cdot 8159}{5} = \frac{16{,}318}{5} \approx 3{,}264 \text{ wore helmets};$$

For those who wore helmets, find the number wearing them correctly:

$\frac{13}{20}$ of helmeted

$$= \frac{13}{20}\cdot 3{,}264 = \frac{13\cdot\cancel{4}\cdot 816}{\cancel{4}\cdot 5} = \frac{10{,}608}{5} \approx 2{,}122$$

2,122 children wore helmets correctly.

Estimating

87. $\frac{11}{5}\cdot\frac{19}{20}$ rounds to $\frac{10}{5}\cdot\frac{20}{20}=2\cdot 1=2$

89. $\frac{16}{5}\cdot\frac{23}{24}$ rounds to $\frac{15}{5}\cdot\frac{24}{24}=3\cdot 1=3$

91. $\frac{1}{8}\cdot\frac{15}{32}$ rounds to $\frac{0}{8}\cdot\frac{16}{32}=0\cdot\frac{1}{2}=0$

Review Problems

93. $6\div 2=3$

95. $8\div 4=2$

97. $15\div 3=5$

99. $18\div 6=3$

Extending the Concepts

101. $1,\frac{1}{3},\frac{1}{9},\ldots$

The first number in the sequence is 1. Note that $1\cdot\frac{1}{3}=\frac{1}{3}$ and $\frac{1}{3}\cdot\frac{1}{3}=\frac{1}{9}$, i.e. each number after 1 comes from multiplying the previous number by $\frac{1}{3}$. The next number in the sequence is $\frac{1}{9}\cdot\frac{1}{3}=\frac{1}{27}$.

103. $\frac{3}{2},1,\frac{2}{3},\frac{4}{9},\ldots$

The first number in the sequence is $\frac{3}{2}$.

Note that $\frac{3}{2}\cdot\frac{2}{3}=1$, $1\cdot\frac{2}{3}=\frac{2}{3}$ and $\frac{2}{3}\cdot\frac{2}{3}=\frac{4}{9}$,

i.e. each number after $\frac{3}{2}$ comes from

multiplying the previous number by $\frac{2}{3}$.

The next number in the sequence is

$\frac{4}{9}\cdot\frac{2}{3}=\frac{8}{27}$.

Improving your Quantitative Literacy

105. $\frac{3}{4}$ of 55 companies [

$=\frac{3}{4}\cdot\frac{55}{1}$

$=\frac{165}{4}$

$=41\text{ R}1$

Since ¾ of 55 is more than 41, at least 42 of the companies will cut spending.

Section 3.4

1. $\frac{3}{4} \div \frac{1}{5} = \frac{3}{4} \cdot \frac{5}{1}$
$= \frac{3 \cdot 5}{4 \cdot 1}$
$= \frac{15}{4}$

3. $-\frac{2}{3} \div \frac{1}{2} = -\frac{2}{3} \cdot \frac{2}{1}$
$= -\frac{4}{3}$

5. $6 \div \left(-\frac{2}{3}\right) = 6 \cdot \left(-\frac{3}{2}\right)$
$= -\frac{6 \cdot 3}{2}$
$= -\frac{3 \cdot \cancel{2} \cdot 3}{\cancel{2}}$
$= -9$

7. $20 \div \frac{1}{10} = 20 \cdot \frac{10}{1}$
$= \frac{20 \cdot 10}{1}$
$= 200$

9. $\frac{3}{4} \div (-2) = \frac{3}{4} \cdot \left(-\frac{1}{2}\right)$
$= -\frac{3 \cdot 1}{4 \cdot 2}$
$= -\frac{3}{8}$

11. $\frac{7}{8} \div \frac{7}{8} = \frac{7}{8} \cdot \frac{8}{7}$
$= \frac{\cancel{7} \cdot \cancel{8}}{\cancel{8} \cdot \cancel{7}}$
$= 1$

13. $\frac{7}{8} \div \frac{8}{7} = \frac{7}{8} \cdot \frac{7}{8}$
$= \frac{7 \cdot 7}{8 \cdot 8}$
$= \frac{49}{64}$

15. $\frac{9}{16} \div \left(-\frac{3}{4}\right) = \frac{9}{16} \cdot \left(-\frac{4}{3}\right)$
$= -\frac{9 \cdot 4}{16 \cdot 3}$
$= -\frac{\cancel{3} \cdot 3 \cdot \cancel{4}}{\cancel{4} \cdot 4 \cdot \cancel{3}}$
$= -\frac{3}{4}$

17. $\frac{25}{46} \div \frac{40}{69} = \frac{25}{46} \cdot \frac{69}{40}$
$= \frac{\cancel{5} \cdot 5 \cdot \cancel{23} \cdot 3}{\cancel{23} \cdot 2 \cdot \cancel{5} \cdot 8}$
$= \frac{15}{16}$

19. $\frac{13}{28} \div \frac{39}{14} = \frac{13}{28} \cdot \frac{14}{39}$
$= \frac{\cancel{13} \cdot \cancel{14}}{\cancel{14} \cdot 2 \cdot \cancel{13} \cdot 3}$
$= \frac{1}{6}$

21. $\frac{27}{196} \div \frac{9}{392} = \frac{27}{196} \cdot \frac{392}{9}$
$= \frac{3 \cdot \cancel{9} \cdot \cancel{196} \cdot 2}{\cancel{196} \cdot \cancel{9}}$
$= 6$

23. $\frac{x^2}{y^3} \div \frac{x}{y^2} = \frac{x^2}{y^3} \cdot \frac{y^2}{x}$

$= \frac{\cancel{x} \cdot x \cdot \cancel{y} \cdot \cancel{y}}{\cancel{y} \cdot \cancel{y} \cdot y \cdot \cancel{x}}$

$= \frac{x}{y}$

25. $\frac{ab^2}{c} \div \frac{b}{c} = \frac{ab^2}{c} \cdot \frac{c}{b}$

$= \frac{a \cdot \cancel{b} \cdot b \cdot \cancel{c}}{\cancel{c} \cdot \cancel{b}}$

$= ab$

27. $\frac{10a^2}{3b} \div \frac{5a}{6b} = \frac{10a^2}{3b} \cdot \frac{6b}{5a}$

$= \frac{2 \cdot \cancel{5} \cdot \cancel{a} \cdot a \cdot \cancel{3} \cdot 2 \cdot \cancel{b}}{\cancel{3} \cdot \cancel{b} \cdot \cancel{5} \cdot \cancel{a}}$

$= 4a$

29. $10 \div \left(\frac{1}{2}\right)^2 = 10 \div \frac{1}{4}$

$= 10 \cdot \frac{4}{1}$

$= 40$

31. $\frac{18}{35} \div \left(\frac{6}{7}\right)^2 = \frac{18}{35} \div \frac{36}{49}$

$= \frac{18}{35} \cdot \frac{49}{36}$

$= \frac{\cancel{18} \cdot \cancel{7} \cdot 7}{\cancel{7} \cdot 5 \cdot \cancel{18} \cdot 2}$

$= \frac{7}{10}$

33. $\frac{4}{5} \div \frac{1}{10} + 5 = \frac{4}{5} \cdot \frac{10}{1} + 5$

$= 8 + 5$

$= 13$

35. $10 + \frac{11}{12} \div \frac{11}{24} = 10 + \frac{11}{12} \cdot \frac{24}{11}$

$= 10 + 2$

$= 12$

37. $24 \div \left(\frac{2}{5}\right)^2 + 25 \div \left(\frac{5}{6}\right)^2 = 24 \div \frac{4}{25} + 25 \div \frac{25}{36}$

$= 24 \cdot \frac{25}{4} + 25 \cdot \frac{36}{25}$

$= 6 \cdot 25 + 1 \cdot 36$

$= 150 + 36$

$= 186$

39. $100 \div \left(\frac{5}{7}\right)^2 + 200 \div \left(\frac{2}{3}\right)^2$

$= 100 \div \frac{25}{49} + 200 \div \frac{4}{9}$

$= 100 \cdot \frac{49}{25} + 200 \cdot \frac{9}{4}$

$= 4 \cdot 49 + 50 \cdot 9$

$= 196 + 450$

$= 646$

41. $\frac{3}{8} \div \frac{5}{8} = \frac{3}{8} \cdot \frac{8}{5}$

$= \frac{3}{5}$

43. $\left(18 \div \frac{3}{5}\right) + 10 = 18 \cdot \frac{5}{3} + 10$

$= 6 \cdot 5 + 10$

$= 30 + 10$

$= 40$

The number is 40.

45. $3 \div \frac{1}{5} = 3 \cdot \frac{5}{1} = 3 \cdot 5$

Applying the Concepts

47. Total yds ÷ yds per blanket = blankets

$$12 \div \frac{6}{7} = 12 \cdot \frac{7}{6}$$
$$= 2 \cdot 7$$
$$= 14$$

14 blankets can be made.

49. Total lbs ÷ lbs per bag = bags

$$12 \div \frac{1}{4} = 12 \cdot \frac{4}{1}$$
$$= 48$$

48 bags can be filled.

51. Total tsp ÷ tsp per spoon = spoons

$$\frac{3}{4} \div \frac{1}{8} = \frac{3}{4} \cdot \frac{8}{1}$$
$$= 3 \cdot 2$$
$$= 6$$

He will need to fill 6 spoons.

53. $\dfrac{\text{Number of females}}{\text{Number of students}} = \dfrac{14}{32}$

$$\frac{14}{32} = \frac{2 \cdot 7}{2 \cdot 16} = \frac{7}{16}$$

55. From problem #53, $\frac{7}{16}$ of the students are female.

$$\frac{7}{16} \text{ of total students} = \frac{7}{16} \cdot 4{,}064$$
$$= \frac{7 \cdot \cancel{16} \cdot 254}{\cancel{16}}$$
$$= 1{,}778$$

1,778 of the students are female.

57. Total pints ÷ pints per carton = cartons

$$14 \div \frac{1}{2} = 14 \cdot \frac{2}{1}$$
$$= 28$$

28 cartons can be filled.

Estimating

59. $\frac{11}{5} \div \frac{19}{20}$ rounds to $\frac{10}{5} \div \frac{20}{20} = 2 \div 1 = 2$

61. $\frac{15}{8} \div \frac{23}{24}$ rounds to $\frac{16}{8} \div \frac{24}{24} = 2 \div 1 = 2$

63. $\frac{1}{2} \div 40 = \frac{1}{2} \cdot \frac{1}{40}$ rounds to $\frac{1}{2} \cdot 0 = 0$

Review Problems

65. $\frac{1}{2} = \frac{1 \cdot 3}{2 \cdot 3} = \frac{3}{6}$

67. $\frac{3}{2} = \frac{3 \cdot 3}{2 \cdot 3} = \frac{9}{6}$

69. $\frac{1}{3} = \frac{1 \cdot 4}{3 \cdot 4} = \frac{4}{12}$

71. $\frac{2}{3} = \frac{2 \cdot 4}{3 \cdot 4} = \frac{8}{12}$

73. $\frac{5}{6} = \frac{5 \cdot 2x}{6 \cdot 2x} = \frac{10x}{12x}$

75. $\frac{2}{3} = \frac{2 \cdot 4x}{3 \cdot 4x} = \frac{8x}{12x}$

Section 3.5

1. $\frac{3}{6}+\frac{1}{6}=\frac{3+1}{6}$
$=\frac{4}{6}$
$=\frac{2}{3}$

3. $\frac{3}{8}-\frac{5}{8}=\frac{3-5}{8}$
$=\frac{-2}{8}$
$=-\frac{1}{4}$

5. $-\frac{1}{4}+\frac{3}{4}=\frac{-1+3}{4}$
$=\frac{2}{4}$
$=\frac{1}{2}$

7. $\frac{x}{3}-\frac{1}{3}=\frac{x-1}{3}$

9. $\frac{1}{4}+\frac{2}{4}+\frac{3}{4}=\frac{1+2+3}{4}$
$=\frac{6}{4}$
$=\frac{3}{2}$

11. $\frac{x+7}{2}-\frac{1}{2}=\frac{x+7-1}{2}$
$=\frac{x+6}{2}$

13. $\frac{1}{10}-\frac{3}{10}-\frac{4}{10}=\frac{1-3-4}{10}$
$=\frac{-6}{10}$
$=-\frac{3}{5}$

15. $\frac{1}{a}+\frac{4}{a}+\frac{5}{a}=\frac{1+4+5}{a}$
$=\frac{10}{a}$

17. The table is as follows:

a	b	$a+b$
$\frac{1}{2}$	$\frac{1}{3}$	$\frac{1}{2}+\frac{1}{3}=\frac{1\cdot3}{2\cdot3}+\frac{1\cdot2}{3\cdot2}$ $=\frac{3}{6}+\frac{2}{6}=\frac{5}{6}$
$\frac{1}{3}$	$\frac{1}{4}$	$\frac{1}{3}+\frac{1}{4}=\frac{1\cdot4}{3\cdot4}+\frac{1\cdot3}{4\cdot3}$ $=\frac{4}{12}+\frac{3}{12}=\frac{7}{12}$
$\frac{1}{4}$	$\frac{1}{5}$	$\frac{1}{4}+\frac{1}{5}=\frac{1\cdot5}{4\cdot5}+\frac{1\cdot4}{5\cdot4}$ $=\frac{5}{20}+\frac{4}{20}=\frac{9}{20}$
$\frac{1}{5}$	$\frac{1}{6}$	$\frac{1}{5}+\frac{1}{6}=\frac{1\cdot6}{5\cdot6}+\frac{1\cdot5}{6\cdot5}$ $=\frac{6}{30}+\frac{5}{30}=\frac{11}{30}$

19. The table is as follows:

a	b	$a+b$
$\frac{1}{12}$	$\frac{1}{2}$	$\frac{1}{12}+\frac{1}{2}=\frac{1}{12}+\frac{1\cdot 6}{2\cdot 6}$ $=\frac{1}{12}+\frac{6}{12}=\frac{7}{12}$
$\frac{1}{12}$	$\frac{1}{3}$	$\frac{1}{12}+\frac{1}{3}=\frac{1}{12}+\frac{1\cdot 4}{3\cdot 4}$ $=\frac{1}{12}+\frac{4}{12}=\frac{5}{12}$
$\frac{1}{12}$	$\frac{1}{4}$	$\frac{1}{12}+\frac{1}{4}=\frac{1}{12}+\frac{1\cdot 3}{4\cdot 3}$ $=\frac{1}{12}+\frac{3}{12}=\frac{4}{12}=\frac{1}{3}$
$\frac{1}{12}$	$\frac{1}{6}$	$\frac{1}{12}+\frac{1}{6}=\frac{1}{12}+\frac{1\cdot 2}{6\cdot 2}$ $=\frac{1}{12}+\frac{2}{12}=\frac{3}{12}=\frac{1}{4}$

21. $\left.\begin{array}{l} 9=3\cdot 3 \\ 3=3 \end{array}\right\} \text{LCD}=3\cdot 3=9$

$$\begin{aligned}\frac{4}{9}+\frac{1}{3}&=\frac{4}{9}+\frac{1\cdot 3}{3\cdot 3}\\&=\frac{4}{9}+\frac{3}{9}\\&=\frac{7}{9}\end{aligned}$$

23. $\text{LCD}=1\cdot 3=3$

$$\begin{aligned}2+\frac{1}{3}&=\frac{2\cdot 3}{1\cdot 3}+\frac{1}{3}\\&=\frac{6}{3}+\frac{1}{3}\\&=\frac{7}{3}\end{aligned}$$

25. $\text{LCD}=4\cdot 1=4$

$$\begin{aligned}-\frac{3}{4}+1&=\frac{-3}{4}+\frac{1\cdot 4}{1\cdot 4}\\&=\frac{-3}{4}+\frac{4}{4}\\&=\frac{1}{4}\end{aligned}$$

27. $\text{LCD}=2\cdot 3=6$

$$\begin{aligned}\frac{1}{2}+\frac{2}{3}&=\frac{1\cdot 3}{2\cdot 3}+\frac{2\cdot 2}{3\cdot 2}\\&=\frac{3}{6}+\frac{4}{6}\\&=\frac{7}{6}\end{aligned}$$

29. $\text{LCD}=4\cdot 5=20$

$$\begin{aligned}\frac{x}{4}+\frac{1}{5}&=\frac{x\cdot 5}{4\cdot 5}+\frac{1\cdot 4}{5\cdot 4}\\&=\frac{5x}{20}+\frac{4}{20}\\&=\frac{5x+4}{20}\end{aligned}$$

31. $\text{LCD}=x\cdot 5=5x$

$$\begin{aligned}\frac{2}{x}+\frac{3}{5}&=\frac{2\cdot 5}{x\cdot 5}+\frac{3\cdot x}{5\cdot x}\\&=\frac{10}{5x}+\frac{3x}{5x}\\&=\frac{10+3x}{5x}\end{aligned}$$

33. $\left.\begin{array}{l} 12=2\cdot 2\cdot 3 \\ 8=2\cdot 2\cdot 2 \end{array}\right\} \text{LCD}=2\cdot 2\cdot 2\cdot 3=24$

$$\begin{aligned}\frac{5}{12}-\left(-\frac{3}{8}\right)&=\frac{5\cdot 2}{12\cdot 2}+\frac{3\cdot 3}{8\cdot 3}\\&=\frac{10}{24}+\frac{9}{24}\\&=\frac{19}{24}\end{aligned}$$

35. $\left.\begin{aligned}20 &= 2\cdot 2\cdot 5\\ 30 &= 2\cdot 3\cdot 5\end{aligned}\right\}\text{LCD} = 2\cdot 2\cdot 3\cdot 5 = 60$

$$\begin{aligned}-\frac{1}{20}+\frac{8}{30} &= \frac{-1\cdot 3}{20\cdot 3}+\frac{8\cdot 2}{30\cdot 2}\\ &= \frac{-3}{60}+\frac{16}{60}\\ &= \frac{13}{60}\end{aligned}$$

37. $\left.\begin{aligned}10 &= 2\cdot 5\\ 100 &= 2\cdot 2\cdot 5\cdot 5\end{aligned}\right\}\text{LCD} = 2\cdot 2\cdot 5\cdot 5 = 100$

$$\begin{aligned}\frac{a}{10}+\frac{1}{100} &= \frac{a\cdot 10}{10\cdot 10}+\frac{1}{100}\\ &= \frac{10a}{100}+\frac{1}{100}\\ &= \frac{10a+1}{100}\end{aligned}$$

39. $\text{LCD} = 7\cdot x = 7x$

$$\begin{aligned}\frac{3}{7}+\frac{4}{x} &= \frac{3\cdot x}{7\cdot x}+\frac{4\cdot 7}{x\cdot 7}\\ &= \frac{3x}{7x}+\frac{28}{7x}\\ &= \frac{3x+28}{7x}\end{aligned}$$

41. $\left.\begin{aligned}30 &= 2\cdot 3\cdot 5\\ 42 &= 2\cdot 3\cdot 7\end{aligned}\right\}\text{LCD} = 2\cdot 3\cdot 5\cdot 7 = 210$

$$\begin{aligned}\frac{17}{30}+\frac{11}{42} &= \frac{17\cdot 7}{30\cdot 7}+\frac{11\cdot 5}{42\cdot 5}\\ &= \frac{119}{210}+\frac{55}{210}\\ &= \frac{174}{210}\\ &= \frac{29}{35}\end{aligned}$$

43. $\left.\begin{aligned}84 &= 2\cdot 2\cdot 3\cdot 7\\ 90 &= 2\cdot 3\cdot 3\cdot 5\end{aligned}\right\}\text{LCD} = 2\cdot 2\cdot 3\cdot 3\cdot 5\cdot 7 = 1{,}260$

$$\begin{aligned}\frac{25}{84}+\frac{41}{90} &= \frac{25\cdot 15}{84\cdot 15}+\frac{41\cdot 14}{90\cdot 14}\\ &= \frac{375}{1{,}260}+\frac{574}{1{,}260}\\ &= \frac{949}{1{,}260}\end{aligned}$$

45. $\left.\begin{aligned}126 &= 2\cdot 3\cdot 3\cdot 7\\ 180 &= 2\cdot 2\cdot 3\cdot 3\cdot 5\end{aligned}\right\}\text{LCD} = 2\cdot 3\cdot 3\cdot 3\cdot 5\cdot 7 = 1{,}260$

$$\begin{aligned}\frac{13}{126}-\frac{13}{180} &= \frac{13\cdot 10}{126\cdot 10}-\frac{13\cdot 7}{180\cdot 7}\\ &= \frac{130}{1{,}260}-\frac{91}{1{,}260}\\ &= \frac{39}{1{,}260}\\ &= \frac{13}{420}\end{aligned}$$

47. $\left.\begin{aligned}4 &= 2\cdot 2\\ 8 &= 2\cdot 2\cdot 2\\ 6 &= 2\cdot 3\end{aligned}\right\}\text{LCD} = 2\cdot 2\cdot 2\cdot 3 = 24$

$$\begin{aligned}\frac{3}{4}+\frac{1}{8}+\frac{5}{6} &= \frac{3\cdot 6}{4\cdot 6}+\frac{1\cdot 3}{8\cdot 3}+\frac{5\cdot 4}{6\cdot 4}\\ &= \frac{18}{24}+\frac{3}{24}+\frac{20}{24}\\ &= \frac{41}{24}\end{aligned}$$

49. $\text{LCD} = 2\cdot 3\cdot y = 6y$

$$\begin{aligned}\frac{4}{y}+\frac{2}{3}+\frac{1}{2} &= \frac{4\cdot 6}{y\cdot 6}+\frac{2\cdot 2y}{3\cdot 2y}+\frac{1\cdot 3y}{2\cdot 3y}\\ &= \frac{24}{6y}+\frac{4y}{6y}+\frac{3y}{6y}\\ &= \frac{24+7y}{6y}\end{aligned}$$

51. $$\left.\begin{aligned}2&=2\\3&=3\\4&=2\cdot 2\\6&=2\cdot 3\end{aligned}\right\}\text{LCD}=2\cdot 2\cdot 3=12$$

$$\begin{aligned}\frac{1}{2}+\frac{1}{3}+\frac{1}{4}+\frac{1}{6}&=\frac{1\cdot 6}{2\cdot 6}+\frac{1\cdot 4}{3\cdot 4}+\frac{1\cdot 3}{4\cdot 3}+\frac{1\cdot 2}{6\cdot 2}\\&=\frac{6}{12}+\frac{4}{12}+\frac{3}{12}+\frac{2}{12}\\&=\frac{15}{12}\\&=\frac{5}{4}\end{aligned}$$

53. $$\left.\begin{aligned}7&=7\\9&=3\cdot 3\end{aligned}\right\}\text{LCD}=7\cdot 3\cdot 3=63$$

$$\begin{aligned}\frac{3}{7}+2+\frac{1}{9}&=\frac{3\cdot 9}{7\cdot 9}+\frac{2\cdot 63}{1\cdot 63}+\frac{1\cdot 7}{9\cdot 7}\\&=\frac{27}{63}+\frac{126}{63}+\frac{7}{63}\\&=\frac{160}{63}\end{aligned}$$

55. $$\left.\begin{aligned}4&=2\cdot 2\\8&=2\cdot 2\cdot 2\end{aligned}\right\}\text{LCD}=2\cdot 2\cdot 2=8$$

$$\begin{aligned}\frac{7}{8}-\frac{1}{4}&=\frac{7}{8}-\frac{1\cdot 2}{4\cdot 2}\\&=\frac{7}{8}-\frac{2}{8}\\&=\frac{5}{8}\end{aligned}$$

57. $$\begin{aligned}\frac{1}{2}x+\frac{1}{6}x&=\left(\frac{1}{2}+\frac{1}{6}\right)x\\&=\left(\frac{3}{6}+\frac{1}{6}\right)x\\&=\frac{4}{6}x\\&=\frac{2}{3}x\end{aligned}$$

59. $$\begin{aligned}\frac{1}{2}x-\frac{3}{4}x&=\left(\frac{1}{2}-\frac{3}{4}\right)x\\&=\left(\frac{2}{4}-\frac{3}{4}\right)x\\&=-\frac{1}{4}x\end{aligned}$$

61. $$\begin{aligned}\frac{1}{3}x+\frac{3}{5}x&=\left(\frac{1}{3}+\frac{3}{5}\right)x\\&=\left(\frac{5}{15}+\frac{9}{15}\right)x\\&=\frac{14}{15}x\end{aligned}$$

63. $$\begin{aligned}\frac{3x}{4}+\frac{x}{6}&=\left(\frac{3}{4}+\frac{1}{6}\right)x\\&=\left(\frac{9}{12}+\frac{2}{12}\right)x\\&=\frac{11}{12}x\end{aligned}$$

65. $$\begin{aligned}\frac{2x}{5}+\frac{5x}{8}&=\left(\frac{2}{5}+\frac{5}{8}\right)x\\&=\left(\frac{16}{40}+\frac{25}{40}\right)x\\&=\frac{41}{40}x\end{aligned}$$

Applying the Concepts

67. Milk in carton 1 + Milk in carton 2 = total

$$\begin{aligned}\frac{1}{2}+4&=\frac{1}{2}+\frac{4\cdot 2}{1\cdot 2}\\&=\frac{1}{2}+\frac{8}{2}\\&=\frac{9}{2}\end{aligned}$$

The total is $\frac{9}{2}$ pints.

69. $\frac{5}{8}$ of monthly income = house payment

$$\frac{5}{8}\cdot 2{,}120 = \frac{5\cdot 2{,}120}{8}$$
$$= \frac{5\cdot \not{8}\cdot 265}{\not{8}}$$
$$= 1{,}325$$

They can spend $1,325 on a house payment.

71. Engineering fraction + business fraction = total fraction

$$\frac{1}{4}+\frac{3}{20} = \frac{5}{20}+\frac{3}{20}$$
$$= \frac{8}{20}$$
$$= \frac{2}{5}$$

73. The table is as follows:

Grade	Number	Fraction
A	$\frac{1}{8}\cdot 40 = 5$	$\frac{1}{8}$
B	$\frac{1}{5}\cdot 40 = 8$	$\frac{1}{5}$
C	$\frac{1}{2}\cdot 40 = 20$	$\frac{1}{2}$
below C	7	$\frac{7 \text{ students}}{40 \text{ total}} = \frac{7}{40}$
Total	40	1

75. Total acres ÷ acres per lot = lots

$$6 \div \frac{3}{5} = \frac{6}{1}\cdot\frac{5}{3}$$
$$= \frac{\not{3}\cdot 2\cdot 5}{1\cdot \not{3}}$$
$$= 10$$

There are 10 lots.

77. $P = 4s$

$$= 4\cdot\left(\frac{3}{8}\text{ in}\right)$$
$$= \frac{4\cdot 3}{8}\text{ in}$$
$$= \frac{\not{4}\cdot 3}{\not{4}\cdot 2}\text{ in}$$
$$= \frac{3}{2}\text{ in}$$

79. $P =$ sum of all sides

$$= \left(\frac{3}{5}+\frac{3}{5}+\frac{3}{10}+\frac{3}{10}\right)\text{ft}$$
$$= \left(\frac{3\cdot 2}{5\cdot 2}+\frac{3\cdot 2}{5\cdot 2}+\frac{3}{10}+\frac{3}{10}\right)\text{ ft}$$
$$= \left(\frac{6}{10}+\frac{6}{10}+\frac{3}{10}+\frac{3}{10}\right)\text{ ft}$$
$$= \frac{18}{10}\text{ ft}$$
$$= \frac{9}{5}\text{ ft}$$

Calculator Problems

81. Since $497 \div 71 = 7$, we have

$$\frac{7}{71}+\frac{17}{497} = \frac{7\cdot 7}{71\cdot 7}+\frac{17}{497}$$
$$= \frac{49}{497}+\frac{17}{497}$$
$$= \frac{66}{497}$$

83. Since
$1{,}247 \div 29 = 43$ and $1{,}247 \div 43 = 19$, we have:

$$\frac{10}{19}+\frac{20}{23} = \frac{10\cdot 23}{19\cdot 23}+\frac{20\cdot 19}{23\cdot 19}$$
$$= \frac{230}{437}+\frac{380}{437}$$
$$= \frac{610}{437}$$

Review Problems

85. $\frac{3}{4} \div \frac{5}{6} = \frac{3}{4} \cdot \frac{6}{5}$

$= \frac{3 \cdot \cancel{2} \cdot 3}{\cancel{2} \cdot 2 \cdot 5}$

$= \frac{9}{10}$

87. $-12 \cdot \frac{2}{3} = -\frac{12 \cdot 2}{3}$

$= -\frac{4 \cdot \cancel{3} \cdot 2}{\cancel{3}}$

$= -8$

89. $4 \cdot \frac{3}{4} = \frac{\cancel{4} \cdot 3}{\cancel{4}} = 3$

91. $-\frac{7}{6} \div \left(-\frac{7}{12}\right) = \frac{7}{6} \cdot \frac{12}{7}$

$= \frac{\cancel{7} \cdot 2 \cdot \cancel{6}}{\cancel{6} \cdot \cancel{7}}$

$= 2$

93. $\frac{x^2}{y} \div \frac{x}{y^3} = \frac{x^2}{y} \cdot \frac{y^3}{x}$

$= \frac{\cancel{x} \cdot x \cdot \cancel{y} \cdot y \cdot y}{\cancel{y} \cdot \cancel{x}}$

$= xy^2$

95. $\frac{2}{3} \cdot \frac{3}{4} \cdot \frac{4}{5} \cdot \frac{5}{6} \cdot \frac{6}{7} = \frac{2 \cdot \cancel{3} \cdot \cancel{4} \cdot \cancel{5} \cdot \cancel{6}}{\cancel{3} \cdot \cancel{4} \cdot \cancel{5} \cdot \cancel{6} \cdot 7}$

$= \frac{2}{7}$

97. $\frac{35}{110} \cdot \frac{80}{63} \div \frac{16}{27} = \frac{35}{110} \cdot \frac{80}{63} \cdot \frac{27}{16}$

$= \frac{\cancel{5} \cdot \cancel{7} \cdot \cancel{16} \cdot 5 \cdot \cancel{9} \cdot 3}{\cancel{5} \cdot 22 \cdot \cancel{9} \cdot \cancel{7} \cdot \cancel{16}}$

$= \frac{15}{22}$

Extending the Concepts

99. $1, \frac{4}{3}, \frac{5}{3}, 2, \ldots$

The first term in the sequence is 1. Note: $1 + \frac{1}{3} = \frac{4}{3}$, $\frac{4}{3} + \frac{1}{3} = \frac{5}{3}$, $\frac{5}{3} + \frac{1}{3} = \frac{6}{3} = 2$, i.e. each term after 1 in the sequence comes from adding $\frac{1}{3}$ to the previous term. The next term in the sequence is $2 + \frac{1}{3} = \frac{2 \cdot 3}{1 \cdot 3} + \frac{1}{3} = \frac{7}{3}$.

101. $\frac{3}{2}, 2, \frac{5}{2}, \ldots$

The first term in the sequence is $\frac{1}{2}$. Note: $\frac{3}{2} + \frac{1}{2} = \frac{4}{2} = 2$ and $\frac{4}{2} + \frac{1}{2} = \frac{5}{2}$, i.e. each term after $\frac{3}{2}$ in the sequence comes from adding $\frac{1}{2}$ to the previous term. The next term in the sequence is $\frac{5}{2} + \frac{1}{2} = \frac{6}{2} = 3$.

Section 3.6

1. $4\frac{2}{3}=\frac{3\cdot 4+2}{3}=\frac{12+2}{3}=\frac{14}{3}$

3. $5\frac{1}{4}=\frac{4\cdot 5+1}{4}=\frac{20+1}{4}=\frac{21}{4}$

5. $1\frac{5}{8}=\frac{8\cdot 1+5}{8}=\frac{8+5}{8}=\frac{13}{8}$

7. $15\frac{2}{3}=\frac{3\cdot 15+2}{3}=\frac{45+2}{3}=\frac{47}{3}$

9. $4\frac{20}{21}=\frac{21\cdot 4+20}{21}=\frac{84+20}{21}=\frac{104}{21}$

11. $12\frac{31}{33}=\frac{33\cdot 12+31}{33}=\frac{396+31}{33}=\frac{427}{33}$

13. $\frac{9}{8}$: $8\overline{)9}$ quotient 1, $\underline{8}$, remainder 1, so $\frac{9}{8}=1+\frac{1}{8}=1\frac{1}{8}$

15. $\frac{19}{4}$: $4\overline{)19}$ quotient 4, $\underline{16}$, remainder 3, so $\frac{19}{4}=4+\frac{3}{4}=4\frac{3}{4}$

17. $\frac{29}{6}$: $6\overline{)29}$ quotient 4, $\underline{24}$, remainder 5, so $\frac{29}{6}=4+\frac{5}{6}=4\frac{5}{6}$

19. $\frac{13}{4}$: $4\overline{)13}$ quotient 3, $\underline{12}$, remainder 1, so $\frac{13}{4}=3+\frac{1}{4}=3\frac{1}{4}$

21. $\frac{109}{27}$: $27\overline{)109}$ quotient 4, $\underline{108}$, remainder 1, so $\frac{109}{27}=4+\frac{1}{27}=4\frac{1}{27}$

23. $\frac{428}{15}$: $15\overline{)428}$ quotient 28, $\underline{30}\downarrow$, 128, $\underline{120}$, remainder 8, so $\frac{428}{15}=28+\frac{8}{15}=28\frac{8}{15}$

25. $\text{LCD}=1\cdot x=x$;

$$8+\frac{3}{x}=\frac{8\cdot x}{x}+\frac{3}{x}$$
$$=\frac{8x+3}{x}$$

27. $\text{LCD}=1\cdot y=y$;

$$2-\frac{5}{y}=\frac{2\cdot y}{y}-\frac{5}{y}$$
$$=\frac{2y-5}{y}$$

29. $\text{LCD}=1\cdot 6=6$;

$$x+\frac{5}{6}=\frac{x\cdot 6}{6}+\frac{5}{6}$$
$$=\frac{6x+5}{6}$$

31. $\text{LCD}=1\cdot 3=3$;

$$a-\frac{1}{3}=\frac{a\cdot 3}{1\cdot 3}-\frac{1}{3}$$
$$=\frac{3a-1}{3}$$

Applying the Concepts

33. a. $\dfrac{\text{Cicadas}}{\text{City traffic}} = \dfrac{90}{80} = \dfrac{9}{8} = 1\dfrac{1}{8}$

b. $\dfrac{\text{Firecracker}}{\text{City traffic}} = \dfrac{120}{80} = \dfrac{12}{8} = 1\dfrac{4}{8} = 1\dfrac{1}{2}$

35. Original price + increase = new price

$$5\frac{1}{4} + \frac{3}{4} = 5 + \frac{1}{4} + \frac{3}{4}$$
$$= 5 + \frac{1+3}{4}$$
$$= 5 + \frac{4}{4}$$
$$= 5 + 1$$
$$= 6$$

The new price is \$6 per share.

37. $5\dfrac{11}{12} = \dfrac{12\cdot 5 + 11}{12} = \dfrac{60+11}{12} = \dfrac{71}{12}$

The man is $\dfrac{71}{12}$ inches tall.

39. Above each age, draw a dot corresponding to the hours of sleep as measured by the vertical axis. To construct the line graph, connect the dots.

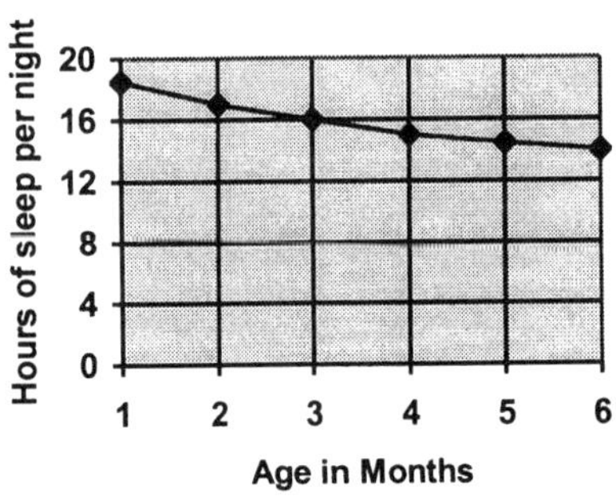

41. $135\dfrac{9}{10} = \dfrac{10\cdot 135 + 9}{10} = \dfrac{1,350+9}{10} = \dfrac{1,359}{10}$

The price is $\dfrac{1,359}{10}$ ¢ per gallon.

Review Problems

43. $\dfrac{3}{8}\cdot\dfrac{3}{5} = \dfrac{3\cdot 3}{8\cdot 5} = \dfrac{9}{40}$

45. $\left(-\dfrac{2}{3}\right)\left(-\dfrac{9}{16}\right) = \dfrac{2\cdot 9}{3\cdot 16}$

$$= \frac{\cancel{2}\cdot\cancel{3}\cdot 3}{\cancel{3}\cdot\cancel{2}\cdot 8}$$
$$= \frac{3}{8}$$

47. $\dfrac{4}{5} \div \left(-\dfrac{7}{8}\right) = \dfrac{4}{5}\cdot\left(-\dfrac{8}{7}\right)$

$$= -\frac{4\cdot 8}{5\cdot 7}$$
$$= -\frac{32}{35}$$

49. $\dfrac{9}{10} \div \dfrac{3}{5} = \dfrac{9}{10}\cdot\dfrac{5}{3}$

$$= \frac{3\cdot\cancel{3}\cdot\cancel{5}}{\cancel{5}\cdot 2\cdot\cancel{3}}$$
$$= \frac{3}{2} \text{ or } 1\frac{1}{2}$$

51. $\dfrac{2}{3}\cdot\dfrac{3}{4} \div \dfrac{5}{8} = \dfrac{2}{3}\cdot\dfrac{3}{4}\cdot\dfrac{8}{5}$

$$= \frac{2\cdot\cancel{3}\cdot 2\cdot\cancel{4}}{\cancel{3}\cdot\cancel{4}\cdot 5}$$
$$= \frac{4}{5}$$

53. $\dfrac{a^2c}{b}\cdot\dfrac{b^2}{ac^2} = \dfrac{\cancel{a}\cdot a\cdot\cancel{c}\cdot\cancel{b}\cdot b}{\cancel{b}\cdot\cancel{a}\cdot\cancel{c}\cdot c}$

$$= \frac{ab}{c}$$

Improving your Quantitative Literacy

55. a. $1\frac{1}{2}\cdot 1{,}405{,}146 = \frac{3}{2}\cdot\frac{1{,}405{,}146}{1}$

$$= \frac{3\cdot \cancel{2}\cdot 702{,}573}{\cancel{2}}$$

$$= 2{,}107{,}719$$

b. $1\frac{1}{3}\cdot 1{,}405{,}146 = \frac{4}{3}\cdot\frac{1{,}405{,}146}{1}$

$$= \frac{4\cdot \cancel{3}\cdot 468{,}382}{\cancel{3}}$$

$$= 1{,}873{,}528$$

(a) is closer to the truth; 2,107,719 is closer to 2,082,487 than 1,873,528 is.

Section 3.7

1. $3\frac{2}{5}\cdot 1\frac{1}{2}=\frac{17}{5}\cdot\frac{3}{2}$

$=\frac{51}{10}$

$=5\frac{1}{10}$

3. $5\frac{1}{8}\cdot 2\frac{2}{3}=\frac{41}{8}\cdot\frac{8}{3}$

$=\frac{41\cdot \cancel{8}}{\cancel{8}\cdot 3}$

$=\frac{41}{3}$

$=13\frac{2}{3}$

5. $2\frac{1}{10}\cdot 3\frac{3}{10}=\frac{21}{10}\cdot\frac{33}{10}$

$=\frac{693}{100}$

$=6\frac{93}{100}$

7. $1\frac{1}{4}\cdot 4\frac{2}{3}=\frac{5}{4}\cdot\frac{14}{3}$

$=\frac{5\cdot \cancel{2}\cdot 7}{\cancel{2}\cdot 2\cdot 3}$

$=\frac{35}{6}$

$=5\frac{5}{6}$

9. $2\cdot 4\frac{7}{8}=\frac{2}{1}\cdot\frac{39}{8}$

$=\frac{\cancel{2}\cdot 39}{1\cdot \cancel{2}\cdot 4}$

$=\frac{39}{4}$

$=9\frac{3}{4}$

11. $\frac{3}{5}\cdot 5\frac{1}{3}=\frac{3}{5}\cdot\frac{16}{3}$

$=\frac{\cancel{3}\cdot 16}{5\cdot \cancel{3}}$

$=\frac{16}{5}$

$=3\frac{1}{5}$

13. $2\frac{1}{2}\cdot 3\frac{1}{3}\cdot 1\frac{1}{2}=\frac{5}{2}\cdot\frac{10}{3}\cdot\frac{3}{2}$

$=\frac{5\cdot \cancel{2}\cdot 5\cdot \cancel{3}}{\cancel{2}\cdot \cancel{3}\cdot 2}$

$=\frac{25}{2}$

$=12\frac{1}{2}$

15. $\frac{3}{4}\cdot 7\cdot 1\frac{4}{5}=\frac{3}{4}\cdot\frac{7}{1}\cdot\frac{9}{5}$

$=\frac{3\cdot 7\cdot 9}{4\cdot 1\cdot 5}$

$=\frac{189}{20}$

$=9\frac{9}{20}$

17. $3\frac{1}{5} \div 4\frac{1}{2} = \frac{16}{5} \div \frac{9}{2}$
$= \frac{16}{5} \cdot \frac{2}{9}$
$= \frac{32}{45}$

19. $6\frac{1}{4} \div 3\frac{3}{4} = \frac{25}{4} \div \frac{15}{4}$
$= \frac{25}{4} \cdot \frac{4}{15}$
$= \frac{\cancel{5} \cdot 5 \cdot \cancel{4}}{\cancel{4} \cdot \cancel{5} \cdot 3}$
$= \frac{5}{3}$
$= 1\frac{2}{3}$

21. $10 \div 2\frac{1}{2} = \frac{10}{1} \div \frac{5}{2}$
$= \frac{10}{1} \cdot \frac{2}{5}$
$= \frac{2 \cdot \cancel{5} \cdot 2}{1 \cdot \cancel{5}}$
$= 4$

23. $8\frac{3}{5} \div 2 = \frac{43}{5} \div \frac{2}{1}$
$= \frac{43}{5} \cdot \frac{1}{2}$
$= \frac{43 \cdot 1}{5 \cdot 2}$
$= \frac{43}{10}$
$= 4\frac{3}{10}$

25. $\left(\frac{3}{4} \div 2\frac{1}{2}\right) \div 3 = \left(\frac{3}{4} \div \frac{5}{2}\right) \div \frac{3}{1}$
$= \left(\frac{3}{4} \cdot \frac{2}{5}\right) \cdot \frac{1}{3}$
$= \frac{\cancel{3} \cdot \cancel{2} \cdot 1}{\cancel{2} \cdot 2 \cdot 5 \cdot \cancel{3}}$
$= \frac{1}{10}$

27. $\left(8 \div 1\frac{1}{4}\right) \div 2 = \left(\frac{8}{1} \div \frac{5}{4}\right) \div \frac{2}{1}$
$= \left(\frac{8}{1} \cdot \frac{4}{5}\right) \cdot \frac{1}{2}$
$= \frac{8 \cdot 2 \cdot \cancel{2} \cdot 1}{1 \cdot 5 \cdot \cancel{2}}$
$= \frac{16}{5}$
$= 3\frac{1}{5}$

29. $2\frac{1}{2} \cdot \left(3\frac{2}{5} \div 4\right) = \frac{5}{2} \cdot \left(\frac{17}{5} \div \frac{4}{1}\right)$
$= \frac{5}{2} \cdot \left(\frac{17}{5} \cdot \frac{1}{4}\right)$
$= \frac{\cancel{5} \cdot 17 \cdot 1}{2 \cdot \cancel{5} \cdot 4}$
$= \frac{17}{8}$
$= 2\frac{1}{8}$

31. $2\frac{1}{2} \cdot 3 = \frac{5}{2} \cdot \frac{3}{1}$
$= \frac{5 \cdot 3}{2 \cdot 1}$
$= \frac{15}{2}$
$= 7\frac{1}{2}$

33. $2\frac{3}{4} \div 3\frac{1}{4} = \frac{11}{4} \div \frac{13}{4}$

$= \frac{11}{4} \cdot \frac{4}{13}$

$= \frac{11 \cdot \cancel{4}}{\cancel{4} \cdot 13}$

$= \frac{11}{13}$

Applying the Concepts

35. $2 \cdot \left(2\frac{3}{4}\right) = \frac{2}{1} \cdot \frac{11}{4}$

$= \frac{\cancel{2} \cdot 11}{1 \cdot \cancel{2} \cdot 2}$

$= \frac{11}{2}$

$= 5\frac{1}{2}$

$5\frac{1}{2}$ cups will be needed.

37. $\frac{1}{3}$ of $2\frac{1}{2}$ cups = amount needed

$\frac{1}{3} \cdot 2\frac{1}{2} = \frac{1}{3} \cdot \frac{5}{2}$

$= \frac{1 \cdot 5}{3 \cdot 2}$

$= \frac{5}{6}$

$\frac{5}{6}$ cup will be needed.

39. $\frac{3}{4}$ of $1\frac{7}{9} = \frac{3}{4} \cdot \frac{16}{9}$

$= \frac{\cancel{3} \cdot \cancel{4} \cdot 4}{\cancel{4} \cdot \cancel{3} \cdot 3}$

$= \frac{4}{3}$

$= 1\frac{1}{3}$

41. 8 gallons × cost per gallon = total cost

$8 \cdot 135\frac{9}{10} = \frac{8}{1} \cdot \frac{1{,}359}{10}$

$= \frac{\cancel{2} \cdot 4 \cdot 1{,}359}{1 \cdot \cancel{2} \cdot 5}$

$= \frac{5{,}436}{5}$

$= 1{,}087\frac{1}{5}$

The total cost is $1{,}087\frac{1}{5}$¢.

43. 5 gallons × miles per gallon = total miles

$5 \cdot 32\frac{3}{4} = \frac{5}{1} \cdot \frac{131}{4}$

$= \frac{5 \cdot 131}{1 \cdot 4}$

$= \frac{655}{4}$

$= 163\frac{3}{4}$

The car will travel $163\frac{3}{4}$ miles.

45. 3 covers × yards per cover = total yards

$$3\cdot 1\frac{1}{2}=\frac{3}{1}\cdot\frac{3}{2}$$
$$=\frac{3\cdot 3}{1\cdot 2}$$
$$=\frac{9}{2}$$
$$=4\frac{1}{2}$$

It will take $4\frac{1}{2}$ yards.

47. $A=\frac{1}{2}bh$

$$=\frac{1}{2}\cdot 3\text{ yd}\cdot 1\frac{1}{2}\text{ yd}$$
$$=\frac{1}{2}\cdot\frac{3}{1}\cdot\frac{3}{2}\text{ yd}^2$$
$$=\frac{9}{4}\text{ yd}^2$$
$$=2\frac{1}{4}\text{ yd}^2$$

49.

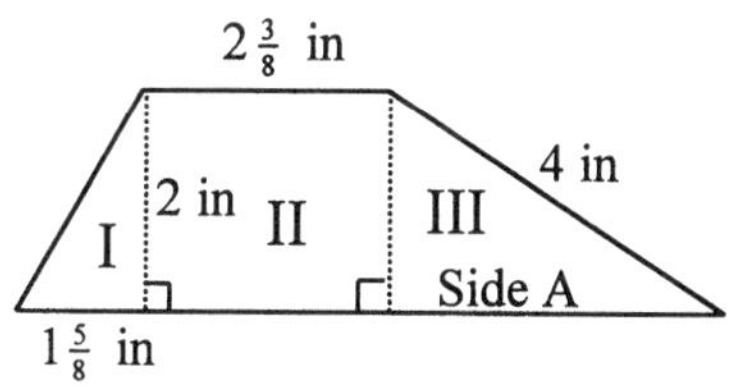

Separate ınto 3 regions. Find the length of the missing side:

$$\text{Side A}=7-1\frac{5}{8}-2\frac{3}{8}$$
$$=\frac{7}{1}-\frac{13}{8}-\frac{19}{8}$$
$$=\frac{7\cdot 8}{1\cdot 8}-\frac{13}{8}-\frac{19}{8}$$
$$=\frac{56}{8}-\frac{13}{8}-\frac{19}{8}$$
$$=\frac{24}{8}=3\text{ in.}$$

49. (continued)

$$\text{Area I}=\frac{1}{2}bh$$
$$=\frac{1}{2}\left(1\frac{5}{8}\right)\cdot 2$$
$$=\frac{1}{2}\cdot\frac{13}{8}\cdot\frac{2}{1}$$
$$=\frac{13\cdot \not{2}}{\not{2}\cdot 8\cdot 1}$$
$$=\frac{13}{8}$$
$$=1\frac{5}{8}\text{ in}^2$$

$$\text{Area II}=lw$$
$$=2\frac{3}{8}\cdot 2$$
$$=\frac{19}{8}\cdot\frac{2}{1}$$
$$=\frac{19\cdot \not{2}}{\not{2}\cdot 4\cdot 1}$$
$$=\frac{19}{4}$$
$$=4\frac{3}{4}\text{ in}^2$$

$$\text{Area III}=\frac{1}{2}bh=\frac{1}{2}\cdot 3\cdot 2$$
$$=\frac{1}{\not{2}}\cdot\frac{3}{1}\cdot\frac{\not{2}}{1}$$
$$=3\text{ in}^2$$

Area = Area I + Area II + Area III

$$A=1\frac{5}{8}+4\frac{3}{4}+3$$
$$=\frac{13}{8}+\frac{19}{4}+\frac{3}{1}$$
$$=\frac{13}{8}+\frac{19\cdot 2}{4\cdot 2}+\frac{3\cdot 8}{1\cdot 8}$$
$$=\frac{13+38+24}{8}$$
$$=\frac{75}{8}\text{ or }9\frac{3}{8}\text{ in}^2$$

51. Can 1 calories – Can 2 calories = difference

$$\begin{aligned}45\cdot 3\frac{1}{2}-25\cdot 3\frac{1}{2}&=(45-25)\cdot 3\frac{1}{2}\\&=20\cdot 3\frac{1}{2}\\&=\frac{20}{1}\cdot\frac{7}{2}\\&=\frac{140}{2}\\&=70\end{aligned}$$

Can 1 has 70 more calories than can 2.

53. Can 1 sodium – Can 2 sodium = difference

$$\begin{aligned}560\cdot 3\frac{1}{2}-300\cdot 3\frac{1}{2}&=(560-300)\cdot 3\frac{1}{2}\\&=260\cdot 3\frac{1}{2}\\&=\frac{260}{1}\cdot\frac{7}{2}\\&=\frac{1{,}820}{2}\\&=910\end{aligned}$$

Can 1 has 910 mg more sodium than Can 2.

Review Problems

55. $\frac{3}{4}+\frac{5}{4}=\frac{3+5}{4}=\frac{8}{4}=2$

57. $\left.\begin{aligned}56&=2\cdot 2\cdot 2\cdot 7\\42&=2\cdot 3\cdot 7\end{aligned}\right\}\text{LCD}=2\cdot 2\cdot 2\cdot 3\cdot 7=168$

$$\begin{aligned}\frac{15}{56}-\frac{5}{42}&=\frac{15\cdot 3}{56\cdot 3}-\frac{5\cdot 4}{42\cdot 4}\\&=\frac{45}{168}-\frac{20}{168}\\&=\frac{25}{168}\end{aligned}$$

59. $\left.\begin{aligned}18&=2\cdot 3\cdot 3\\12&=2\cdot 2\cdot 3\\4&=2\cdot 2\end{aligned}\right\}\text{LCD}=2\cdot 2\cdot 3\cdot 3=36$

$$\begin{aligned}\frac{5}{18}+\frac{1}{12}+\frac{3}{4}&=\frac{5\cdot 2}{18\cdot 2}+\frac{1\cdot 3}{12\cdot 3}+\frac{3\cdot 9}{4\cdot 9}\\&=\frac{10}{36}+\frac{3}{36}+\frac{27}{36}\\&=\frac{40}{36}\\&=\frac{10}{9}\\&=1\frac{1}{9}\end{aligned}$$

Extending the Concepts

61. $\left(1\frac{1}{2}\right)^2=\left(\frac{3}{2}\right)^2=\frac{3}{2}\cdot\frac{3}{2}=\frac{9}{4}=2\frac{1}{4}$

63. $\left(1\frac{3}{4}\right)^2=\left(\frac{7}{4}\right)^2=\frac{7}{4}\cdot\frac{7}{4}=\frac{49}{16}=3\frac{1}{16}$

65. $A=2\frac{1}{4}\text{ ft}^2=\frac{9}{4}\text{ ft}^2$.

Because $A=s^2$, we need to find s so that $s^2=s\cdot s=\frac{9}{4}\text{ ft}^2$. By inspection, $s=\frac{3}{2}\text{ ft}=1\frac{1}{2}\text{ ft}$.

67. $V=3\frac{3}{8}\text{ ft}^3=\frac{27}{8}\text{ ft}^3$.

Because $V=s^3$, we need to find s so that $s^3=s\cdot s\cdot s=\frac{27}{8}\text{ ft}^3$. By inspection, $s=\frac{3}{2}\text{ ft}=1\frac{1}{2}\text{ ft}$.

Section 3.8

1. $$\begin{array}{r} 2\frac{1}{5} \\ +\ 3\frac{3}{5} \\ \hline 5\frac{4}{5} \end{array}$$

3. $$\begin{array}{r} 4\frac{3}{10} \\ +\ 8\frac{1}{10} \\ \hline 12\frac{4}{10} = 12\frac{2}{5} \end{array}$$

5. $$\begin{array}{r} 6\frac{8}{9} \\ -\ 3\frac{4}{9} \\ \hline 3\frac{4}{9} \end{array}$$

7. $$\begin{array}{r} 9\frac{1}{6} \\ +\ 2\frac{5}{6} \\ \hline 11\frac{6}{6} = 11+1 = 12 \end{array}$$

9. $$\begin{array}{rcrcr} 3\frac{5}{8} & = & 3\frac{5}{8} & = & 3\frac{5}{8} \\ -\ 2\frac{1}{4} & = & -\ 2\frac{1\cdot 2}{4\cdot 2} & = & -\ 2\frac{2}{8} \\ \hline & & & & 1\frac{3}{8} \end{array}$$

11. $$\begin{array}{rcrcr} 11\frac{1}{3} & = & 11\frac{1\cdot 2}{3\cdot 2} & = & 11\frac{2}{6} \\ +\ 2\frac{5}{6} & = & +\ 2\frac{5}{6} & = & +\ 2\frac{5}{6} \\ \hline & & & & 13\frac{7}{6} = 14\frac{1}{6} \end{array}$$

13. $$\begin{array}{rcrcr} 7\frac{5}{12} & = & 7\frac{5}{12} & = & 7\frac{5}{12} \\ -\ 3\frac{1}{3} & = & -\ 3\frac{1\cdot 4}{3\cdot 4} & = & -\ 3\frac{4}{12} \\ \hline & & & & 4\frac{1}{12} \end{array}$$

15. $$\begin{array}{rcrcr} 6\frac{1}{3} & = & 6\frac{1\cdot 4}{3\cdot 4} & = & 6\frac{4}{12} \\ -\ 4\frac{1}{4} & = & -\ 4\frac{1\cdot 3}{4\cdot 3} & = & -\ 4\frac{3}{12} \\ \hline & & & & 2\frac{1}{12} \end{array}$$

17. $$\begin{array}{rcrcr} 10\frac{5}{6} & = & 10\frac{5\cdot 2}{6\cdot 2} & = & 10\frac{10}{12} \\ +\ 15\frac{3}{4} & = & +\ 15\frac{3\cdot 3}{4\cdot 3} & = & +\ 15\frac{9}{12} \\ \hline & & & & 25\frac{19}{12} = 26\frac{7}{12} \end{array}$$

19. $$\begin{array}{r} 5\frac{2}{3} \\ +\ 6\frac{1}{3} \\ \hline 11\frac{3}{3} = 12 \end{array}$$

21.

$$\begin{array}{r} 10\frac{13}{16} \\ -\ 8\frac{5}{16} \\ \hline 2\frac{8}{16} = 2\frac{1}{2} \end{array}$$

23.

$$\begin{array}{rcrcr} 6\frac{1}{2} & = & 6\frac{1\cdot 7}{2\cdot 7} & = & 6\frac{7}{14} \\ +\ 2\frac{5}{14} & = & +\ 2\frac{5}{14} & = & +\ 2\frac{5}{14} \\ \hline & & & & 8\frac{12}{14} = 8\frac{6}{7} \end{array}$$

25.

$$\begin{array}{rcrcr} 1\frac{5}{8} & = & 1\frac{5}{8} & = & 1\frac{5}{8} \\ +\ 1\frac{3}{4} & = & +\ 1\frac{3\cdot 2}{4\cdot 2} & = & +\ 1\frac{6}{8} \\ \hline & & & & 2\frac{11}{8} = 3\frac{3}{8} \end{array}$$

27.

$$\begin{array}{rcrcr} 4\frac{2}{3} & = & 4\frac{2\cdot 5}{3\cdot 5} & = & 4\frac{10}{15} \\ +\ 5\frac{3}{5} & = & +\ 5\frac{3\cdot 3}{5\cdot 3} & = & +\ 5\frac{9}{15} \\ \hline & & & & 9\frac{19}{15} = 10\frac{4}{15} \end{array}$$

29.

$$\begin{array}{rcrcr} 5\frac{4}{10} & = & 5\frac{4\cdot 3}{10\cdot 3} & = & 5\frac{12}{30} \\ -\ 3\frac{1}{3} & = & -\ 3\frac{1\cdot 10}{3\cdot 10} & = & -\ 3\frac{10}{30} \\ \hline & & & & 2\frac{2}{30} = 2\frac{1}{15} \end{array}$$

31.

$$\begin{array}{r} 1\frac{1}{4} \\ 2\frac{3}{4} \\ +\ 5 \\ \hline 8\frac{4}{4} = 9 \end{array}$$

33.

$$\begin{array}{r} 7\frac{1}{10} \\ 8\frac{3}{10} \\ +\ 2\frac{7}{10} \\ \hline 17\frac{11}{10} = 18\frac{1}{10} \end{array}$$

35.

$$\begin{array}{r} \frac{3}{4} \\ 8\frac{1}{4} \\ +\ 5 \\ \hline 13\frac{4}{4} = 14 \end{array}$$

37.

$$\begin{array}{rcrcr} 3\frac{1}{2} & = & 3\frac{1\cdot 3}{2\cdot 3} & = & 3\frac{3}{6} \\ 8\frac{1}{3} & = & 8\frac{1\cdot 2}{3\cdot 2} & = & 8\frac{2}{6} \\ +\ 5\frac{1}{6} & = & +\ 5\frac{1}{6} & = & +\ 5\frac{1}{6} \\ \hline & & & & 16\frac{6}{6} = 17 \end{array}$$

39.

$$\begin{array}{rcrcr}
8\frac{2}{3} & = & 8\frac{2\cdot 8}{3\cdot 8} & = & 8\frac{16}{24} \\
9\frac{1}{8} & = & 9\frac{1\cdot 3}{8\cdot 3} & = & 9\frac{3}{24} \\
\underline{+\ 6\frac{1}{4}} & = & \underline{+\ 6\frac{1\cdot 6}{4\cdot 6}} & = & \underline{+\ 6\frac{6}{24}} \\
 & & & & 23\frac{25}{24} = 24\frac{1}{24}
\end{array}$$

41.

$$\begin{array}{rcrcr}
6\frac{1}{7} & = & 6\frac{1\cdot 2}{7\cdot 2} & = & 6\frac{2}{14} \\
9\frac{3}{14} & = & 9\frac{3}{14} & = & 9\frac{3}{14} \\
\underline{+\ 12\frac{1}{2}} & = & \underline{+\ 12\frac{1\cdot 7}{2\cdot 7}} & = & \underline{+\ 12\frac{7}{14}} \\
 & & & & 27\frac{12}{14} = 27\frac{6}{7}
\end{array}$$

43.

$$\begin{array}{rcrcr}
10\frac{1}{20} & = & 10\frac{1}{20} & = & 10\frac{1}{20} \\
11\frac{4}{5} & = & 11\frac{4\cdot 4}{5\cdot 4} & = & 11\frac{16}{20} \\
\underline{+\ 15\frac{3}{10}} & = & \underline{+\ 15\frac{3\cdot 2}{10\cdot 2}} & = & \underline{+\ 15\frac{6}{20}} \\
 & & & & 36\frac{23}{20} = 37\frac{3}{20}
\end{array}$$

45.

$$\begin{array}{rcr}
8 & = & 7\frac{4}{4} \\
\underline{-\ 1\frac{3}{4}} & = & \underline{-\ 1\frac{3}{4}} \\
 & & 6\frac{1}{4}
\end{array}$$

47.

$$\begin{array}{rcr}
15 & = & 14\frac{10}{10} \\
\underline{-\ 5\frac{3}{10}} & = & \underline{-\ 5\frac{3}{10}} \\
 & & 9\frac{7}{10}
\end{array}$$

49.

$$\begin{array}{rcr}
8\frac{1}{4} & = & 7\frac{5}{4} \\
\underline{-\ 2\frac{3}{4}} & = & \underline{-\ 2\frac{3}{4}} \\
 & & 5\frac{2}{4} = 5\frac{1}{2}
\end{array}$$

51.

$$\begin{array}{rcr}
9\frac{1}{3} & = & 8\frac{4}{3} \\
\underline{-\ 8\frac{2}{3}} & = & \underline{-\ 8\frac{2}{3}} \\
 & & \frac{2}{3}
\end{array}$$

53.

$$\begin{array}{rcrcrcr}
4\frac{1}{4} & = & 4\frac{1\cdot 3}{4\cdot 3} & = & 4\frac{3}{12} & = & 3\frac{15}{12} \\
\underline{-\ 2\frac{1}{3}} & = & \underline{-\ 2\frac{1\cdot 4}{3\cdot 4}} & = & \underline{-\ 2\frac{4}{12}} & = & \underline{-\ 2\frac{4}{12}} \\
 & & & & & & 1\frac{11}{12}
\end{array}$$

55.

$$\begin{array}{rcrcrcr}
9\frac{2}{3} & = & 9\frac{2\cdot 4}{3\cdot 4} & = & 9\frac{8}{12} & = & 8\frac{20}{12} \\
\underline{-\ 5\frac{3}{4}} & = & \underline{-\ 5\frac{3\cdot 3}{4\cdot 3}} & = & \underline{-\ 5\frac{9}{12}} & = & \underline{-\ 5\frac{9}{12}} \\
 & & & & & & 3\frac{11}{12}
\end{array}$$

57.

$$\begin{array}{rcrcrcr}
16\frac{3}{4} & = & 16\frac{3\cdot 5}{4\cdot 5} & = & 16\frac{15}{20} & = & 15\frac{35}{20} \\
\underline{-\ 10\frac{4}{5}} & = & \underline{-\ 10\frac{4\cdot 4}{5\cdot 4}} & = & \underline{-\ 10\frac{16}{20}} & = & \underline{-\ 10\frac{16}{20}} \\
 & & & & & & 5\frac{19}{20}
\end{array}$$

59.
$$\begin{array}{rcrcrcr} 10\dfrac{3}{10} & = & 10\dfrac{3}{10} & = & 10\dfrac{3}{10} & = & 9\dfrac{13}{10} \\ -\,4\dfrac{4}{5} & = & -\,4\dfrac{4\cdot 2}{5\cdot 2} & = & -\,4\dfrac{8}{10} & = & -\,4\dfrac{8}{10} \\ \hline & & & & & & 5\dfrac{5}{10}=5\dfrac{1}{2} \end{array}$$

61.
$$\begin{array}{rcrcrcr} 13\dfrac{1}{6} & = & 13\dfrac{1\cdot 4}{6\cdot 4} & = & 13\dfrac{4}{24} & = & 12\dfrac{28}{24} \\ -\,12\dfrac{5}{8} & = & -\,12\dfrac{5\cdot 3}{8\cdot 3} & = & -\,12\dfrac{15}{24} & = & -\,12\dfrac{15}{24} \\ \hline & & & & & & \dfrac{13}{24} \end{array}$$

63.
$$\begin{array}{rcrcrcr} 6\dfrac{1}{5} & = & 6\dfrac{1\cdot 2}{5\cdot 2} & = & 6\dfrac{2}{10} & = & 5\dfrac{12}{10} \\ -\,2\dfrac{7}{10} & = & -\,2\dfrac{7}{10} & = & -\,2\dfrac{7}{10} & = & -\,2\dfrac{7}{10} \\ \hline & & & & & & 3\dfrac{5}{10}=3\dfrac{1}{2} \end{array}$$

65.
$$\begin{array}{rcrcr} 3\dfrac{1}{8} & = & 3\dfrac{1\cdot 5}{8\cdot 5} & = & 3\dfrac{5}{40} \\ +\,2\dfrac{3}{5} & = & +\,2\dfrac{3\cdot 8}{5\cdot 8} & = & +\,2\dfrac{24}{40} \\ \hline & & & & 5\dfrac{29}{40} \end{array}$$

Applying the Concepts

67. Add the two lengths:
$$\begin{array}{r} 5\dfrac{7}{8} \\ +\,6\dfrac{3}{8} \\ \hline 11\dfrac{10}{8}=12\dfrac{2}{8}=12\dfrac{1}{4} \end{array}$$

The total length is $12\frac{1}{4}$ inches.

69. Find the difference between the distances:
$$\begin{array}{rcrcr} 1\dfrac{1}{2} & = & 1\dfrac{1\cdot 4}{2\cdot 4} & = & 1\dfrac{4}{8} \\ -\,1\dfrac{3}{8} & = & -\,1\dfrac{3}{8} & = & -\,1\dfrac{3}{8} \\ \hline & & & & \dfrac{1}{8} \end{array}$$

The horse ran $\frac{1}{8}$ mile further.

71. Original length – shrinkage = new length
$$\begin{array}{rcrcr} 32\dfrac{1}{2} & = & 32\dfrac{1\cdot 3}{2\cdot 3} & = & 32\dfrac{3}{6} \\ -\,1\dfrac{1}{3} & = & -\,1\dfrac{1\cdot 2}{3\cdot 2} & = & -\,1\dfrac{2}{6} \\ \hline & & & & 31\dfrac{1}{6} \end{array}$$

The new length is $31\frac{1}{6}$ inches.

73. NFL: $P = 2l + 2w$

$$= 2 \cdot 100 \text{ yd} + 2 \cdot 53\frac{1}{3} \text{ yd}$$
$$= \frac{200}{1} \text{ yd} + \frac{2}{1} \cdot \frac{160}{3} \text{ yd}$$
$$= \frac{600}{3} \text{ yd} + \frac{320}{3} \text{ yd}$$
$$= \frac{920}{3} \text{ yd} = 306\frac{2}{3} \text{ yd}$$

Canadian: $P = 2l + 2w$

$$= 2 \cdot 110 \text{ yd} + 2 \cdot 65 \text{ yd}$$
$$= 220 \text{ yd} + 130 \text{ yd}$$
$$= 350 \text{ yd}$$

Arena: $P = 2l + 2w$

$$= 2 \cdot 50 \text{ yd} + 2 \cdot 28\frac{1}{3} \text{ yd}$$
$$= \frac{100}{1} \text{ yd} + \frac{2}{1} \cdot \frac{85}{3} \text{ yd}$$
$$= \frac{300}{3} \text{ yd} + \frac{170}{3} \text{ yd}$$
$$= \frac{470}{3} \text{ yd} = 156\frac{2}{3} \text{ yd}$$

75. a. 3/13 price – 3/8 price = difference

$$\begin{array}{rcr} 77\frac{3}{8} & = & 76\frac{11}{8} \\ -\ 74\frac{7}{8} & = & -\ 74\frac{7}{8} \\ \hline & & 2\frac{4}{8} = 2\frac{1}{2} \end{array}$$

The difference is $\$2\frac{1}{2}$.

b. 100 shares × difference in price = net gain

$$100 \cdot \$2\frac{1}{2} = 100 \cdot \frac{5}{2} = \$250$$

The stock is worth $250 more.

77. 200 shares × difference in price = net gain/loss

$$200\left(34\frac{3}{8} - 32\frac{7}{8}\right) = 200\left(33\frac{11}{8} - 32\frac{7}{8}\right)$$
$$= 200\left(1\frac{4}{8}\right)$$
$$= 200\left(1\frac{1}{2}\right)$$
$$= 200 \cdot \frac{3}{2}$$
$$= 300$$

The stock would be worth $300 more.

Review Problems

81. $3 + 2 \cdot 7 = 3 + 14$
$= 17$

83. $4 \cdot 5 - 3 \cdot 2 = 20 - 6$
$= 14$

85. $3 \cdot 2^3 + 5 \cdot 4^2 = 3 \cdot 8 + 5 \cdot 16$
$= 24 + 80$
$= 104$

87. $3[2 + 5(6)] = 3[2 + 30]$
$= 3[32]$
$= 96$

89. $(7-3)(8+2) = (4)(10)$
$= 40$

Improving Your Quantitative Literacy

91. Halfbridle's speed:

$$\begin{array}{rcr} 35\frac{1}{5} & = & 34\frac{6}{5} \\ -\frac{2}{5} & = & -\frac{2}{5} \\ \hline & & 34\frac{4}{5} \text{ sec} \end{array}$$

Section 3.9

1. $3+\left(1\frac{1}{2}\right)\left(2\frac{2}{3}\right)=3+\left(\frac{3}{2}\right)\left(\frac{8}{3}\right)$
$=3+4$
$=7$

3. $8-\left(\frac{6}{11}\right)\left(1\frac{5}{6}\right)=8-\left(\frac{6}{11}\right)\left(\frac{11}{6}\right)$
$=8-1$
$=7$

5. $\frac{2}{3}\left(1\frac{1}{2}\right)+\frac{3}{4}\left(1\frac{1}{3}\right)=\left(\frac{2}{3}\right)\left(\frac{3}{2}\right)+\left(\frac{3}{4}\right)\left(\frac{4}{3}\right)$
$=1+1$
$=2$

7. $2\left(1\frac{1}{2}\right)+5\left(6\frac{2}{5}\right)=2\left(\frac{3}{2}\right)+5\left(\frac{32}{5}\right)$
$=3+32$
$=35$

9. $\left(\frac{3}{5}+\frac{1}{10}\right)\left(\frac{1}{2}+\frac{3}{4}\right)=\left(\frac{3\cdot 2}{5\cdot 2}+\frac{1}{10}\right)\left(\frac{1\cdot 2}{2\cdot 2}+\frac{3}{4}\right)$
$=\left(\frac{6}{10}+\frac{1}{10}\right)\left(\frac{2}{4}+\frac{3}{4}\right)$
$=\left(\frac{7}{10}\right)\left(\frac{5}{4}\right)$
$=\frac{7}{8}$

11. $\left(2+\frac{2}{3}\right)\left(3+\frac{1}{8}\right)=\left(2\frac{2}{3}\right)\left(3\frac{1}{8}\right)$
$=\left(\frac{8}{3}\right)\left(\frac{25}{8}\right)$
$=\frac{25}{3}$
$=8\frac{1}{3}$

13. $\left(1+\frac{5}{6}\right)\left(1-\frac{5}{6}\right)=\left(\frac{6}{6}+\frac{5}{6}\right)\left(\frac{6}{6}-\frac{5}{6}\right)$
$=\left(\frac{11}{6}\right)\left(\frac{1}{6}\right)$
$=\frac{11}{36}$

15. $\frac{2}{3}+\frac{1}{3}\left(2\frac{1}{2}+\frac{1}{2}\right)^2=\frac{2}{3}+\frac{1}{3}(3)^2$
$=\frac{2}{3}+\frac{1}{3}(9)$
$=\frac{2}{3}+3$
$=3\frac{2}{3}$

17. $2\frac{3}{8}+\frac{1}{2}\left(\frac{1}{3}+\frac{5}{3}\right)^3=2\frac{3}{8}+\frac{1}{2}\left(\frac{6}{3}\right)^3$
$=2\frac{3}{8}+\frac{1}{2}(2)^3$
$=2\frac{3}{8}+\frac{1}{2}(8)$
$=2\frac{3}{8}+4$
$=6\frac{3}{8}$

19. $2\left(\frac{1}{2}+\frac{1}{3}\right)+3\left(\frac{2}{3}+\frac{1}{4}\right)$

$=2\cdot\frac{1}{2}+2\cdot\frac{1}{3}+3\cdot\frac{2}{3}+3\cdot\frac{1}{4}$

$=1+\frac{2}{3}+2+\frac{3}{4}$

$=(1+2)+\left(\frac{2\cdot4}{3\cdot4}+\frac{3\cdot3}{4\cdot3}\right)$

$=3+\left(\frac{8}{12}+\frac{9}{12}\right)$

$=3\frac{17}{12}$

$=4\frac{5}{12}$

21. $\frac{\frac{2}{3}}{\frac{3}{4}}=\frac{2}{3}\div\frac{3}{4}$

$=\frac{2}{3}\cdot\frac{4}{3}$

$=\frac{8}{9}$

23. $\frac{\frac{2}{3}}{\frac{4}{3}}=\frac{2}{3}\div\frac{4}{3}$

$=\frac{2}{3}\cdot\frac{3}{4}$

$=\frac{2}{4}$

$=\frac{1}{2}$

25. $\frac{\frac{11}{20}}{\frac{5}{10}}=\frac{11}{20}\div\frac{5}{10}$

$=\frac{11}{20}\cdot\frac{10}{5}$

$=\frac{11}{10}$

$=1\frac{1}{10}$

27. $\frac{\frac{1}{2}+\frac{1}{3}}{\frac{1}{2}-\frac{1}{3}}=\frac{6\left(\frac{1}{2}+\frac{1}{3}\right)}{6\left(\frac{1}{2}-\frac{1}{3}\right)}$

$=\frac{6\cdot\frac{1}{2}+6\cdot\frac{1}{3}}{6\cdot\frac{1}{2}-6\cdot\frac{1}{3}}$

$=\frac{3+2}{3-2}$

$=5$

29. $\frac{\frac{5}{8}-\frac{1}{4}}{\frac{1}{8}+\frac{1}{2}}=\frac{8\left(\frac{5}{8}-\frac{1}{4}\right)}{8\left(\frac{1}{8}+\frac{1}{2}\right)}$

$=\frac{8\cdot\frac{5}{8}-8\frac{1}{4}}{8\cdot\frac{1}{8}+8\cdot\frac{1}{2}}$

$=\frac{5-2}{1+4}$

$=\frac{3}{5}$

31. $$\frac{\frac{9}{20}-\frac{1}{10}}{\frac{1}{10}+\frac{9}{20}}=\frac{20\left(\frac{9}{20}-\frac{1}{10}\right)}{20\left(\frac{1}{10}+\frac{9}{20}\right)}$$
$$=\frac{20\cdot\frac{9}{20}-20\cdot\frac{1}{10}}{20\cdot\frac{1}{10}+20\cdot\frac{9}{20}}$$
$$=\frac{9-2}{2+9}$$
$$=\frac{7}{11}$$

33. $$\frac{1+\frac{2}{3}}{1-\frac{2}{3}}=\frac{3\left(1+\frac{2}{3}\right)}{3\left(1-\frac{2}{3}\right)}$$
$$=\frac{3\cdot1+3\cdot\frac{2}{3}}{3\cdot1-3\cdot\frac{2}{3}}$$
$$=\frac{3+2}{3-2}$$
$$=5$$

35. $$\frac{2+\frac{5}{6}}{5-\frac{1}{3}}=\frac{6\left(2+\frac{5}{6}\right)}{6\left(5-\frac{1}{3}\right)}$$
$$=\frac{6\cdot2+6\cdot\frac{5}{6}}{6\cdot5-6\cdot\frac{1}{3}}$$
$$=\frac{12+5}{30-2}$$
$$=\frac{17}{28}$$

37. $$\frac{3+\frac{5}{6}}{1+\frac{5}{3}}=\frac{6\left(3+\frac{5}{6}\right)}{6\left(1+\frac{5}{3}\right)}$$
$$=\frac{6\cdot3+6\cdot\frac{5}{6}}{6\cdot1+6\cdot\frac{5}{3}}$$
$$=\frac{18+5}{6+10}$$
$$=\frac{23}{16}$$
$$=1\frac{7}{16}$$

39. $$\frac{\frac{1}{3}+\frac{3}{4}}{2-\frac{1}{6}}=\frac{12\left(\frac{1}{3}+\frac{3}{4}\right)}{12\left(2-\frac{1}{6}\right)}$$
$$=\frac{12\cdot\frac{1}{3}+12\cdot\frac{3}{4}}{12\cdot2-12\cdot\frac{1}{6}}$$
$$=\frac{4+9}{24-2}$$
$$=\frac{13}{22}$$

41. $$\frac{\frac{5}{6}}{3+\frac{2}{3}}=\frac{6\left(\frac{5}{6}\right)}{6\left(3+\frac{2}{3}\right)}$$
$$=\frac{6\cdot\frac{5}{6}}{6\cdot3+6\cdot\frac{2}{3}}$$
$$=\frac{5}{18+4}$$
$$=\frac{5}{22}$$

43. $\dfrac{2\frac{1}{2}+\frac{1}{2}}{3\frac{3}{5}-\frac{2}{5}}=\dfrac{\frac{5}{2}+\frac{1}{2}}{\frac{18}{5}-\frac{2}{5}}$

$$=\dfrac{\frac{6}{2}}{\frac{16}{5}}=\dfrac{\frac{3}{1}}{\frac{16}{5}}$$

$$=3\div\frac{16}{5}$$

$$=3\cdot\frac{5}{16}$$

$$=\frac{15}{16}$$

45. $\dfrac{2+1\frac{2}{3}}{3\frac{5}{6}-1}=\dfrac{2+\frac{5}{3}}{\frac{23}{6}-1}$

$$=\frac{6\left(2+\frac{5}{3}\right)}{6\left(\frac{23}{6}-1\right)}$$

$$=\frac{12+10}{23-6}$$

$$=\frac{22}{17}$$

$$=1\frac{5}{17}$$

47. $\dfrac{3\frac{1}{4}-2\frac{1}{2}}{5\frac{3}{4}+1\frac{1}{2}}=\dfrac{\frac{13}{4}-\frac{5}{2}}{\frac{23}{4}+\frac{3}{2}}$

$$=\frac{4\left(\frac{13}{4}-\frac{5}{2}\right)}{4\left(\frac{23}{4}+\frac{3}{2}\right)}$$

$$=\frac{13-10}{23+6}$$

$$=\frac{3}{29}$$

49. $\dfrac{3\frac{1}{4}+5\frac{1}{6}}{2\frac{1}{3}+3\frac{1}{4}}=\dfrac{\frac{13}{4}+\frac{31}{6}}{\frac{7}{3}+\frac{13}{4}}$

$$=\frac{12\left(\frac{13}{4}+\frac{31}{6}\right)}{12\left(\frac{7}{3}+\frac{13}{4}\right)}$$

$$=\frac{39+62}{28+39}$$

$$=\frac{101}{67}$$

$$=1\frac{34}{67}$$

51. $\dfrac{6\frac{2}{3}+7\frac{3}{4}}{8\frac{1}{2}+9\frac{7}{8}}=\dfrac{\frac{20}{3}+\frac{31}{4}}{\frac{17}{2}+\frac{79}{8}}$

$$=\frac{24\left(\frac{20}{3}+\frac{31}{4}\right)}{24\left(\frac{17}{2}+\frac{79}{8}\right)}$$

$$=\frac{160+186}{204+237}$$

$$=\frac{346}{441}$$

53. $2\left(2\frac{1}{5}+\frac{3}{6}\right)=2\left(\frac{11}{5}+\frac{3}{6}\right)$

$$=2\left(\frac{11\cdot 6}{5\cdot 6}+\frac{3\cdot 5}{6\cdot 5}\right)$$

$$=2\left(\frac{66}{30}+\frac{15}{30}\right)$$

$$=2\cdot\frac{81}{30}$$

$$=\frac{81}{15}$$

$$=5\frac{6}{15}$$

$$=5\frac{2}{5}$$

55. $\left(\frac{3}{4}+2\right)+5\frac{1}{4}=\frac{3}{4}+\frac{2\cdot 4}{1\cdot 4}+\frac{21}{4}$
$=\frac{3+8+21}{4}$
$=\frac{32}{4}$
$=8$

57. $\frac{1}{3}\left(6x+3\frac{3}{5}\right)=\frac{1}{3}(6x)+\frac{1}{3}\left(\frac{18}{5}\right)$
$=2x+\frac{6}{5}$
$=2x+1\frac{1}{5}$

59. $\frac{2}{7}(7a-4)+3\frac{1}{7}=\frac{2}{7}(7a)-\frac{2}{7}(4)+3\frac{1}{7}$
$=2a-\frac{8}{7}+\frac{22}{7}$
$=2a+\frac{14}{7}$
$=2a+2$

61. $2\frac{1}{2}(4x+2)=\frac{5}{2}(4x+2)$
$=\frac{5}{2}(4x)+\frac{5}{2}(2)$
$=10x+5$

63. $5\left(\frac{1}{2}x-\frac{1}{3}\right)+2\frac{2}{3}=5\left(\frac{1}{2}x\right)-5\left(\frac{1}{3}\right)+2\frac{2}{3}$
$=\frac{5}{2}x-\frac{5}{3}+\frac{8}{3}$
$=\frac{5}{2}x-\frac{3}{3}$
$=2\frac{1}{2}x-1$

Applying the Concepts

65. Usual purchase:

$$\begin{array}{rcrcr} 32\frac{1}{2} & = & 32\frac{1\cdot 3}{2\cdot 3} & = & 32\frac{3}{6} \\ +\ 25\frac{1}{3} & = & +\ 25\frac{1\cdot 2}{3\cdot 2} & = & +\ 25\frac{2}{6} \\ \hline & & & & 57\frac{5}{6}\text{ yds} \end{array}$$

Double the order:

$2\left(57\frac{5}{6}\right)=2\left(\frac{347}{6}\right)=\frac{347}{3}=115\frac{2}{3}$ yds

67. $\text{Mean}=\frac{\text{Sum of the gains/losses}}{\text{Number of gains/losses}}$
$=\frac{\frac{3}{4}+\frac{9}{16}+\frac{3}{32}+\frac{7}{32}+\frac{1}{16}}{5}$
$=\frac{\frac{24}{32}+\frac{18}{32}+\frac{3}{32}+\frac{7}{32}+\frac{2}{32}}{5}$
$=\frac{\frac{54}{32}}{5}$
$=\frac{54}{32}\cdot\frac{1}{5}$
$=\frac{27}{16}\cdot\frac{1}{5}$
$=\frac{27}{80}$

The mean gain for the week is $\$\frac{27}{80}$.

Review Problems

69. $\frac{3}{4}\cdot\frac{8}{9}=\frac{\not{3}\cdot\not{4}\cdot 2}{\not{4}\cdot\not{3}\cdot 3}=\frac{2}{3}$

71. $\frac{2}{3}\div(-4)=\frac{2}{3}\cdot\left(-\frac{1}{4}\right)=-\frac{\not{2}\cdot 1}{3\cdot\not{2}\cdot 2}=-\frac{1}{6}$

73. $-\frac{3}{7}+\frac{2}{7}=\frac{-3+2}{7}=\frac{-1}{7}=-\frac{1}{7}$

75. $10-\frac{2}{9}=\frac{10\cdot 9}{1\cdot 9}-\frac{2}{9}=\frac{90}{9}-\frac{2}{9}=\frac{88}{9}=9\frac{7}{9}$

77. $\frac{x}{3}+\frac{4}{5}=\frac{x\cdot 5}{3\cdot 5}+\frac{4\cdot 3}{5\cdot 3}=\frac{5x}{15}+\frac{12}{15}=\frac{5x+12}{15}$

79. $4-\frac{2}{x}=\frac{4\cdot x}{1\cdot x}-\frac{2}{x}=\frac{4x-2}{x}$

81. $\frac{7}{a}+\frac{2}{3}=\frac{7\cdot 3}{a\cdot 3}+\frac{2\cdot a}{3\cdot a}=\frac{21}{3a}+\frac{2a}{3a}=\frac{21+2a}{3a}$

Chapter 3 Review

1. $\frac{6}{8}=\frac{\cancel{2}\cdot 3}{\cancel{2}\cdot 4}=\frac{3}{4}$

3. $\frac{110a^3}{70a}=\frac{\cancel{10}\cdot 11\cdot \cancel{a}\cdot a\cdot a}{\cancel{10}\cdot 7\cdot \cancel{a}}=\frac{11a^2}{7}$

5. $\frac{1}{5}(5x)=\left(\frac{1}{5}\cdot 5\right)x=\left(\frac{\cancel{1}}{\cancel{5}}\cdot\frac{\cancel{5}}{\cancel{1}}\right)x=x$

7.
$$\begin{aligned}&\frac{96}{25}\cdot\frac{15}{98}\cdot\frac{35}{54}\\&=\frac{(\cancel{2}\cdot\cancel{2}\cdot 2\cdot 2\cdot 2\cdot\cancel{3})\cdot(\cancel{3}\cdot\cancel{5})\cdot(\cancel{5}\cdot\cancel{7})}{(\cancel{5}\cdot\cancel{5})(\cancel{2}\cdot\cancel{7}\cdot 7)(\cancel{2}\cdot\cancel{3}\cdot\cancel{3}\cdot 3)}\\&=\frac{8}{21}\end{aligned}$$

9. $\frac{8}{9}\div\frac{4}{3}=\frac{8}{9}\cdot\frac{3}{4}=\frac{2\cdot\cancel{4}\cdot\cancel{3}}{3\cdot\cancel{3}\cdot\cancel{4}}=\frac{2}{3}$

11. $\frac{a^3}{b}\div\frac{a}{b^2}=\frac{a^3}{b}\cdot\frac{b^2}{a}=\frac{\cancel{a}\cdot a\cdot a\cdot\cancel{b}\cdot b}{\cancel{b}\cdot\cancel{a}}=a^2b$

13. $\frac{6}{8}-\frac{2}{8}=\frac{6-2}{8}=\frac{4}{8}=\frac{1}{2}$

15.
$$\begin{aligned}-3-\frac{1}{2}&=\frac{-3\cdot 2}{1\cdot 2}-\frac{1}{2}\\&=\frac{-6}{2}-\frac{1}{2}\\&=\frac{-6-1}{2}\\&=\frac{-7}{2}\\&=-\frac{7}{2}\end{aligned}$$

17. $\left.\begin{aligned}126&=2\cdot 3\cdot 3\cdot 7\\84&=2\cdot 2\cdot 3\cdot 7\end{aligned}\right\}\text{LCD}=2\cdot 2\cdot 3\cdot 3\cdot 7=252$

$$\begin{aligned}\frac{11}{126}-\frac{5}{84}&=\frac{11\cdot 2}{126\cdot 2}-\frac{5\cdot 3}{84\cdot 3}\\&=\frac{22}{252}-\frac{15}{252}\\&=\frac{7}{252}\\&=\frac{\cancel{7}}{\cancel{7}\cdot 36}\\&=\frac{1}{36}\end{aligned}$$

19. $3\frac{5}{8}=\frac{8\cdot 3+5}{8}=\frac{29}{8}$

21. $x-\frac{3}{4}=\frac{x\cdot 4}{1\cdot 4}-\frac{3}{4}=\frac{4x-3}{4}$

23. $2\div 3\frac{1}{4}=2\div\frac{13}{4}=2\cdot\frac{4}{13}=\frac{2\cdot 4}{1\cdot 13}=\frac{8}{13}$

25. $6\cdot 2\frac{1}{2}\cdot\frac{4}{5}=\frac{6}{1}\cdot\frac{5}{2}\cdot\frac{4}{5}=\frac{3\cdot\cancel{2}\cdot\cancel{5}\cdot 4}{1\cdot\cancel{2}\cdot\cancel{5}}=12$

27.
$$\begin{array}{rcrcr}8\frac{2}{3}&=&8\frac{2\cdot 4}{3\cdot 4}&=&8\frac{8}{12}\\+\;9\frac{1}{4}&=&+\;9\frac{1\cdot 3}{4\cdot 3}&=&+\;9\frac{3}{12}\\\hline&&&&17\frac{11}{12}\end{array}$$

29. $3+2\left(4\frac{1}{3}\right)=3+2\left(\frac{13}{3}\right)$

$=3+\frac{26}{3}$

$=\frac{3\cdot 3}{1\cdot 3}+\frac{26}{3}$

$=\frac{9}{3}+\frac{26}{3}$

$=\frac{35}{3}$

$=11\frac{2}{3}$

31. $\dfrac{1+\frac{2}{3}}{1-\frac{2}{3}}=\dfrac{3\left(1+\frac{2}{3}\right)}{3\left(1-\frac{2}{3}\right)}=\dfrac{3+2}{3-2}=\dfrac{5}{1}=5$

33. $\dfrac{\frac{7}{8}-\frac{1}{2}}{\frac{1}{4}+\frac{1}{2}}=\dfrac{8\left(\frac{7}{8}-\frac{1}{2}\right)}{8\left(\frac{1}{4}+\frac{1}{2}\right)}=\dfrac{7-4}{2+4}=\dfrac{3}{6}=\dfrac{1}{2}$

35. $\frac{1}{10}$ of 200 = number of defective items

$\frac{1}{10}\cdot 200=\frac{1\cdot 200}{10\cdot 1}=\frac{1\cdot \cancel{10}\cdot 20}{\cancel{10}\cdot 1}=\frac{20}{1}=20$

There are 20 defective items.

37. $3\left(2\frac{1}{4}+\frac{3}{4}\right)=3\left(2+\frac{1}{4}+\frac{3}{4}\right)$

$=3\left(2+\frac{4}{4}\right)$

$=3(2+1)$

$=3(3)$

$=9$

39. Find the fraction of the recipe to be made:

$\frac{\text{Cookies needed}}{\text{Total cookies}}=\frac{36}{48}=\frac{3}{4}$ of recipe

$\frac{3}{4}$ of $2\frac{1}{2}$ cups = flour needed

$\frac{3}{4}\cdot 2\frac{1}{2}=\frac{3}{4}\cdot\frac{5}{2}=\frac{3\cdot 5}{4\cdot 2}=\frac{15}{8}=1\frac{7}{8}$ cups

41. $3\cdot 3\frac{1}{2}=3\cdot\frac{7}{2}=\frac{3\cdot 7}{1\cdot 2}=\frac{21}{2}=10\frac{1}{2}$

$10\frac{1}{2}$ tbsp will be used.

43. $A=\frac{1}{2}bh$

$=\frac{1}{2}\cdot 12\frac{3}{5}\cdot 4$

$=\frac{1}{2}\cdot\frac{63}{5}\cdot\frac{4}{1}$

$=\frac{1\cdot 63\cdot \cancel{2}\cdot 2}{\cancel{2}\cdot 5\cdot 1}$

$=\frac{126}{5}$

$=25\frac{1}{5}$ ft^2

$P=10\frac{2}{5}$

$$\begin{array}{r} 12\frac{3}{5} \\ +\ 5 \\ \hline 27\frac{5}{5} \end{array}$$

$27\frac{5}{5}=28$ ft

Chapters 1-3 Cumulative Review

1. $(6+2)+(3+6)=8+9=17$

3.
$$\begin{array}{r} 1,985 \\ -\ 141 \\ \hline 1,844 \end{array}$$

5. $13x-19x=(13-19)x=-6x$

7. $\frac{3}{4}\left(\frac{8}{3}x\right)=\left(\frac{\cancel{3}}{\cancel{4}}\cdot\frac{\cancel{4}\cdot 2}{\cancel{3}}\right)x=2x$

9.
$$\begin{array}{r} 5,280 \\ \times\ 26 \\ \hline 31,680 \\ 105,600 \\ \hline 137,280 \end{array}$$

11.
$$\begin{aligned} 2(-2)2-5(-3)&=-4(2)-(-15)\\ &=-8+15\\ &=7 \end{aligned}$$

13. $\frac{2}{3}\cdot 6=\frac{2}{3}\cdot\frac{6}{1}=\frac{2\cdot 2\cdot\cancel{3}}{\cancel{3}\cdot 1}=\frac{4}{1}=4$

15. $\frac{3}{4}\cdot 4=\frac{3}{4}\cdot\frac{4}{1}=\frac{3\cdot\cancel{4}}{\cancel{4}\cdot 1}=\frac{3}{1}=3$

17. $8\div 4(2+6)=8\div 4(8)=2(8)=16$

19. $\frac{11}{77}=\frac{\cancel{11}}{\cancel{11}\cdot 7}=\frac{1}{7}$

21. $121\div 11\div 11=(121\div 11)\div 11=11\div 11=1$

23.
$$\begin{aligned} \left(\frac{3}{5}+\frac{2}{25}\right)-\frac{4}{125}&=\frac{3\cdot 25}{5\cdot 25}+\frac{2\cdot 5}{25\cdot 5}-\frac{4}{125}\\ &=\frac{75}{125}+\frac{10}{125}-\frac{4}{125}\\ &=\frac{81}{125} \end{aligned}$$

25.
$$\begin{aligned} 2\frac{1}{3}\left(1\frac{1}{4}\div\frac{1}{5}\right)&=\frac{7}{3}\cdot\frac{5}{4}\div\frac{1}{5}\\ &=\frac{7}{3}\cdot\frac{5}{4}\cdot\frac{5}{1}\\ &=\frac{175}{12}\\ &=14\frac{7}{12} \end{aligned}$$

27.
$$\begin{aligned} \frac{2}{3}+\frac{1}{9}+\frac{3}{4}&=\frac{2\cdot 12}{3\cdot 12}+\frac{1\cdot 4}{9\cdot 4}+\frac{3\cdot 9}{4\cdot 9}\\ &=\frac{24}{36}+\frac{4}{36}+\frac{27}{36}\\ &=\frac{55}{36}\\ &=1\frac{19}{36} \end{aligned}$$

29. $\frac{7}{9}$ of $3\frac{1}{4}=\frac{7}{9}\cdot 3\frac{1}{4}=\frac{7}{9}\cdot\frac{13}{4}=\frac{91}{36}=2\frac{19}{36}$

31.

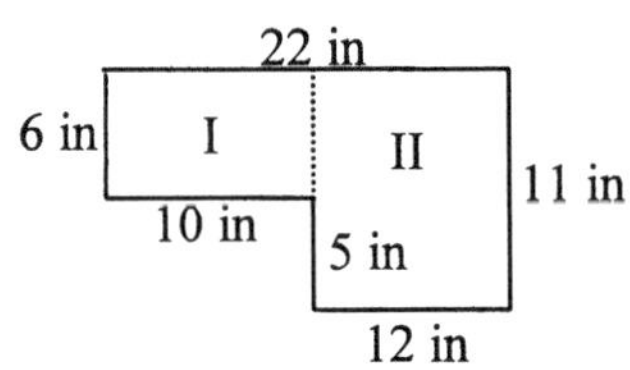

Separate into 2 regions.

Area I = $lw=10\cdot 6=60\text{ in}^2$

Area II = $lw=12\cdot 11=132\text{ in}^2$

Total Area = Area I + Area II

$=60\text{ in}^2+132\text{ in}^2=192\text{ in}^2$

33. $\frac{14x^2y}{49x^3y^4} = \frac{\cancel{7} \cdot 2 \cdot \cancel{x} \cdot \cancel{x} \cdot \cancel{y}}{\cancel{7} \cdot 7 \cdot \cancel{x} \cdot \cancel{x} \cdot x \cdot \cancel{y} \cdot y \cdot y \cdot y}$

$= \frac{2}{7xy^3}$

35. $\frac{1}{2} \cdot \frac{4}{9} + \left(-\frac{4}{5}\right) = \frac{1 \cdot \cancel{2} \cdot 2}{\cancel{2} \cdot 9} - \frac{4}{5}$

$= \frac{2}{9} - \frac{4}{5}$

$= \frac{2 \cdot 5}{9 \cdot 5} - \frac{4 \cdot 9}{5 \cdot 9}$

$= \frac{10}{45} - \frac{36}{45}$

$= -\frac{26}{45}$

37. $\frac{5}{6}$ of 36 exposures + $\frac{5}{6}$ of 2 rolls of 24 exposures = pictures taken

$\frac{5}{6} \cdot 36 + \frac{5}{6} \cdot 2 \cdot 24 = \frac{5}{6} \cdot \frac{36}{1} + \frac{5}{6} \cdot \frac{48}{1}$

$= 30 + 40$

$= 70$

She took 70 pictures.

Chapter 3 Test

1. Answers will vary.

$\frac{1}{8}$: 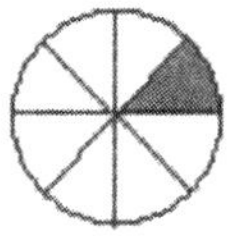 $\frac{3}{8}$: 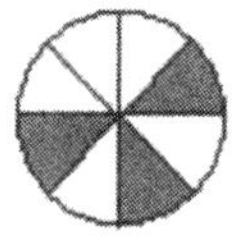

$\frac{5}{8}$: 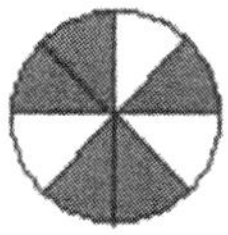 $\frac{7}{8}$: 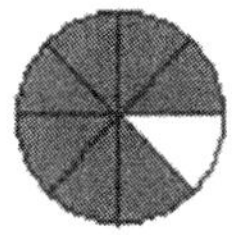

3. $\frac{3}{5}(30x) = \left(\frac{3}{5} \cdot \frac{30}{1}\right)x = \left(\frac{3 \cdot \cancel{5} \cdot 6}{\cancel{5} \cdot 1}\right)x = 18x$

5. $\frac{5}{18} \div \frac{15}{16} = \frac{5}{18} \cdot \frac{16}{15} = \frac{\cancel{5} \cdot \cancel{2} \cdot 8}{\cancel{2} \cdot 9 \cdot \cancel{5} \cdot 3} = \frac{8}{27}$

7. $\frac{3}{10} + \frac{1}{10} = \frac{3+1}{10} = \frac{4}{10} = \frac{2}{5}$

9.
$$\begin{aligned} -4 - \frac{3}{5} &= \frac{-4 \cdot 5}{1 \cdot 5} - \frac{3}{5} \\ &= \frac{-20}{5} - \frac{3}{5} \\ &= \frac{-20-3}{5} \\ &= -\frac{23}{5} \end{aligned}$$

11.
$$\left.\begin{aligned} 6 &= 2 \cdot 3 \\ 9 &= 3 \cdot 3 \\ 4 &= 2 \cdot 2 \end{aligned}\right\} \text{LCD} = 2 \cdot 2 \cdot 3 \cdot 3 = 36$$

$$\begin{aligned} \frac{5}{6} + \frac{2}{9} + \frac{1}{4} &= \frac{5 \cdot 6}{6 \cdot 6} + \frac{2 \cdot 4}{9 \cdot 4} + \frac{1 \cdot 9}{4 \cdot 9} \\ &= \frac{30}{36} + \frac{8}{36} + \frac{9}{36} \\ &= \frac{47}{36} \text{ or } 1\frac{11}{36} \end{aligned}$$

13.
$$\begin{array}{r} 8 \\ 5\overline{)43} \\ \underline{40} \\ 3 \end{array} \quad \text{so} \quad \frac{43}{5} = 8 + \frac{3}{5} = 8\frac{3}{5}$$

15. $6 \div 1\frac{1}{3} = 6 \div \frac{4}{3} = \frac{6}{1} \cdot \frac{3}{4} = \frac{\cancel{2} \cdot 3 \cdot 3}{1 \cdot \cancel{2} \cdot 2} = \frac{9}{2}$

17.
$$\begin{array}{rcrcrcr} 5\frac{1}{6} & = & 5\frac{1}{6} & = & 5\frac{1}{6} & = & 4\frac{7}{6} \\ \underline{-\ 1\frac{1}{2}} & = & \underline{-\ 1\frac{1 \cdot 3}{2 \cdot 3}} & = & \underline{-\ 1\frac{3}{6}} & = & \underline{-\ 1\frac{3}{6}} \\ & & & & & & 3\frac{4}{6} = 3\frac{2}{3} \end{array}$$

19.
$$\begin{aligned} \left(2\frac{1}{3} + \frac{1}{2}\right)\left(3\frac{2}{3} - \frac{1}{6}\right) &= \left(\frac{7}{3} + \frac{1}{2}\right)\left(\frac{11}{3} - \frac{1}{6}\right) \\ &= \left(\frac{14}{6} + \frac{3}{6}\right)\left(\frac{22}{6} - \frac{1}{6}\right) \\ &= \left(\frac{17}{6}\right)\left(\frac{21}{6}\right) \\ &= \frac{17 \cdot \cancel{3} \cdot 7}{2 \cdot \cancel{3} \cdot 2 \cdot 3} \\ &= \frac{119}{12} \\ &= 9\frac{11}{12} \end{aligned}$$

21. $\frac{1}{3}$ of 120 = number spoiled

$\frac{1}{3} \cdot 120 = \frac{1 \cdot \cancel{3} \cdot 40}{\cancel{3} \cdot 1} = \frac{40}{1} = 40$

40 grapefruit are spoiled.

23. $2\left(4\frac{2}{3}\right) = \left(\frac{2}{1}\right)\left(\frac{14}{3}\right) = \frac{28}{3} = 9\frac{1}{3}$

$9\frac{1}{3}$ cups will be used.

25. $A = \frac{1}{2}bh$

$= \frac{1}{2} \cdot 9\frac{1}{3} \cdot 5$

$= \frac{1}{2} \cdot \frac{28}{3} \cdot \frac{5}{1}$

$= \frac{1 \cdot \cancel{2} \cdot 14 \cdot 5}{\cancel{2} \cdot 3 \cdot 1}$

$= \frac{70}{3} = 23\frac{1}{3}\ \text{ft}^2$

$P = 8\frac{1}{3}$

$5\frac{2}{3}$

$+\ 9\frac{1}{3}$

$22\frac{4}{3} = 23\frac{1}{3}$ ft

Chapter 4 Preview

1. $-2+7=5$

3. $-2+(-4)=-6$

5. $(-4)(5)=-20$

7. $\frac{3}{2}(12)=\frac{3}{2}\cdot\frac{12}{1}=\frac{3\cdot\cancel{2}\cdot 6}{\cancel{2}\cdot 1}=18$

9. $\left(-\frac{5}{4}\right)\left(\frac{8}{15}\right)=-\frac{\cancel{5}\cdot 2\cdot\cancel{4}}{\cancel{4}\cdot 3\cdot\cancel{5}}=-\frac{2}{3}$

11. $\frac{1}{3}(15)+2=5+2=7$

13. $3x+7x=(3+7)x=10x$

15. $4(x-5)=4\cdot x-4\cdot 5=4x-20$

17. $6x-x=(6-1)x=5x$

19.
$$\begin{aligned} P &= 2l+2w \\ &= 2(3x)+2(x) \\ &= 6x+2x \\ &= 8x \end{aligned} \qquad \begin{aligned} A &= lw \\ &= (3x)(x) \\ &= 3x^2 \end{aligned}$$

Chapter 4 Pretest

1. $$\begin{aligned}4a-1-5a+8&=4a-5a-1+8\\&=(4-5)a-1+8\\&=-a+7\end{aligned}$$

3. When $x=-2$:

$$\begin{aligned}3x-4&=3(-2)-4\\&=-6-4\\&=-10\end{aligned}$$

5. $$\begin{aligned}a+4&=-2\\a+4+(-4)&=-2+(-4)\\a&=-6\end{aligned}$$

7. $$\begin{aligned}\frac{1}{5}x&=3\\5\left(\frac{1}{5}x\right)&=5(3)\\x&=15\end{aligned}$$

9. $$\begin{aligned}2a+1&=5(a-2)-1\\2a+1&=5a-10-1\\2a+1&=5a-11\\2a+1+(-2a)&=5a-11+(-2a)\\1&=3a-11\\1+11&=3a-11+11\\12&=3a\\4&=a\end{aligned}$$

11. Let x = number asked for

The sum of a number and 6 is -17

↓ ↓ ↓

$x+6 \quad = \quad -17$

$x+6=-17$

$x=-23$

The number is -23.

13. Let w = width.
Then $w+4$ = length.

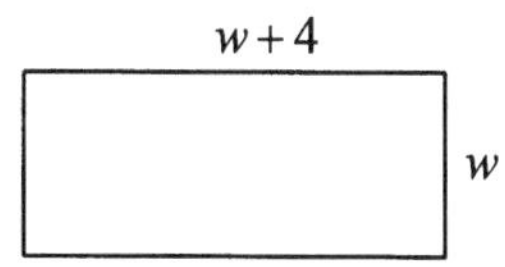

$$\begin{aligned}P&=2l+2w\\44&=2(w+4)+2w\\44&=2w+8+2w\\44&=4w+8\\36&=4w\\9&=w\end{aligned}$$

The width is $w=9$ cm.
The length is $w+4=9+4=13$ cm.

15. When $y=-3$:

$$\begin{aligned}5x+2y&=4\\5x+2(-3)&=4\\5x-6&=4\\5x&=10\\x&=2\end{aligned}$$

17.

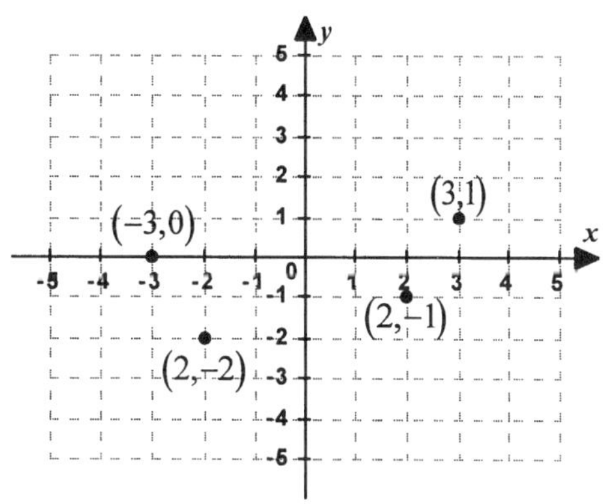

19. The equation $x = 2$ is equivalent to the equation $x + 0y = 2$. Solutions are any ordered pairs that have an x-coordinate of 2; y can be any number. Some solutions are: $(2,0),(2,3),(2,-2)$

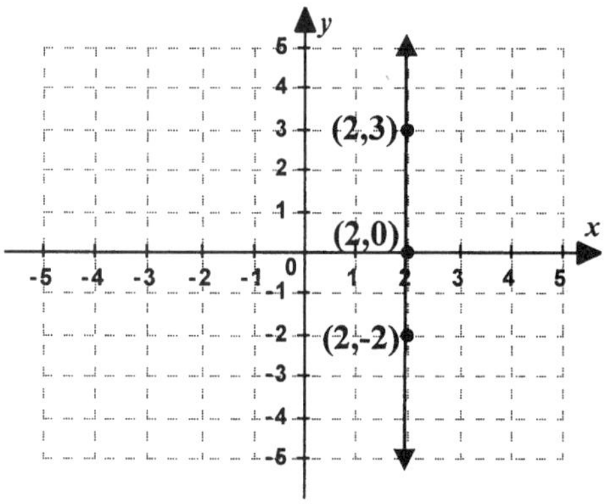

Section 4.1

1. $2x + 8x = (2+8)x = 10x$

3. $6a - 2a = (6-2)a = 4a$

5. $-4y + 5y = (-4+5)y = 1y = y$

7. $-6x - 2x = (-6-2)x = -8x$

9. $4a - a = 4a - 1a = (4-1)a = 3a$

11. $x - 6x = 1x - 6x = (1-6)x = -5x$

13. $4x + 2x + 3 + 8 = (4+2)x + 3 + 8$
$= 6x + 11$

15. $7x - 5x + 6 - 4 = (7-5)x + 6 - 4$
$= 2x + 2$

17. $-2a + a + 7 + 5 = (-2+1)a + 7 + 5$
$= -1a + 12$
$= -a + 12$

19. $6y - 2y - 5 + 1 = (6-2)y - 5 + 1$
$= 4y - 4$

21. $4x + 2x - 8x + 4 = (4+2-8)x + 4$
$= -2x + 4$

23. $9x - x - 5 - 1 = (9-1)x - 5 - 1$
$= 8x - 6$

25. $2a + 4 + 3a + 5 = 2a + 3a + 4 + 5$
$= (2+3)a + 9$
$= 5a + 9$

27. $3x + 2 - 4x + 1 = 3x - 4x + 2 + 1$
$= (3-4)x + 2 + 1$
$= -1x + 3$
$= -x + 3$

29. $12y + 3 + 5y = 12y + 5y + 3$
$= (12+5)y + 3$
$= 17y + 3$

31. $4a - 3 - 5a + 2a = 4a - 5a + 2a - 3$
$= (4-5+2)a - 3$
$= 1a - 3$
$= a - 3$

33. $2(3x+4) + 8 = 2(3x) + 2(4) + 8$
$= 6x + 8 + 8$
$= 6x + 16$

35. $5(2x-3) + 4 = 5(2x) - 5(3) + 4$
$= 10x - 15 + 4$
$= 10x - 11$

37. $8(2y+4) + 3y = 8(2y) + 8(4) + 3y$
$= 16y + 32 + 3y$
$= 16y + 3y + 32$
$= 19y + 32$

39. $6(4y-3) + 6y = 6(4y) - 6(3) + 6y$
$= 24y - 18 + 6y$
$= 24y + 6y - 18$
$= 30y - 18$

41. $2(x+3) + 4(x+2)$
$= 2(x) + 2(3) + 4(x) + 4(2)$
$= 2x + 6 + 4x + 8$
$= 2x + 4x + 6 + 8$
$= 6x + 14$

43. $3(2a+4)+7(3a-1)$

$=3(2a)+3(4)+7(3a)+7(-1)$

$=6a+12+21a-7$

$=6a+21a+12-7$

$=27a+5$

45. When $x=5$:

$2x+4=2(5)+4$

$=10+4$

$=14$

47. When $x=5$:

$7x-8=7(5)-8$

$=35-8$

$=27$

49. When $x=5$:

$-4x+1=-4(5)+1$

$=-20+1$

$=-19$

51. When $x=5$:

$-8+3x=-8+3(5)$

$=-8+15$

$=7$

53. When $a=-2$:

$2a+5=2(-2)+5$

$=-4+5$

$=1$

55. When $a=-2$:

$-7a+4=-7(-2)+4$

$=14+4$

$=18$

57. When $a=-2$:

$-a+10=-(-2)+10$

$=2+10$

$=12$

59. When $a=-2$:

$-4+3a=-4+3(-2)$

$=-4+(-6)$

$=-10$

61. When $x=3$:

$3x+5x+4=3(3)+5(3)+4$

$=9+15+4$

$=28$

63. When $x=3$:

$9x+x+3+7=9(3)+(3)+3+7$

$=27+3+3+7$

$=40$

65. When $x=3$:

$4x+3+2x+5=4(3)+3+2(3)+5$

$=12+3+6+5$

$=26$

67. When $x=3$:

$3x-8+2x-3=3(3)-8+2(3)-3$

$=9-8+6-3$

$=4$

69. $\text{Area}=6(x)+6(4)$ $\quad$ $\text{Area}=6(x+4)$

$=6x+24$ $\quad$ $=6x+24$

71. $P=4s=4(x+1)=4x+4$

73. $P=2a+2b$

$=2(3x+1)+2(2x-3)$

$=6x+2+4x-6$

$=10x-4$

Applying the Concepts

75. Complement = $90° - 25° = 65°$
Supplement = $180° - 25° = 155°$
25° is an acute angle, because it is between 0° and 90°.

77. a. When $A = 12,000$ ft:

$$\begin{aligned} 72 - \frac{A}{300} &= 72 - \frac{12,000}{300} \\ &= 72 - 40 \\ &= 32 \end{aligned}$$

The temperature is 32°F.

b. When $A = 15,000$ ft:

$$\begin{aligned} 72 - \frac{A}{300} &= 72 - \frac{15,000}{300} \\ &= 72 - 50 \\ &= 22 \end{aligned}$$

The temperature is 22°F.

c. When $A = 27,000$ ft:

$$\begin{aligned} 72 - \frac{A}{300} &= 72 - \frac{27,000}{300} \\ &= 72 - 90 \\ &= -18 \end{aligned}$$

The temperature is −18°F.

79. a. When $g = 10$ gallons:

$$\begin{aligned} 7 + 2g &= 7 + 2(10) \\ &= 7 + 20 \\ &= 27 \end{aligned}$$

The monthly bill is $27.

b. When $g = 20$ gallons:

$$\begin{aligned} 7 + 2g &= 7 + 2(20) \\ &= 7 + 40 \\ &= 47 \end{aligned}$$

The monthly bill is $47.

Review Problems

81. The opposite of 9 is −9.

83. The opposite of −6 is $-(-6) = 6$.

85. $2(6+5) = 2(6) + 2(5)$
Distributive Property

87. $x + 5 = 5 + x$
Commutative Property of Addition

89. $(x+5) + 1 = 1 + (x+5)$
Commutative Property of Addition

Section 4.2

1. When $x = 2$:

$$\begin{aligned} 2x+1&=5 \\ 2(2)+1&=5 \\ 4+1&=5 \\ 5&=5 \quad \text{true} \end{aligned}$$

$x = 2$ is a solution.

3. When $x = 5$:

$$\begin{aligned} 3x+4&=19 \\ 3(5)+4&=19 \\ 15+4&=19 \\ 19&=19 \quad \text{true} \end{aligned}$$

$x = 5$ is a solution.

5. When $x = 4$:

$$\begin{aligned} 2x-4&=2 \\ 2(4)-4&=2 \\ 8-4&=2 \\ 4&=2 \quad \text{false} \end{aligned}$$

$x = 4$ is not a solution.

7. When $x = -2$:

$$\begin{aligned} 2x+1&=3x+3 \\ 2(-2)+1&=3(-2)+3 \\ -4+1&=-6+3 \\ -3&=-3 \quad \text{true} \end{aligned}$$

$x = -2$ is a solution.

9. When $x = -4$:

$$\begin{aligned} x-4&=2x+1 \\ (-4)-4&=2(-4)+1 \\ -8&=-8+1 \\ -8&=-7 \quad \text{false} \end{aligned}$$

$x = -4$ is not a solution.

11.

$$\begin{aligned} x+2&=8 \\ x+2+(-2)&=8+(-2) \\ x+0&=6 \\ x&=6 \end{aligned}$$

13.

$$\begin{aligned} x-4&=7 \\ x-4+4&=7+4 \\ x-0&=11 \\ x&=11 \end{aligned}$$

15.

$$\begin{aligned} a+9&=-6 \\ a+9+(-9)&=-6+(-9) \\ a+0&=-15 \\ a&=-15 \end{aligned}$$

17.

$$\begin{aligned} x-5&=-4 \\ x-5+5&=-4+5 \\ x-0&=1 \\ x&=1 \end{aligned}$$

19.

$$\begin{aligned} y-3&=-6 \\ y-3+3&=-6+3 \\ y-0&=-3 \\ y&=-3 \end{aligned}$$

21.

$$\begin{aligned} a+\frac{1}{3}&=-\frac{2}{3} \\ a+\frac{1}{3}+\left(-\frac{1}{3}\right)&=-\frac{2}{3}+\left(-\frac{1}{3}\right) \\ a+0&=-\frac{3}{3} \\ a&=-1 \end{aligned}$$

23. $$x-\frac{3}{5}=\frac{4}{5}$$
$$x-\frac{3}{5}+\frac{3}{5}=\frac{4}{5}+\frac{3}{5}$$
$$x-0=\frac{7}{5}$$
$$x=\frac{7}{5}$$

25. $$a-\frac{5}{6}=\frac{1}{12}$$
$$a-\frac{5}{6}+\frac{5}{6}=\frac{1}{12}+\frac{5}{6}$$
$$a-0=\frac{1}{12}+\frac{10}{12}$$
$$a=\frac{11}{12}$$

27. $$x+4-7=3-10$$
$$x-3=-7$$
$$x-3+3=-7+3$$
$$x=-4$$

29. $$x-6+4=-3-2$$
$$x-2=-5$$
$$x-2+2=-5+2$$
$$x=-3$$

31. $$3-5=a-4$$
$$-2=a-4$$
$$-2+4=a-4+4$$
$$2=a$$

33. $$3a+7-2a=1$$
$$a+7=1$$
$$a+7+(-7)=1+(-7)$$
$$a=-6$$

35. $$6a-2-5a=-9+1$$
$$a-2=-8$$
$$a-2+2=-8+2$$
$$a=-6$$

37. $$8-5=3x-2x+4$$
$$3=x+4$$
$$3+(-4)=x+4+(-4)$$
$$-1=x$$

39. $$2(x+3)-x=4$$
$$2x+6-x=4$$
$$x+6=4$$
$$x+6+(-6)=4+(-6)$$
$$x=-2$$

41. $$-3(x-4)+4x=3-7$$
$$-3x+12+4x=-4$$
$$x+12=-4$$
$$x+12+(-12)=-4+(-12)$$
$$x=-16$$

43. $$5(2a+1)-9a=8-6$$
$$10a+5-9a=2$$
$$a+5=2$$
$$a+5+(-5)=2+(-5)$$
$$a=-3$$

45. $$-(x+3)+2x-1=6$$
$$-x-3+2x-1=6$$
$$x-4=6$$
$$x-4+4=6+4$$
$$x=10$$

47. $P = \text{sum of sides}$

$36 = 10 + 10 + x + 12$

$36 = x + 32$

$36 + (-32) = x + 32 + (-32)$

$4 = x$

49. $P = \text{sum of sides}$

$16 = 5 + 5 + x - 6$

$16 = x + 4$

$12 = x$

51.
$$x + 23° = 90°$$
$$x + 23° + (-23°) = 90° + (-23°)$$
$$x = 67°$$

53. a.
$$x + 86 + 89 = 400$$
$$x + 175 = 400$$
$$x + 175 + (-175) = 400 + (-175)$$
$$x = 225$$

There are 225 on the stage level.

b. \$30 × stage seats + \$25 × other seats = maximum amount of money

$30(225) + 25(86 + 89)$

$= 30(225) + 25(175)$

$= 6{,}750 + 4{,}375$

$= 11{,}125$

The maximum amount of money is \$11,125.

55.
$$x + 12 = 30$$
$$x + 12 + (-12) = 30 + (-12)$$
$$x = 18$$

57. $\underbrace{\text{The difference of 8 and 5}}\ \underbrace{\text{is equal to}}\ \underbrace{\text{the sum of } x \text{ and } 7}$

$$8 - 5 \quad = \quad x + 7$$
$$8 - 5 = x + 7$$
$$3 = x + 7$$
$$3 + (-7) = x + 7 + (-7)$$
$$-4 = x$$

Review Problems

59. $4(5x) = (4 \cdot 5)x = 20x$

61. $-2(3y) = (-2 \cdot 3)y = -6y$

63. $4(-3a) = (4 \cdot -3)a = -12a$

65. $\frac{1}{2}(2x) = \left(\frac{1}{2} \cdot 2\right)x = 1x = x$

67. $\frac{1}{5}(5y) = \left(\frac{1}{5} \cdot 5\right)y = 1y = y$

69. $3\left(\frac{1}{3}a\right) = \left(3 \cdot \frac{1}{3}\right)a = 1a = a$

Section 4.3

1. $\frac{1}{4}x = 2$

$4 \cdot \frac{1}{4}x = 4 \cdot 2$

$x = 8$

3. $\frac{1}{2}x = -3$

$2\left(\frac{1}{2}x\right) = 2(-3)$

$x = -6$

5. $-\frac{1}{3}x = 2$

$-3\left(-\frac{1}{3}x\right) = -3(2)$

$x = -6$

7. $-\frac{1}{6}x = -1$

$-6\left(-\frac{1}{6}x\right) = -6(-1)$

$x = 6$

9. $\frac{3}{4}y = 12$

$\frac{4}{3} \cdot \frac{3}{4}y = \frac{4}{3} \cdot 12$

$y = 16$

11. $3a = 48$

$\frac{3a}{3} = \frac{48}{3}$

$a = 16$

13. $-\frac{3}{5}x = \frac{9}{10}$

$-\frac{5}{3}\left(-\frac{3}{5}x\right) = -\frac{5}{3}\left(\frac{9}{10}\right)$

$x = -\frac{3}{2}$

15. $5x = -35$

$\frac{5x}{5} = \frac{-35}{5}$

$x = -7$

17. $-8y = 64$

$\frac{-8y}{-8} = \frac{64}{-8}$

$y = -8$

19. $-7x = -42$

$\frac{-7x}{-7} = \frac{-42}{-7}$

$x = 6$

21. $3x - 1 = 5$

$3x - 1 + 1 = 5 + 1$

$3x = 6$

$\frac{3x}{3} = \frac{6}{3}$

$x = 2$

23. $-4a + 3 = -9$

$-4a + 3 + (-3) = -9 + (-3)$

$-4a = -12$

$\frac{-4a}{-4} = \frac{-12}{-4}$

$a = 3$

25.
$$\begin{aligned} 6x-5&=19\\ 6x-5+5&=19+5\\ 6x&=24\\ \frac{6x}{6}&=\frac{24}{6}\\ x&=4 \end{aligned}$$

27.
$$\begin{aligned} \frac{1}{3}a+3&=-5\\ \frac{1}{3}a+3+(-3)&=-5+(-3)\\ \frac{1}{3}a&=-8\\ 3\left(\frac{1}{3}a\right)&=3(-8)\\ a&=-24 \end{aligned}$$

29.
$$\begin{aligned} -\frac{1}{4}a+5&=2\\ -\frac{1}{4}a+5+(-5)&=2+(-5)\\ -\frac{1}{4}a&=-3\\ -4\left(-\frac{1}{4}a\right)&=-4(-3)\\ a&=12 \end{aligned}$$

31.
$$\begin{aligned} 2x-4&=-20\\ 2x-4+4&=-20+4\\ 2x&=-16\\ \frac{2x}{2}&=\frac{-16}{2}\\ x&=-8 \end{aligned}$$

33.
$$\begin{aligned} \frac{2}{3}x-4&=6\\ \frac{2}{3}x-4+4&=6+4\\ \frac{2}{3}x&=10\\ \frac{3}{2}\cdot\frac{2}{3}x&=\frac{3}{2}\cdot 10\\ x&=15 \end{aligned}$$

35.
$$\begin{aligned} -11a+4&=-29\\ -11a+4+(-4)&=-29+(-4)\\ -11a&=-33\\ \frac{-11a}{-11}&=\frac{-33}{-11}\\ a&=3 \end{aligned}$$

37.
$$\begin{aligned} -3y-2&=1\\ -3y-2+2&=1+2\\ -3y&=3\\ \frac{-3y}{-3}&=\frac{3}{-3}\\ y&=-1 \end{aligned}$$

39.
$$\begin{aligned} -2x-5&=-7\\ -2x-5+5&=-7+5\\ -2x&=-2\\ \frac{-2x}{-2}&=\frac{-2}{-2}\\ x&=1 \end{aligned}$$

41.
$$\begin{aligned} 2x+3x-5&=7+3\\ 5x-5&=10\\ 5x-5+5&=10+5\\ 5x&=15\\ \frac{5x}{5}&=\frac{15}{5}\\ x&=3 \end{aligned}$$

43. $4x-7+2x=9-10$
$$6x-7=-1$$
$$6x-7+7=-1+7$$
$$6x=6$$
$$\frac{6x}{6}=\frac{6}{6}$$
$$x=1$$

45. $3a+2a+a=7-13$
$$6a=-6$$
$$\frac{6a}{6}=\frac{-6}{6}$$
$$a=-1$$

47. $5x+4x+3x=4-8$
$$12x=-4$$
$$\frac{12x}{12}=\frac{-4}{12}$$
$$x=-\frac{1}{3}$$

49. $5-18=3y-2y+1$
$$-13=y+1$$
$$-13+(-1)=y+1+(-1)$$
$$-14=y$$

51. $P=4s$
$$72=4(2x)$$
$$72=8x$$
$$\frac{72}{8}=\frac{8x}{8}$$
$$9=x$$

53. $P=2l+2w$
$$80=2(3x)+2(2x)$$
$$80=6x+4x$$
$$80=10x$$
$$\frac{80}{10}=\frac{10x}{10}$$
$$8=x$$

55. $1+2(4)+3x=21$
$$9+3x=21$$
$$9+(-9)+3x=21+(-9)$$
$$3x=12$$
$$\frac{3x}{3}=\frac{12}{3}$$
$$x=4$$
She made 4 three point shots.

57. $32x-25x-18=2,082$
$$7x-18=2,082$$
$$7x-18+18=2,082+18$$
$$7x=2,100$$
$$\frac{7x}{7}=\frac{2,100}{7}$$
$$x=300$$
You originally purchased 300 shares.

59. $\underbrace{\text{The sum of } 2x \text{ and } 5}$ is 19.

$\downarrow$ $\downarrow\downarrow$

$2x+5 \quad = 19$
$$2x+5=19$$
$$2x+5+(-5)=19+(-5)$$
$$2x=14$$
$$\frac{2x}{2}=\frac{14}{2}$$
$$x=7$$

61. $\underbrace{\text{The difference of } 5x \text{ and } 6}$ is -9.

$$\begin{array}{ccc} \downarrow & & \downarrow\ \downarrow \\ 5x-6 & & =\ -9 \end{array}$$

$$5x-6=-9$$
$$5x-6+6=-9+6$$
$$5x=-3$$
$$\frac{5x}{5}=\frac{-3}{5}$$
$$x=-\frac{3}{5}$$

Review Problems

63. $2(3a-8)=2\cdot 3a-2\cdot 8$
$=6a-16$

65. $-3(5x-1)=-3(5x)+(-3)(-1)$
$=-15x+3$

67. $3(y-5)+6=3\cdot y-3\cdot 5+6$
$=3y-15+6$
$=3y-9$

69. $6(2x-1)+4x=6\cdot 2x-6\cdot 1+4x$
$=12x-6+4x$
$=12x+4x-6$
$=16x-6$

Section 4.4

1.
$$5(x+1)=20$$
$$5x+5=20$$
$$5x+5+(-5)=20+(-5)$$
$$5x=15$$
$$\frac{5x}{5}=\frac{15}{5}$$
$$x=3$$

3.
$$6(x-3)=-6$$
$$6x-18=-6$$
$$6x-18+18=-6+18$$
$$6x=12$$
$$\frac{6x}{6}=\frac{12}{6}$$
$$x=2$$

5.
$$2x+4=3x+7$$
$$2x+4+(-3x)=3x+7+(-3x)$$
$$-1x+4=7$$
$$-1x+4+(-4)=7+(-4)$$
$$-1x=3$$
$$\frac{-1x}{-1}=\frac{3}{-1}$$
$$x=-3$$

7.
$$7y-3=4y-15$$
$$7y-3+(-4y)=4y-15+(-4y)$$
$$3y-3=-15$$
$$3y-3+3=-15+3$$
$$3y=-12$$
$$\frac{3y}{3}=\frac{-12}{3}$$
$$y=-4$$

9.
$$12x+3=-2x+17$$
$$12x+3+2x=-2x+17+2x$$
$$14x+3=17$$
$$14x+3+(-3)=17+(-3)$$
$$14x=14$$
$$\frac{14x}{14}=\frac{14}{14}$$
$$x=1$$

11.
$$6x-8=-x-8$$
$$6x-8+x=-x-8+x$$
$$7x-8=-8$$
$$7x-8+8=-8+8$$
$$7x=0$$
$$\frac{7x}{7}=\frac{0}{7}$$
$$x=0$$

13.
$$7(a-1)+4=11$$
$$7a-7+4=11$$
$$7a-3=11$$
$$7a-3+3=11+3$$
$$7a=14$$
$$\frac{7a}{7}=\frac{14}{7}$$
$$a=2$$

15.
$$8(x+5)-6=18$$
$$8x+40-6=18$$
$$8x+34=18$$
$$8x+34+(-34)=18+(-34)$$
$$8x=-16$$
$$\frac{8x}{8}=\frac{-16}{8}$$
$$x=-2$$

17. $2(3x-6)+1=7$

$$6x-12+1=7$$
$$6x-11=7$$
$$6x-11+11=7+11$$
$$6x=18$$
$$\frac{6x}{6}=\frac{18}{6}$$
$$x=3$$

19. $10(y+1)+4=3y+7$

$$10y+10+4=3y+7$$
$$10y+14=3y+7$$
$$10y+14+(-3y)=3y+7+(-3y)$$
$$7y+14=7$$
$$7y+14+(-14)=7+(-14)$$
$$7y=-7$$
$$\frac{7y}{7}=\frac{-7}{7}$$
$$y=-1$$

21. $4(x-6)+1=2x-9$

$$4x-24+1=2x-9$$
$$4x-23=2x-9$$
$$4x-23+(-2x)=2x-9+(-2x)$$
$$2x-23=-9$$
$$2x-23+23=-9+23$$
$$2x=14$$
$$\frac{2x}{2}=\frac{14}{2}$$
$$x=7$$

23. $2(3x+1)=4(x-1)$

$$6x+2=4x-4$$
$$6x+2+(-4x)=4x-4+(-4x)$$
$$2x+2=-4$$
$$2x+2+(-2)=-4+(-2)$$
$$2x=-6$$
$$\frac{2x}{2}=\frac{-6}{2}$$
$$x=-3$$

25. $3a+4=2(a-5)+15$

$$3a+4=2a-10+15$$
$$3a+4=2a+5$$
$$3a+4+(-2a)=2a+5+(-2a)$$
$$a+4=5$$
$$a+4+(-4)=5+(-4)$$
$$a=1$$

27. $9x-6=-3(x+2)-24$

$$9x-6=-3x-6-24$$
$$9x-6=-3x-30$$
$$9x-6+3x=-3x-30+3x$$
$$12x-6=-30$$
$$12x-6+6=-30+6$$
$$12x=-24$$
$$\frac{12x}{12}=\frac{-24}{12}$$
$$x=-2$$

29.
$$\begin{aligned} 3x-5&=11+2(x-6) \\ 3x-5&=11+2x-12 \\ 3x-5&=-1+2x \\ 3x-5+(-2x)&=-1+2x+(-2x) \\ x-5&=-1 \\ x-5+5&=-1+5 \\ x&=4 \end{aligned}$$

31. $\frac{x}{3}+\frac{x}{6}=5$ LCD = 6
$$\begin{aligned} 6\left(\frac{x}{3}+\frac{x}{6}\right)&=6(5) \\ 6\left(\frac{x}{3}\right)+6\left(\frac{x}{6}\right)&=6(5) \\ 2x+x&=30 \\ 3x&=30 \\ x&=10 \end{aligned}$$

33. $\frac{x}{5}-x=4$ LCD = 5
$$\begin{aligned} 5\left(\frac{x}{5}-x\right)&=5(4) \\ 5\left(\frac{x}{5}\right)-5(x)&=5(4) \\ x-5x&=20 \\ -4x&=20 \\ x&=-5 \end{aligned}$$

35. $3x+\frac{1}{2}=\frac{1}{4}$ LCD = 4
$$\begin{aligned} 4\left(3x+\frac{1}{2}\right)&=4\left(\frac{1}{4}\right) \\ 4(3x)+4\left(\frac{1}{2}\right)&=4\left(\frac{1}{4}\right) \\ 12x+2&=1 \\ 12x&=-1 \\ x&=-\frac{1}{12} \end{aligned}$$

37. $\frac{x}{3}+\frac{1}{2}=-\frac{1}{2}$ LCD = 6
$$\begin{aligned} 6\left(\frac{x}{3}+\frac{1}{2}\right)&=6\left(-\frac{1}{2}\right) \\ 6\left(\frac{x}{3}\right)+6\left(\frac{1}{2}\right)&=6\left(-\frac{1}{2}\right) \\ 2x+3&=-3 \\ 2x&=-6 \\ x&=-3 \end{aligned}$$

39. $\frac{4}{x}=\frac{1}{5}$ LCD = $5x$
$$\begin{aligned} 5x\left(\frac{4}{x}\right)&=5x\left(\frac{1}{5}\right) \\ 20&=x \end{aligned}$$

41. $\frac{3}{x}+1=\frac{2}{x}$ LCD = x
$$\begin{aligned} x\left(\frac{3}{x}+1\right)&=x\left(\frac{2}{x}\right) \\ x\left(\frac{3}{x}\right)+x(1)&=x\left(\frac{2}{x}\right) \\ 3+x&=2 \\ x&=-1 \end{aligned}$$

43. $$\frac{3}{x}-\frac{2}{x}=\frac{1}{5} \qquad \text{LCD} = 5x$$

$$5x\left(\frac{3}{x}-\frac{2}{x}\right)=5x\left(\frac{1}{5}\right)$$

$$5x\left(\frac{3}{x}\right)-5x\left(\frac{2}{x}\right)=5x\left(\frac{1}{5}\right)$$

$$15-10=x$$

$$5=x$$

45. $$\frac{1}{x}-\frac{1}{2}=-\frac{1}{4} \qquad \text{LCD} = 4x$$

$$4x\left(\frac{1}{x}-\frac{1}{2}\right)=4x\left(-\frac{1}{4}\right)$$

$$4x\left(\frac{1}{x}\right)-4x\left(\frac{1}{2}\right)=4x\left(-\frac{1}{4}\right)$$

$$4-2x=-x$$

$$4-2x+2x=-x+2x$$

$$4=x$$

47. P = sum of sides:

$$x+x+x+6=36$$

$$3x+6=36$$

$$3x=30$$

$$x=10$$

49. P = sum of sides:

$$x+x+x+1=16$$

$$3x+1=16$$

$$3x=15$$

$$x=5$$

Applying the Concepts

51. $$58-x=45$$

$$58-x+(-58)=45+(-58)$$

$$-x=-13$$

$$\frac{-1x}{-1}=\frac{-13}{-1}$$

$$x=13$$

$$45+x=58$$

$$45+x+(-45)=58+(-45)$$

$$x=13$$

Solving either equation, the angle is 13°.

53. $$x+\frac{1}{4}x=15 \qquad \text{LCD} = 4$$

$$4\left(x+\frac{1}{4}x\right)=4(15)$$

$$4(x)+4\left(\frac{1}{4}x\right)=4(15)$$

$$4x+x=60$$

$$5x=60$$

$$x=12$$

The quantity is 12.

Review Problems

55. The sum of a number and 2

$$x + 2$$

57. Twice a number

$$2\cdot x = 2x$$

59. Twice the sum of a number and 6

$$2\cdot (x+6) = 2(x+6)$$

61. The difference of

$\underbrace{x \text{ and } 4}$

$x - 4$

63. The sum of

$\underbrace{\text{twice a number}}$ and 5

$2x \quad + \ 5$

Extending the Concepts

65. a. Let $x =$ number of general tickets.
Then $x + 100 =$ number of student tickets.
Student sales + general sales = \$2,400

$\$4(x+100) \quad + \quad \$6x \quad = \$2,400$

b. $4(x+100) + 6x = 2{,}400$

$$4x + 400 + 6x = 2{,}400$$
$$10x + 400 = 2{,}400$$
$$10x = 2{,}000$$
$$x = 200$$

There were $x = 200$ general tickets, and $x + 100 = 200 + 100 = 300$ student tickets.

c. Total tickets: $300 + 200 = 500$.

Section 4.5

1. The sum of $\underbrace{x \text{ and } 3}$

$x+3$

3. The sum of $\underbrace{\text{twice } x}$ and 1

$2x \quad + \quad 1$

5. $\underbrace{\text{Five } x}$ $\underbrace{\text{decreased by}}$ 6

$5x \quad - \quad 6$

7. $\underbrace{\text{Three times}}$ $\underbrace{\text{the sum of } x \text{ and } 1}$

$3\times \quad (x+1) \quad = 3(x+1)$

9. $\underbrace{\text{Five times}}$ $\underbrace{\text{the sum of three } x \text{ and } 4}$

$5\times \quad (3x+4) \quad = 5(3x+4)$

Applying the Concepts

11. Step 1: *Read and list*
    Known items: the numbers 3 and 5.
    Unknown items: the number in question
    Step 2: *Assign a variable and translate*
    Let x = the number asked for
    Then $x+3$ ="the sum of a number and 3".
    Step 3: *Reread and write an equation*
    $\underbrace{\text{The sum of } x \text{ and } 3}$ is 5.

    $\downarrow \qquad \downarrow\downarrow$

    $x+3 \quad = 5$

    Step 4: *Solve the equation*
    $x+3=5$
    $x=2$
    Step 5: *Write your answer*
    The number is 2.
    Step 6: *Reread and check*
    The sum of 2 and 3 is 5.

13. Step 1: *Read and list*

Known items: the numbers 1 and –3, twice a number.

Unknown items: the number in question

Step 2: *Assign a variable and translate*

Let x = the number asked for

Then $2x+1$ = "the sum of twice a number and 1"

Step 3: *Reread and write an equation*

The sum of twice a number and 1 is -3.

$$\underbrace{\text{twice a number and 1}}$$

$$\downarrow \qquad \downarrow \downarrow$$

$$2x+1 \quad = -3$$

Step 4: *Solve the equation*

$$2x+1=-3$$
$$2x=-4$$
$$x=-2$$

Step 5: *Write your answer*

The number is –2.

Step 6: *Reread and check*

Twice –2 is –4. The sum of –4 and 1 is –3.

15. Step 1: *Read and list*

Known items: the numbers 6 and 9, five times a number.

Unknown items: the number in question

Step 2: *Assign a variable and translate*

Let x = the number asked for

Then $5x$ = five times the number

Step 3: *Reread and write and equation*

$\underbrace{\text{6 subtracted from}}$ $\underbrace{\text{five times a number}}$ is 9.

$$\downarrow \qquad \downarrow \downarrow$$

$$5x \quad -6 \quad = 9$$

Step 4: *Solve the equation*

$$5x-6=9$$
$$5x=15$$
$$x=3$$

Step 5: *Write your answer*

The number is 3.

Step 6: *Reread and check*

Five times 3 is 15. 6 subtracted from 15 is 9.

17. Step 1: *Read and list*

Known items: the numbers 1 and 18, three times the sum

Unknown items: the number in question

Step 2: *Assign a variable and translate*

Let x = the number asked for

Then $3(x+1)$ = "three times the sum of a number and 1".

Step 3: *Reread and write an equation*

Three times	the sum of a number and 1	is 18.
↓	↓	↓ ↓
$3\times$	$(x+1)$	$= 18$

Step 4: *Solve the equation*

$$3(x+1)=18$$
$$3x+3=18$$
$$3x=15$$
$$x=5$$

Step 5: *Write your answer*

The number is 5.

Step 6: *Reread and check*

The sum of 5 and 1 is 6. Three times that sum is 18.

19. Step 1: *Read and list*

Known items: the numbers 4 and –10, three times a number, five times the sum

Unknown items: the number in question

Step 2: *Assign a variable and translate*

Let x = the number asked for

Then $3x$ = "three times the number".

Step 3: *Reread and write an equation*

Five times	the sum of three times a number and 4	is -10.
↓	↓	↓ ↓
$5\times$	$(3x+4)$	$=-10$

Step 4: *Solve the equation*

$$5(3x+4)=-10$$
$$15x+20=-10$$
$$15x=-30$$
$$x=-2$$

Step 5: *Write your answer*

The number is –2.

Step 6: *Reread and check*

Three times –2 is –6. The sum of –6 and 4 is –2. Five times –2 is –10.

Geometry Problems

21. Step 1: *Read and list*

Known items: Perimeter is 30 m, the length is twice the width

Unknown items: the length and width

Step 2: *Assign a variable and translate*

Let x = width (in meters)

Then $2x$ = length (in meters)

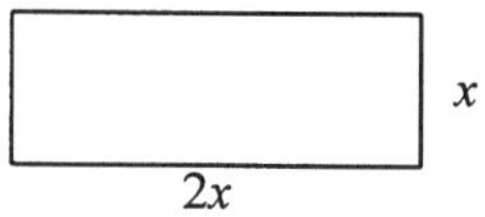

Step 3: *Reread and write an equation*

Perimeter = sum of all sides

$$30 = x + x + 2x + 2x$$

Step 4: *Solve the equation*

$$30 = x + x + 2x + 2x$$

$$30 = 6x$$

$$5 = x$$

Step 5: *Write your answer*

The width is $x = 5$ meters.

The length is $2x = 2(5) = 10$ meters.

Step 6: *Reread and check*

The length of 10 m is twice the width of 5 m. The perimeter is 5 + 5 + 10 + 10 = 30 m.

23. Step 1: *Read and list*

Known items: Perimeter is 32 cm

Unknown items: the length of the square

Step 2: *Assign a variable and translate*

Let x = length (in cm) of one side of the square

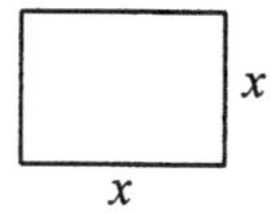

Step 3: *Reread and write an equation*

Perimeter = $4x$

$$32 = 4x$$

Step 4: *Solve the equation*

$$32 = 4x$$

$$8 = x$$

Step 5: *Write your answer*

The length of one side is 8 cm.

Step 6: *Reread and check*

The perimeter is 4(8) cm = 32 cm.

25. Step 1: *Read and list*

Known items: Two equal angles, their sum equals the 3rd angle

Unknown items: the measure of all three angles

Step 2: *Assign a variable and translate*

Let x = measure of 1st angle

Then x = measure of the 2nd angle

and $x+x=2x=$ measure of the 3rd angle

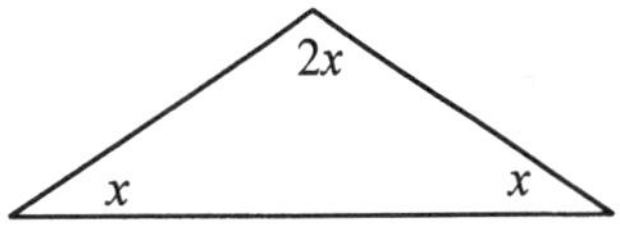

Step 3: *Reread and write an equation*

Sum of all angles = 180°

$$x+x+2x=180°$$

Step 4: *Solve the equation*

$$x+x+2x=180°$$
$$4x=180°$$
$$x=45°$$

Step 5: *Write your answer*

The 1st and 2nd angles equal $x=45°$.

The 3rd angle is $2x=2(45°)=90°$.

Step 6: *Reread and check*

The 3rd angle of 90° is the sum of the two 45° angles.

The sum of the three angles is $45°+45°+90°=180°$.

27. Step 1: *Read and list*

Known items: Smallest angle is 1/3 the largest angle, third angle is twice the smallest

Unknown items: the measure of all three angles

Step 2: *Assign a variable and translate*

Let x = measure of largest angle

Then $\frac{1}{3}x$ = measure of smallest angle

and $2\left(\frac{1}{3}x\right)$ = measure of the 3rd angle

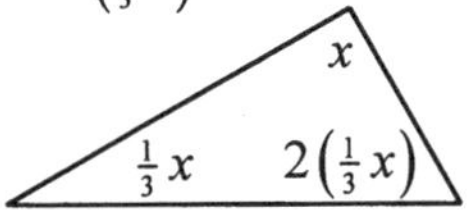

Step 3: *Reread and write an equation*

Sum of all angles = 180°

$$x+\frac{1}{3}x+2\left(\frac{1}{3}x\right)=180°$$

Step 4: *Solve the equation*

$$x+\frac{1}{3}x+2\left(\frac{1}{3}x\right)=180°$$
$$x+\frac{1}{3}x+\frac{2}{3}x=180° \qquad \text{LCD}=3$$
$$3\left(x+\frac{1}{3}x+\frac{2}{3}x\right)=3(180°)$$
$$3x+x+2x=540$$
$$5x=540$$
$$x=90$$

Step 5: *Write your answer*

The largest angle is $x=90°$.

The smallest angle is

$\frac{1}{3}x=\frac{1}{3}(90°)=30°$.

The third angle is

$2\left(\frac{1}{3}x\right)=2\left(\frac{1}{3}\cdot 90°\right)=60°$.

Step 6: *Reread and check*

The smallest angle is 30°, which is 1/3 if 90°. The third angle is 60°, which is twice the smallest angle.

The sum of the three angles is $30°+60°+90°=180°$.

Age Problems

29. Step 1: *Read and list*

Known items: Pat is 20 years older than Patrick. In 2 years, the sum of the their ages will be 90.

Unknown items: Their ages now

Step 2: *Assign a variable and translate*

Let x = Patrick's age

Then $x+20=$ Pat's age.

Step 3: *Reread and write an equation*

	Now	In 2 years
Patrick	x	$x+2$
Pat	$x+20$	$(x+20)+2$ $=x+22$

In 2 years:

Patrick's age + Pat's age = 90

$$(x+2) \quad + (x+22) = 90$$

Step 4: *Solve the equation*

$$(x+2)+(x+22)=90$$
$$2x+24=90$$
$$2x=66$$
$$x=33$$

Step 5: *Write your answer*

Patrick's age is $x=33$ years.

Pat's age is $x+20=33+20=53$ years.

Step 6: *Reread and check*

Pat is 53, twenty years older than Patrick, who is 33. In two years, Pat will be 55 and Patrick will be 35, and the sum of those ages is 90.

31. Step 1: *Read and list*

Known items: Dale is 4 years older than Sue. Five years ago the sum of their ages was 64.

Unknown items: Their ages now

Step 2: *Assign a variable and translate*

Let x = Sue's age

Then $x+4=$ Dale's age

Step 3: *Reread and write an equation*

	Now	5 years ago
Sue	x	$x-5$
Dale	$x+4$	$(x+4)-5=x-1$

5 years ago:

Sue's age + Dale's age = 64

$$(x-5) \quad + \quad (x-1) \quad = 64$$

Step 4: *Solve the equation*

$$(x-5)+(x-1)=64$$
$$2x-6=64$$
$$2x=70$$
$$x=35$$

Step 5: *Write your answer*

Sue's age is $x=35$ years.

Dales's age is $x+4=35+4=39$ years.

Step 6: *Reread and check*

Dale is 39 which is 4 years older than Sue, who is 35. Five years ago Dale was 34, Sue was 30, and the sum of those ages is 64.

Miscellaneous Problems

33. Using the 1st column:

$x+3+4=15$

$x+7=15$

$x=8$

Using the 2nd column:

$1+5+z=15$

$6+z=15$

$z=9$

Using the 3rd column:

$y+7+2=15$

$y+9=15$

$y=6$

35. *Known*: \$14/hr for 1st 35 hours, \$21/hr over 35 hours, made \$574 in 1 week.
Unknown: how many hours she worked
Let $x=$ hours worked
Note: Because \$14/hr × 35 hours = \$490, and she made \$574, she must have worked more than 35 hours. So x is more than 35. If x is the total hours worked, then $x-35$ is the number of overtime hours.

	Regular	Overtime
Hours worked	35	$x-35$
Wage per hour	\$14	\$21
Total wage	$\$14(35)$	$\$21(x-35)$

Regular wages + overtime wages = \$574

$\$14(35) \quad + \$21(x-35) \quad = \$574$

$14(35)+21(x-35)=574$

$490+21x-735=574$

$21x-245=574$

$21x=819$

$x=39$

She worked 39 hours.

37. The least amount of money would come from all 36 people being members.
\$3 per member $\times 36$ members = total receipts

$\$3(36)=\$108$

Ike and Nancy receive half the money:

$\frac{1}{2}(\$108)=\$54$

39. *Known*: members pay \$3, nonmembers pay \$5, 36 people attended, Ike and Nancy received \$80.
Unknown: How many of the 36 people were members and how many were nonmembers.
Let $x=$ number of members
Then $36-x=$ number of nonmembers

	Members	Nonmembers
Number	x	$36-x$
Cost per lesson	\$3	\$5
Total money	$\$3(x)$	$\$5(36-x)$

Because Ike and Nancy received \$80 which was half the receipts, the total receipts were \$160.

Member money + nonmember money = total

$\$3(x) \quad + \quad \$5(36-x) \quad = \$160$

$3x+5(36-x)=160$

$3x+180-5x=160$

$-2x+180=160$

$-2x=-20$

$x=10$

The number of members was $x=10$.
The number of nonmembers was $36-x=36-10=26$.
Yes, it is possible.

Review Problems

41. $\dfrac{3}{4}+\dfrac{2}{3}=\dfrac{3\cdot 3}{4\cdot 3}+\dfrac{2\cdot 4}{3\cdot 4}$

$$=\frac{9}{12}+\frac{8}{12}$$

$$=\frac{17}{12}$$

$$=1\frac{5}{12}$$

43. $\dfrac{3}{4}\left(\dfrac{2}{3}\right)=\dfrac{3\cdot 2}{4\cdot 3}=\dfrac{\cancel{3}\cdot\cancel{2}}{\cancel{2}\cdot 2\cdot\cancel{3}}=\dfrac{1}{2}$

45. $2+\dfrac{4}{5}=2\dfrac{4}{5}$

Note: Mixed numbers imply addition

47. $2\left(\dfrac{4}{5}\right)=\dfrac{2}{1}\cdot\dfrac{4}{5}=\dfrac{8}{5}=1\dfrac{3}{5}$

49.

$$\begin{array}{rcrcr} 3\dfrac{1}{3} & = & 3\dfrac{1\cdot 2}{3\cdot 2} & = & 3\dfrac{2}{6} \\ +\ 2\dfrac{1}{6} & = & +\ 2\dfrac{1}{6} & = & +\ 2\dfrac{1}{6} \\ \hline & & & & 5\dfrac{3}{6}=5\dfrac{1}{2} \end{array}$$

51. $3\dfrac{1}{3}\cdot 2\dfrac{1}{6}=\dfrac{10}{3}\cdot\dfrac{13}{6}=\dfrac{\cancel{2}\cdot 5\cdot 13}{3\cdot\cancel{2}\cdot 3}=\dfrac{65}{9}=7\dfrac{2}{9}$

Section 4.6

1. $A = lw$ *let* $l = 32,\ w = 22$

$= 32 \cdot 22$

$= 704$

The area is 704 square feet.

3. $A = lw$ *let* $l = \frac{3}{2}, w = \frac{3}{4}$

$= \frac{3}{2} \cdot \frac{3}{4}$

$= \frac{9}{8}$

The area is $\frac{9}{8}$ square inches.

5. $G = H \cdot R$ *let* $H = 40, R = 6$

$= 40 \cdot 6$

$= 240$

The gross pay is \$240.

7. $G = H \cdot R$ *let* $H = 30, R = 9\frac{1}{2}$

$= 30 \cdot 9\frac{1}{2}$

$= \frac{30}{1} \cdot \frac{19}{2}$

$= 15 \cdot 19$

$= 285$

The gross pay is \$285.

9. $F = 3 \cdot Y$ *let* $Y = 4$

$= 3 \cdot 4$

$= 12$

4 yards is equivalent to 12 feet.

11. $F = 3 \cdot Y$ *let* $Y = 2\frac{2}{3}$

$= 3 \cdot 2\frac{2}{3}$

$= 3 \cdot \frac{8}{3}$

$= 8$

$2\frac{2}{3}$ yards is equivalent to 8 feet.

13. $I = P \cdot R \cdot T$ *let* $P = 1000, R = \frac{7}{100}, T = 2$

$= 1000 \cdot \frac{7}{100} \cdot 2$

$= \frac{14,000}{100}$

$= 140$

The interest is \$140.

15. $P = 2w + 2l$ *let* $w = 10, l = 19$

$= 2(10) + 2(19)$

$= 20 + 38$

$= 58$

The perimeter is 58 inches.

17. $P = 2w + 2l$ *let* $w = \frac{3}{4}, l = \frac{7}{8}$

$= 2\left(\frac{3}{4}\right) + 2\left(\frac{7}{8}\right)$

$= \frac{3}{2} + \frac{7}{4}$

$= \frac{6}{4} + \frac{7}{4}$

$= \frac{13}{4}$

The perimeter is $\frac{13}{4}$ feet.

19. $C = \frac{5}{9}(F - 32)$ *let* $F = 212$

$= \frac{5}{9}(212 - 32)$

$= \frac{5}{9}(180)$

$= 5 \cdot 20$

$= 100$

212°F is equivalent to 100°C (same as in the table).

21. $C = \frac{5}{9}(F-32) \quad let\ F = 68$

$= \frac{5}{9}(68-32)$

$= \frac{5}{9}(36)$

$= 5 \cdot 4$

$= 20$

68°F is equivalent to 20°C (same as in the table).

23. Since you want to find C, use the formula for C:

$C = \frac{5}{9}(F-32) \quad let\ F = 32$

$= \frac{5}{9}(32-32)$

$= \frac{5}{9}(0)$

$= 0$

32°F is equivalent to 0°C.

25. Since you want to find F, use the formula for F:

$F = \frac{9}{5}C + 32 \qquad let\ C = -15$

$= \frac{9}{5}(-15) + 32$

$= -27 + 32$

$- 5$

−15°C is equivalent to 5°F.

27. The table is as follows:

Age A (yrs)	Max heart rate $M = 220 - A$ (beats per minute)
18	$M = 220 - 18 = 202$
19	$M = 220 - 19 = 201$
20	$M = 220 - 20 = 200$
21	$M = 220 - 21 = 199$
22	$M = 220 - 22 = 198$
23	$M = 220 - 23 = 197$

29. For a 20 year old, the maximum heart rate is $M = 200$ (from problem 27). Therefore, the training heart rate is

$$T = R + \frac{3}{5}(200 - R) = R + 120 - \frac{3}{5}R = \frac{2}{5}R + 120$$

Resting Heart rate, R	Training heart rate $T = \frac{2}{5}R + 120$
60	$T = \frac{2}{5}(60) + 120 = 24 + 120 = 144$
62	$T = \frac{2}{5}(62) + 120 = \frac{124}{5} + 120$ $= 144\frac{4}{5} \approx 145$
64	$T = \frac{2}{5}(64) + 120 = \frac{128}{5} + 120$ $= 145\frac{3}{5} \approx 146$
68	$T = \frac{2}{5}(68) + 120 = \frac{136}{5} + 120$ $= 147\frac{1}{5} \approx 147$
70	$T = \frac{2}{5}(70) + 120 = 28 + 120 = 148$
72	$T = \frac{2}{5}(72) + 120 = \frac{144}{5} + 120$ $= 148\frac{4}{5} \approx 149$

31. a. $6:30 - 2:30 = 4$ hrs

b. $d = rt = \frac{55\text{ mi}}{\text{hr}} \cdot 4\text{ hrs} = 220\text{ mi}$

33. a. $6:30 - 2:30 = 4$ hrs

b. $r = \frac{d}{t} = \frac{260\text{ mi}}{4\text{ hrs}} = \frac{65\text{ mi}}{\text{hr}}$

35. $V = lwh$ *let* $l = 6, w = 12, h = 5$

$$= 6 \cdot 12 \cdot 5$$
$$= 360$$

The volume is 360 in^3.

37. $V = lwh$ *let* $l = 6, w = \frac{1}{2}, h = \frac{1}{3}$

$$= 6\left(\frac{1}{2}\right)\left(\frac{1}{3}\right)$$
$$= \frac{6}{6}$$
$$= 1$$

The volume is 1 yd^3.

39. $y = 3x - 2$ *let* $x = 3$

$$= 3(3) - 2$$
$$= 9 - 2$$
$$= 7$$

41. $y = 3x - 2$ *let* $x = -\frac{1}{3}$

$$= 3\left(-\frac{1}{3}\right) - 2$$
$$= -1 - 2$$
$$= -3$$

43. $y = 3x - 2$ *let* $x = 0$

$$= 3(0) - 2$$
$$= 0 - 2$$
$$= -2$$

45. $x + y = 5$ *let* $y = 2$

$$x + 2 = 5$$
$$x = 3$$

47. $x + y = 5$ *let* $y = 0$

$$x + 0 = 5$$
$$x = 5$$

49. $x + y = 5$ *let* $y = -3$

$$x + (-3) = 5$$
$$x = 8$$

51. $x + y = 3$ *let* $x = 2$

$$2 + y = 3$$
$$y = 1$$

53. $x + y = 3$ *let* $x = 0$

$$0 + y = 3$$
$$y = 3$$

55. $x + y = 3$ *let* $x = \frac{1}{2}$

$$\frac{1}{2} + y = 3$$
$$y = 3 + \left(-\frac{1}{2}\right)$$
$$y = \frac{5}{2}$$

57. $4x + 3y = 12$ *let* $x = 3$

$$4(3) + 3y = 12$$
$$12 + 3y = 12$$
$$3y = 0$$
$$y = 0$$

59. $4x + 3y = 12$ *let* $x = -\frac{1}{4}$

$$4\left(-\frac{1}{4}\right) + 3y = 12$$
$$-1 + 3y = 12$$
$$3y = 13$$
$$y = \frac{13}{3}$$

61. $4x+3y=12$ *let* $x=0$

$$\begin{aligned}4(0)+3y&=12\\0+3y&=12\\3y&=12\\y&=4\end{aligned}$$

63. $4x+3y=12$ *let* $y=4$

$$\begin{aligned}4x+3(4)&=12\\4x+12&=12\\4x&=0\\x&=0\end{aligned}$$

65. $4x+3y=12$ *let* $y=-\frac{1}{3}$

$$\begin{aligned}4x+3\left(-\frac{1}{3}\right)&=12\\4x+(-1)&=12\\4x&=13\\x&=\frac{13}{4}\end{aligned}$$

67. $4x+3y=12$ *let* $y=0$

$$\begin{aligned}4x+3(0)&=12\\4x+0&=12\\4x&=12\\x&=3\end{aligned}$$

69. The complement of 45° is $90°-45°=45°$.
The supplement of 45° is $180°-45°=135°$.

71. The complement of 31° is $90°-31°=59°$.
The supplement of 31° is $180°-31°=149°$.

73. a. $$\begin{aligned}S&=\frac{h\cdot w\cdot fps\cdot t}{35{,}000}\\&=\frac{480\cdot216\cdot30\cdot150}{35{,}000}\\&=\frac{466{,}560{,}000}{35{,}000}\\&=\frac{466{,}560}{35}=13{,}330\frac{10}{35}\end{aligned}$$

The trailer is a little over 13,330 Kb.

b. $$\begin{aligned}S&=\frac{h\cdot w\cdot fps\cdot t}{35{,}000}\\&=\frac{320\cdot144\cdot15\cdot150}{35{,}000}\\&=\frac{103{,}680{,}000}{35{,}000}\\&=\frac{103{,}680}{35}=2{,}962\frac{10}{35}\end{aligned}$$

The trailer is a little over 2,962 Kb.

75. $A=lw=6\cdot5=30$

$P=30 \qquad N=4$

$$\begin{aligned}W&=\frac{APN}{2{,}000}\\&=\frac{30\cdot30\cdot4}{2{,}000}\\&=\frac{3{,}600}{2{,}000}\\&=\frac{36}{20}\\&=\frac{9}{5}\text{ or }1\frac{4}{5}\end{aligned}$$

The weight is $1\frac{4}{5}$ tons.

Note: $\frac{9}{5}\text{tons}\cdot\frac{2000\text{ lb}}{\text{ton}}=3{,}600$ lbs.

Review Problems

77. $\dfrac{\frac{3}{5}}{\frac{4}{5}}=\dfrac{3}{5}\div\dfrac{4}{5}=\dfrac{3}{5}\cdot\dfrac{5}{4}=\dfrac{3}{4}$

79. $$\begin{aligned}\frac{1+\frac{1}{2}}{1-\frac{1}{2}}&=\frac{2\left(1+\frac{1}{2}\right)}{2\left(1-\frac{1}{2}\right)}\\&=\frac{2(1)+2\left(\frac{1}{2}\right)}{2(1)-2\left(\frac{1}{2}\right)}\\&=\frac{2+1}{2-1}\\&=\frac{3}{1}\\&=3\end{aligned}$$

81. $$\begin{aligned}\frac{\frac{1}{2}+\frac{1}{4}}{\frac{1}{4}+\frac{1}{8}}&=\frac{8\left(\frac{1}{2}+\frac{1}{4}\right)}{8\left(\frac{1}{4}+\frac{1}{8}\right)}\\&=\frac{4+2}{2+1}\\&=\frac{6}{3}\\&=2\end{aligned}$$

83. $$\begin{aligned}\frac{\frac{3}{5}+\frac{3}{7}}{\frac{3}{5}-\frac{3}{7}}&=\frac{35\left(\frac{3}{5}+\frac{3}{7}\right)}{35\left(\frac{3}{5}-\frac{3}{7}\right)}\\&=\frac{21+15}{21-15}\\&=\frac{36}{6}\\&=6\end{aligned}$$

Section 4.7

1. When $x = 0$:

$$\begin{aligned} x + y &= 4 \\ 0 + y &= 4 \\ y &= 4 \end{aligned}$$

solution: $(0, 4)$

When $x = 3$:

$$\begin{aligned} x + y &= 4 \\ 3 + y &= 4 \\ y &= 1 \end{aligned}$$

solution: $(3,1)$

When $x = -2$:

$$\begin{aligned} x + y &= 4 \\ -2 + y &= 4 \\ y &= 6 \end{aligned}$$

solution: $(-2, 6)$

3. When $x = 0$:

$$\begin{aligned} x + 2y &= 6 \\ 0 + 2y &= 6 \\ 2y &= 6 \\ y &= 3 \end{aligned}$$

solution: $(0,3)$

When $x = 2$:

$$\begin{aligned} x + 2y &= 6 \\ 2 + 2y &= 6 \\ 2y &= 4 \\ y &= 2 \end{aligned}$$

solution: $(2,2)$

When $y = -6$:

$$\begin{aligned} x + 2y &= 6 \\ x + 2(-6) &= 6 \\ x + (-12) &= 6 \\ x &= 18 \end{aligned}$$

solution: $(18, -6)$

5. When $x = 0$:

$$\begin{aligned} 4x + 3y &= 12 \\ 4(0) + 3y &= 12 \\ 0 + 3y &= 12 \\ 3y &= 12 \\ y &= 4 \end{aligned}$$

solution: $(0, 4)$

When $y = 0$:

$$\begin{aligned} 4x + 3y &= 12 \\ 4x + 3(0) &= 12 \\ 4x + 0 &= 12 \\ 4x &= 12 \\ x &= 3 \end{aligned}$$

solution: $(3, 0)$

When $x = -3$:

$$\begin{aligned} 4x + 3y &= 12 \\ 4(-3) + 3y &= 12 \\ -12 + 3y &= 12 \\ 3y &= 24 \\ y &= 8 \end{aligned}$$

solution: $(-3, 8)$

7. When $x = 1$:

$$\begin{aligned} y &= 4x - 3 \\ y &= 4(1) - 3 \\ y &= 4 - 3 \\ y &= 1 \end{aligned}$$

solution: $(1,1)$

When $y = -3$:

$$\begin{aligned} y &= 4x - 3 \\ -3 &= 4x - 3 \\ 0 &= 4x \\ 0 &= x \end{aligned}$$

solution: $(0, -3)$

When $x = 5$:

$$\begin{aligned} y &= 4x - 3 \\ y &= 4(5) - 3 \\ y &= 20 - 3 \\ y &= 17 \end{aligned}$$

solution: $(5,17)$

9. When $x = 0$:

$$y = 2x + 3$$
$$y = 2(0) + 3$$
$$y = 0 + 3$$
$$y = 3$$

solution: $(0, 3)$

When $x = 2$:

$$y = 2x + 3$$
$$y = 2(2) + 3$$
$$y = 4 + 3$$
$$y = 7$$

solution: $(2, 7)$

When $x = -2$:

$$y = 2x + 3$$
$$y = 2(-2) + 3$$
$$y = -4 + 3$$
$$y = -1$$

solution: $(-2, -1)$

11. When $x = 2$:

$$y = 7x$$
$$y = 7(2)$$
$$y = 14$$

solution: $(2, 14)$

When $y = 6$:

$$y = 7x$$
$$6 = 7x$$
$$\frac{6}{7} = x$$

solution: $\left(\frac{6}{7}, 6\right)$

When $x = 0$:

$$y = 7x$$
$$y = 7(0)$$
$$y = 0$$

solution: $(0, 0)$

13. When $x = 0$:

$$y = -2x$$
$$y = -2(0)$$
$$y = 0$$

solution: $(0, 0)$

When $x = -2$:

$$y = -2x$$
$$y = -2(-2)$$
$$y = 4$$

solution: $(-2, 4)$

When $x = 2$:

$$y = -2x$$
$$y = -2(2)$$
$$y = -4$$

solution: $(2, -4)$

15. When $x = 0$:

$$y = \frac{1}{2}x$$
$$y = \frac{1}{2}(0)$$
$$y = 0$$

solution: $(0, 0)$

When $x = 2$:

$$y = \frac{1}{2}x$$
$$y = \frac{1}{2}(2)$$
$$y = 1$$

solution: $(2, 1)$

When $x = 4$:

$$y = \frac{1}{2}x$$
$$y = \frac{1}{2}(4)$$
$$y = 2$$

solution: $(4, 2)$

17. When $x=-2$:

$$y=-\frac{1}{2}x+2$$
$$y=-\frac{1}{2}(-2)+2$$
$$y=1+2$$
$$y=3$$

solution: $(-2,3)$

When $x=0$:

$$y=-\frac{1}{2}x+2$$
$$y=-\frac{1}{2}(0)+2$$
$$y=0+2$$
$$y=2$$

solution: $(0,2)$

When $x=2$:

$$y=-\frac{1}{2}x+2$$
$$y=-\frac{1}{2}(2)+2$$
$$y=-1+2$$
$$y=1$$

solution: $(2,1)$

19. When $x=2$:

$$x+y=5$$
$$2+y=5$$
$$y=3$$

When $x=3$:

$$x+y=5$$
$$3+y=5$$
$$y=2$$

When $y=4$:

$$x+y=5$$
$$x+4=5$$
$$x=1$$

When $y=5$:

$$x+y=5$$
$$x+5=5$$
$$x=0$$

x	y
2	3
3	2
1	4
0	5

21. When $x=0$:

$$y=-4x$$
$$y=-4(0)$$
$$y=0$$

When $y=4$:

$$y=-4x$$
$$4=-4x$$
$$-1=x$$

When $y=8$:

$$y=-4x$$
$$8=-4x$$
$$-2=x$$

When $x=1$:

$$y=-4x$$
$$y=-4(1)$$
$$y=-4$$

x	y
0	0
–1	4
–2	8
1	–4

23. When $x=-2$:

$$3x-2y=6$$
$$3(-2)-2y=6$$
$$-6-2y=6$$
$$-2y=12$$
$$y=-6$$

When $y=0$:

$$3x-2y=6$$
$$3x-2(0)=6$$
$$3x-0=6$$
$$3x=6$$
$$x=2$$

When $x=4$:

$$3x-2y=6$$
$$3(4)-2y=6$$
$$12-2y=6$$
$$-2y=-6$$
$$y=3$$

When $y=-3$:

$$3x-2y=6$$
$$3x-2(-3)=6$$
$$3x+6=6$$
$$3x=0$$
$$x=0$$

x	y
–2	–6
2	0
4	3
0	–3

25. When $x=-1$:

$$y=6x-1$$
$$y=6(-1)-1$$
$$y=-6-1$$
$$y=-7$$

When $y=5$:

$$y=6x-1$$
$$5=6x-1$$
$$6=6x$$
$$1=x$$

When $y=-13$:

$$y=6x-1$$
$$-13=6x-1$$
$$-12=6x$$
$$-2=x$$

When $x=0$:

$$y=6x-1$$
$$y=6(0)-1$$
$$y=0-1$$
$$y=-1$$

x	y
−1	−7
1	5
−2	−13
0	−1

27. Substitute $(2,3)$:

$$2x-5y=10$$
$$2(2)-5(3)=10$$
$$4-15=10$$
$$-11=10$$
false

Substitute $(0,-2)$:

$$2x-5y=10$$
$$2(0)-5(-2)=10$$
$$0+10=10$$
$$10=10$$
true

Substitute $\left(\frac{5}{2},1\right)$:

$$2x-5y=10$$
$$2\left(\frac{5}{2}\right)-5(1)=10$$
$$5-5=10$$
$$0=10$$
false

$(0,-2)$ is a solution.

29. Substitute $(0,3)$:

$$y=2x+3$$
$$3=2(0)+3$$
$$3=0+3$$
$$3=3$$
true

Substitute $(5,4)$:

$$y=2x+3$$
$$4=2(5)+3$$
$$4=10+3$$
$$4=13$$
false

Substitute $(2,0)$:

$$y=2x+3$$
$$0=2(2)+3$$
$$0=4+3$$
$$0=7$$
false

$(0,3)$ is a solution.

31. Substitute $(0,0)$:

$$x+y=0$$
$$0+0=0$$
$$0=0$$
true

Substitute $(5,-5)$:

$$x+y=0$$
$$5+(-5)=0$$
$$0=0$$
true

Substitute $(-3,3)$:

$$x+y=0$$
$$-3+3=0$$
$$0=0$$
true

All three of the pairs are solutions.

33. Substitute $(0,5)$:

$y = 5x$

$5 = 5(0)$

$5 = 0$

false

Substitute $(1,5)$:

$y = 5x$

$5 = 5(1)$

$5 = 5$

true

Substitute $\left(2, \frac{5}{2}\right)$:

$y = 5x$

$\frac{5}{2} = 5(2)$

$\frac{5}{2} = 10$

false

$(1,5)$ is a solution.

Applying the Concepts

35. The table is as follows:

α	θ
$0°$	$\alpha + \theta = 90°$ $0° + \theta = 90°$ $\theta = 90°$
$30°$	$\alpha + \theta = 90°$ $30° + \theta = 90°$ $\theta = 60°$
$\alpha + \theta = 90°$ $\alpha + 45° - 90°$ $\alpha = 45°$	$45°$
$60°$	$\alpha + \theta = 90°$ $60° + \theta = 90°$ $\theta = 30°$
$\alpha + \theta = 90°$ $\alpha + 15° = 90°$ $\alpha = 75°$	$15°$
$90°$	$\alpha + \theta = 90°$ $90° + \theta = 90°$ $\theta = 0°$

37. From table (a):

Substitute $(0,6)$:

$y = 2x - 6$

$6 = 2(0) - 6$

$6 = 0 - 6$

$6 = -6$

Because $(0,6)$ gives a false statement, table (a) does not represent solutions of the equation.

From table (b):

Substitute $(0,-6)$:

$y = 2x - 6$

$-6 = 2(0) - 6$

$-6 = 0 - 6$

$-6 = -6$

true

Substitute $(1,-4)$:

$y = 2x - 6$

$-4 = 2(1) - 6$

$-4 = 2 - 6$

$-4 = -4$

true

Substitute $(2,-2)$:

$y = 2x - 6$

$-2 = 2(2) - 6$

$-2 = 4 - 6$

$-2 = -2$

true

Substitute $(3,0)$:

$y = 2x - 6$

$0 = 2(3) - 6$

$0 = 6 - 6$

$0 = 0$

true

Since all 4 ordered pairs give true statements, the table does represent solutions to the equation.

From table (c):

Substitute $(0,-6)$:

$y = 2x - 6$

$-6 = 2(0) - 6$

$-6 = 0 - 6$

$-6 = -6$

true

Substitute $(1,-5)$:

$y = 2x - 6$

$-5 = 2(1) - 6$

$-5 = 2 - 6$

$-5 = -4$

false

Since $(1,-5)$ gives a false statement, table (e) does not represent solutions to the equation.

Review Problems

39. $P = \text{sum of all sides}$

$= 5 \text{ ft} + 3 \text{ ft} + 4 \text{ ft}$

$= 12 \text{ ft}$

$$A = \frac{1}{2}bh$$
$$= \frac{1}{2}(4 \text{ ft})(3 \text{ ft})$$
$$= 6 \text{ ft}^2$$

41. $P = \text{sum of all sides}$

$= 5 \text{ in} + 5 \text{ in} + 8 \text{ in}$

$= 18 \text{ in}$

$$A = \frac{1}{2}bh$$
$$= \frac{1}{2}(8 \text{ in})(3 \text{ in})$$
$$= 12 \text{ in}^2$$

Section 4.8

1. $(4,2): x=4, y=2$
   From the origin, move 4 units right, 2 units up. See graph next page.

3. $(-4,2): x=-4, y=2$
   From the origin, move 4 units left, 2 units up. See graph next page.

5. $(3,4): x=3, y=4$
   From the origin, move 3 units right, 4 units up. See graph next page.

7. $(-3,-4): x=-3, y=-4$
   From the origin, move 3 units left, 4 units down. See graph next page.

9. $(4,3): x=4, y=3$
   From the origin, move 4 units right, 3 units up. See graph next page.

11. $\left(5,\frac{1}{2}\right): x=5, y=\frac{1}{2}$
    From the origin, move 5 units right, $\frac{1}{2}$ units up. See graph next page.

13. $\left(1,-\frac{3}{2}\right): x=1, y=-\frac{3}{2}$
    From the origin, move 1 units right, $-\frac{3}{2}$ units down. See graph next page.

15. $(2,0): x=2, y=0$
    From the origin, move 2 units right, 0 units up. See graph next page.

17. $(-2,0): x=-2, y=0$
    From the origin, move 2 units left, 0 units up. See graph next page.

Graph for Problems 1 – 17 odds:

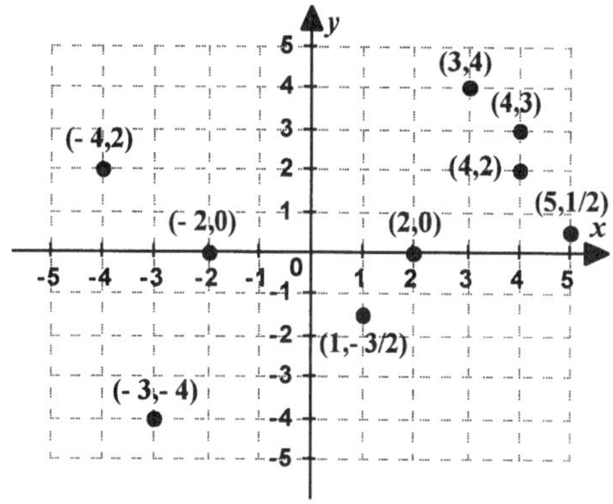

19. The point is 2 units right in the x direction, 2 units up in the y direction. $(2,2)$

21. The point is 3 units left in the x direction, 2 units up in the y direction. $(-3,2)$

23. The point is 3 units right in the x direction, 3 units down in the y direction. $(3,-3)$

25. The point is 4 units left in the x direction, 0 units up in the y direction. $(-4,0)$

27. First plot $(1,3)$ and $(2,4)$ and draw a line through the 2 points. Then plot $(-1,1)$, which does lie on the line.

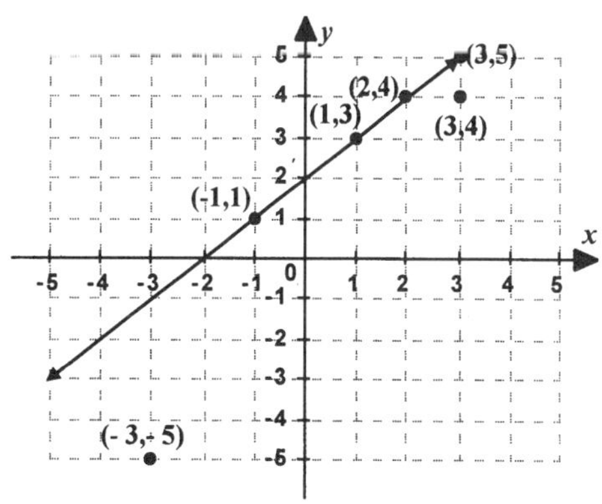

29. Plot $(-3,-5)$ on the graph shown above. It does not lie on the line.

31. First plot $(-3,2)$ and $(1,3)$ and draw a line through the 2 points. Then plot $(5,4)$, which does lie on the line.

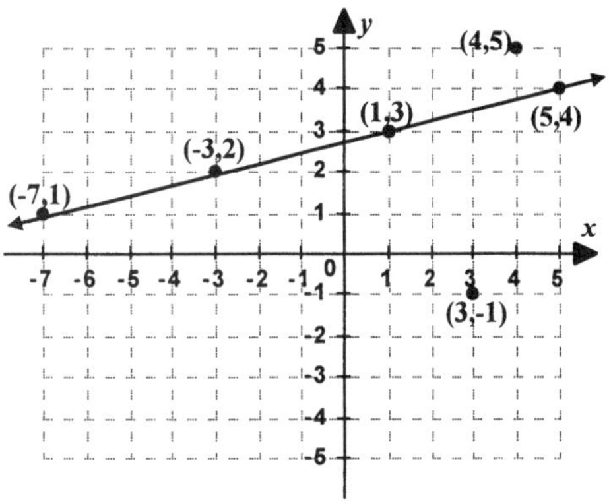

33. Plot $(4,5)$ on the graph shown above. It does not lie on the line.

Applying the Concepts

35. On the given axes, plot the ordered pairs: $(1996,386),(1997,518)$ $(1998,573),(1999,819)$

To construct the line graph, connect the points.

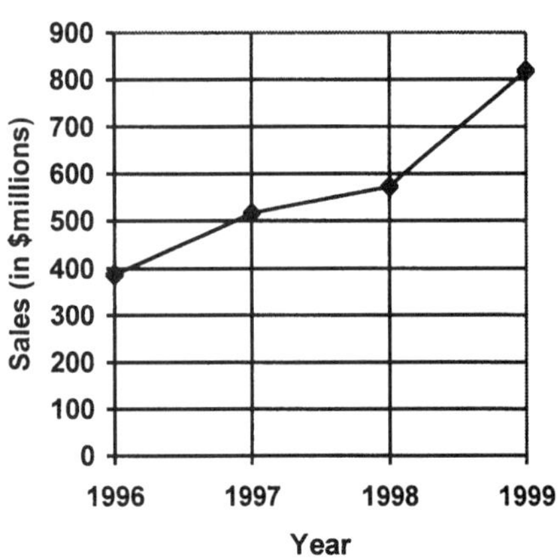

37. a. Answers will vary depending on points chosen. Choosing the 1st three points from left to right:

Hours: 0 Amount earned: \$0 $(0,0)$

Hours: 5 Amount earned: \$40 $(5,40)$

Hours: 10 Amount earned: \$80 $(10,80)$

b. From 40 hours on the x axis, locate the point on the line above it; the y value is 320, or \$320.

c. From \$240 on the y axis, locate the point on the line to right of it; the x value is 30, or 30 hours.

d. No. For 35 hours, the corresponding amount earned is \$280. $(35,280)$

Review Problems

39. $$\frac{x^2}{y}\cdot\frac{y^3}{x}=\frac{\cancel{x}\cdot x\cdot\cancel{y}\cdot y\cdot y}{\cancel{y}\cdot\cancel{x}}=xy^2$$

41. $$\frac{x^2}{y^3}\div\frac{x^3}{y^2}=\frac{x^2}{y^3}\cdot\frac{y^2}{x^3}=\frac{\cancel{x}\cdot\cancel{x}\cdot\cancel{y}\cdot\cancel{y}}{y\cdot\cancel{y}\cdot\cancel{y}\cdot\cancel{x}\cdot\cancel{x}\cdot x}=\frac{1}{xy}$$

43. $$\frac{x}{5}+\frac{3}{4}=\frac{x\cdot4}{5\cdot4}+\frac{3\cdot5}{4\cdot5}=\frac{4x}{20}+\frac{15}{20}=\frac{4x+15}{20}$$

45. $$3-\frac{1}{x}=\frac{3\cdot x}{1\cdot x}-\frac{1}{x}=\frac{3x}{x}-\frac{1}{x}=\frac{3x-1}{x}$$

Extending the Concepts

47. From point C:

5 units left is A $(1,2)$

5 units up is B $(6,7)$

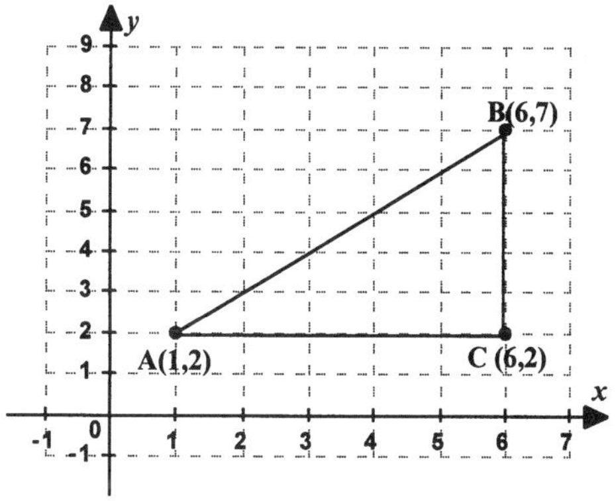

49. From point D:

3 units up is C$(7,5)$

5 units left is A$(2,2)$

From point A:

3 units up is B$(2,5)$

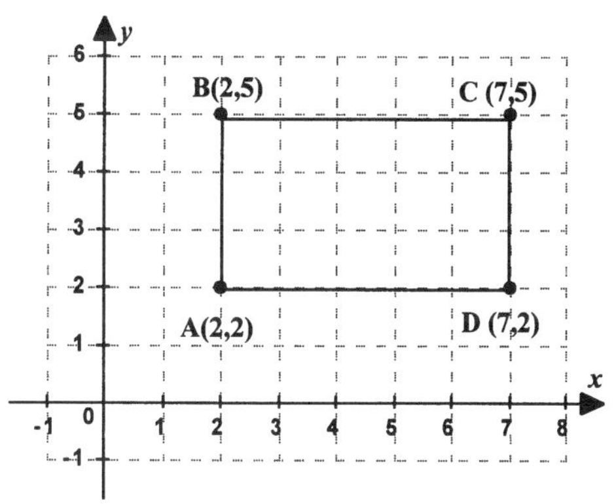

Improving Your Quantitative Literacy

51. a. For 2003, the corresponding health costs are a little less than $6,000.

b. For 2009, the corresponding health costs are a little more than $8,000.

Section 4.9

1.

Let $x = 0$:	Let $x = 2$:	Let $x = -1$:
$x + y = 2$	$x + y = 2$	$x + y = 2$
$0 + y = 2$	$2 + y = 2$	$-1 + y = 2$
$y = 2$	$y = 0$	$y = 3$

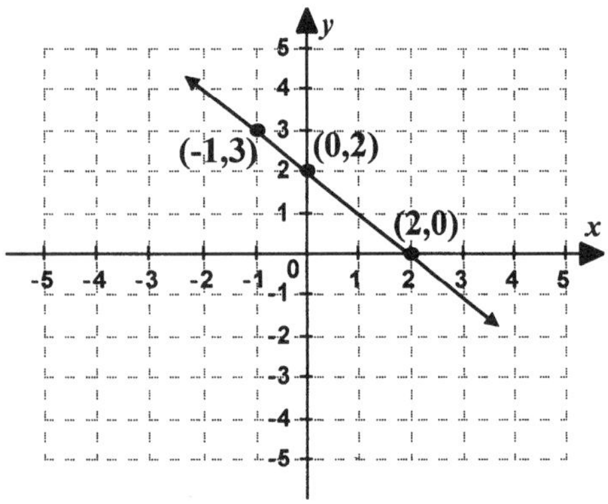

3.

Let $x = 0$:	Let $x = 1$:	Let $x = 2$:
$y = 2x - 4$	$y = 2x - 4$	$y = 2x - 4$
$y = 2(0) - 4$	$y = 2(1) - 4$	$y = 2(2) - 4$
$y = 0 - 4$	$y = 2 - 4$	$y = 4 - 4$
$y = -4$	$y = -2$	$y = 0$

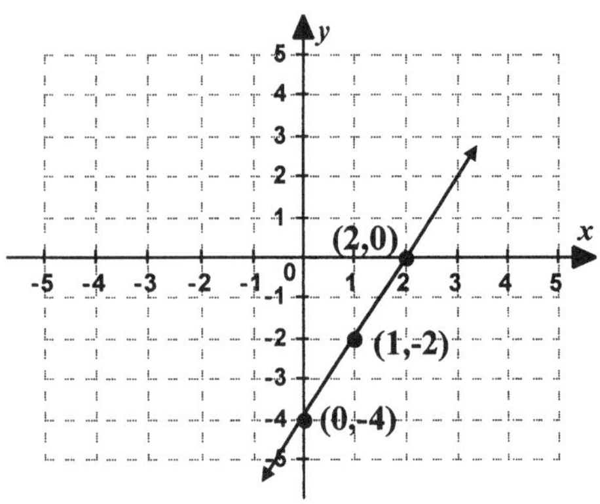

5.

Let $x = 0$:	Let $y = 0$:	Let $x = 3$:
$3x + y = 3$	$3x + y = 3$	$3x + y = 3$
$3(0) + y = 3$	$3x + 0 = 3$	$3(3) + y = 3$
$0 + y = 3$	$3x = 3$	$9 + y = 3$
$y = 3$	$x = 1$	$y = -6$

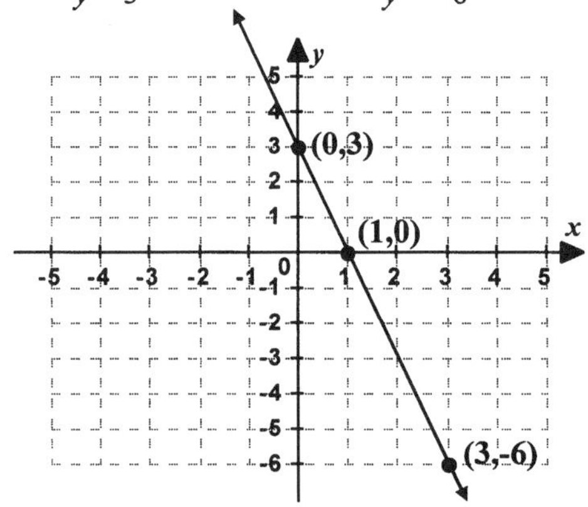

7.

Let $x = 0$:	Let $y = 0$:	Let $x = -4$:
$3x + 4y = 12$	$3x + 4y = 12$	$3x + 4y = 12$
$3 \cdot 0 + 4y = 12$	$3x + 4 \cdot 0 = 12$	$3(-4) + 4y = 12$
$4y = 12$	$3x = 12$	$-12 + 4y = 12$
$y = 3$	$x = 4$	$4y = 24$
		$y = 6$

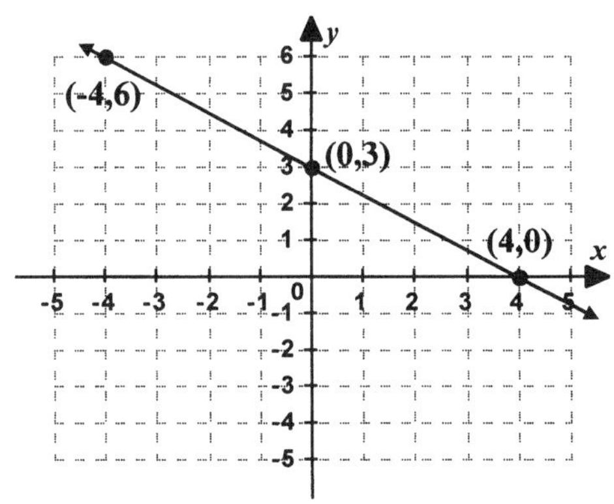

9. Let $x = 0$: Let $y = 0$: Let $x = -4$:

$$\begin{array}{lll} 3x-4y=12 & 3x-4y=12 & 3x-4y=12 \\ 3\cdot 0-4y=12 & 3x-4\cdot 0=12 & 3(-4)-4y=12 \\ 0-4y=12 & 3x-0=12 & -12-4y=12 \\ -4y=12 & 3x=12 & -4y=24 \\ y=-3 & x=4 & y=-6 \end{array}$$

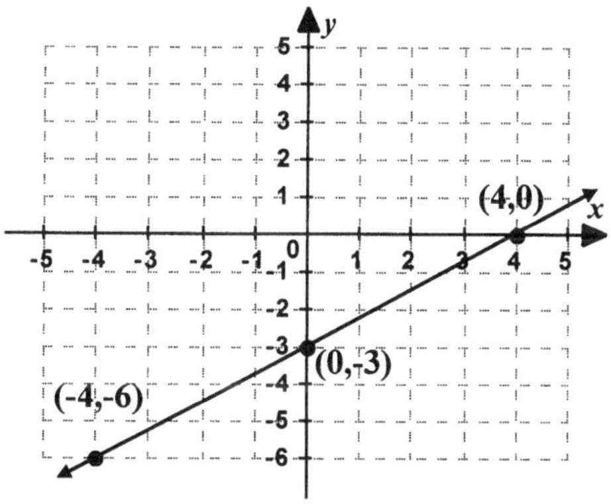

11. Let $x = 0$: Let $x = 2$: Let $x = -2$:

$$\begin{array}{lll} y=\frac{1}{2}x & y=\frac{1}{2}x & y=\frac{1}{2}x \\ y=\frac{1}{2}\cdot 0 & y=\frac{1}{2}\cdot 2 & y=\frac{1}{2}(-2) \\ y=0 & y=1 & y=-1 \end{array}$$

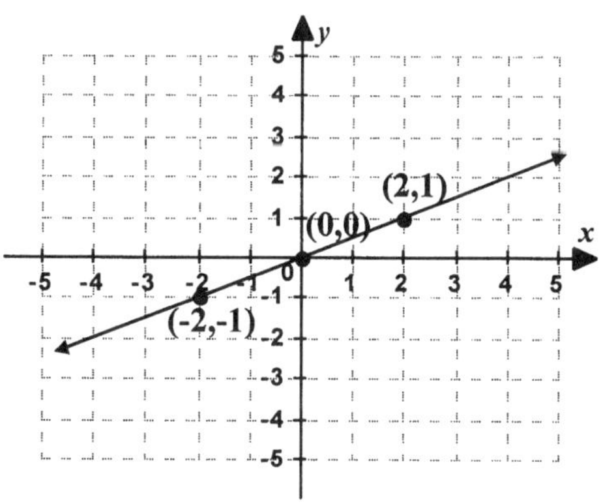

13. Let $x = -2$ Let $x = 0$ Let $x = 2$

$$\begin{array}{lll} y=\frac{1}{2}x+2 & y=\frac{1}{2}x+2 & y=\frac{1}{2}x+2 \\ y=\frac{1}{2}(-2)+2 & y=\frac{1}{2}\cdot 0+2 & y=\frac{1}{2}(2)+2 \\ y=-1+2 & y=0+2 & y=1+2 \\ y=1 & y=2 & y=3 \end{array}$$

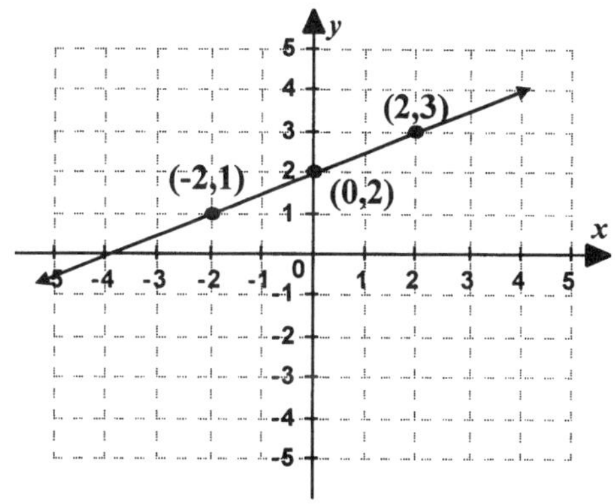

15. The equation $x = 2$ is the same as the equation $x + 0y = 2$. You can replace y with any value and x will still be 2. Some solutions could be: $(2,0),(2,3),(2,-3)$.

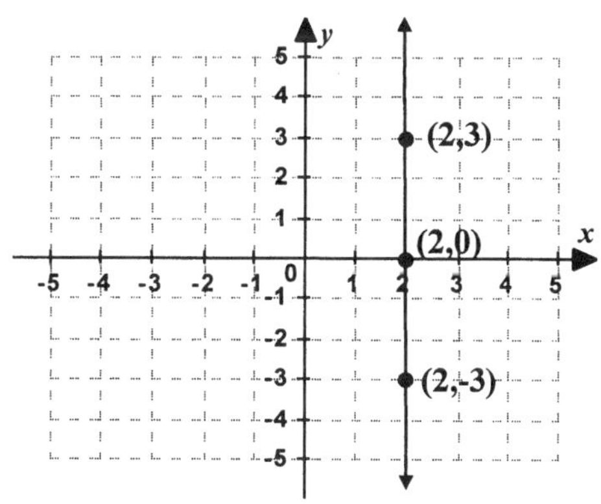

17. Choose any 3 values for x or y:

Let $x = 0$:	Let $y = 0$:	Let $x = 1$:
$x + y = 5$	$x + y = 5$	$x + y = 5$
$0 + y = 5$	$x + 0 = 5$	$1 + y = 5$
$y = 5$	$x = 5$	$y = 4$

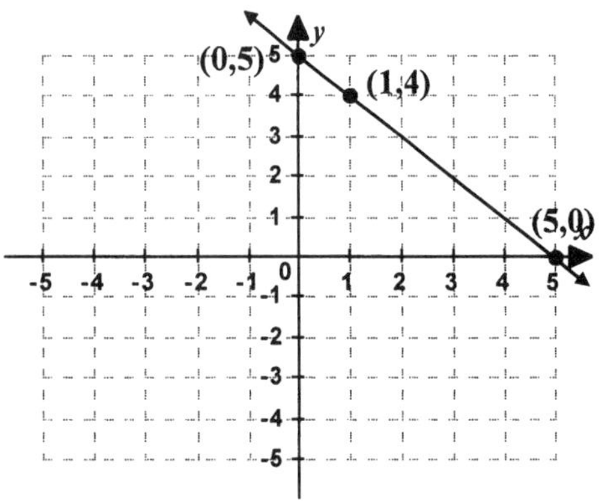

19. Choose any 3 values for x or y:

Let $x = 0$:	Let $y = 0$:	Let $x = 2$:
$x - y = 5$	$x - y = 5$	$x - y = 5$
$0 - y = 5$	$x - 0 = 5$	$2 - y = 5$
$-y = 5$	$x = 5$	$-y = 3$
$y = -5$		$y = -3$

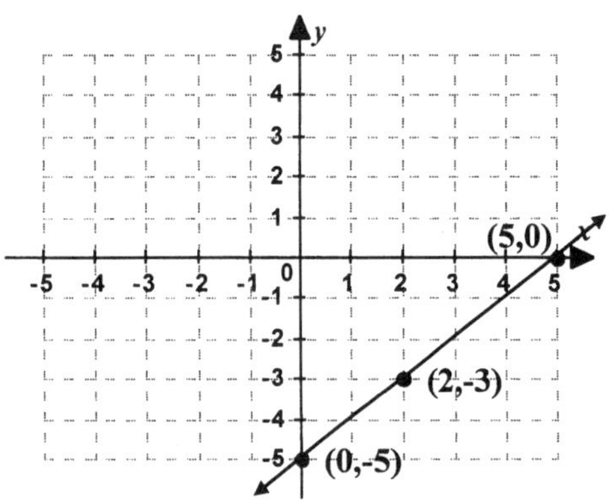

21. Choose any 3 values for x or y:

Let $x = 0$:	Let $y = 0$:	Let $x = 4$:
$2x - 4y = 4$	$2x - 4y = 4$	$2x - 4y = 4$
$2 \cdot 0 - 4y = 4$	$2x - 4 \cdot 0 = 4$	$2 \cdot 4 - 4y = 4$
$-4y = 4$	$2x = 4$	$-4y = -4$
$y = -1$	$x = 2$	$y = 1$

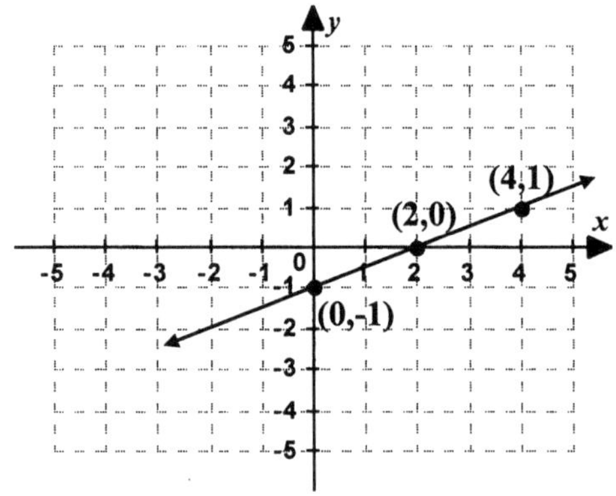

23. Choose any 3 values for x or y:

Let $x = 0$:	Let $y = 0$:	Let $x = 4$:
$2x + 4y = 4$	$2x + 4y = 4$	$2x + 4y = 4$
$2 \cdot 0 + 4y = 4$	$2x + 4 \cdot 0 = 4$	$2 \cdot 4 + 4y = 4$
$4y = 4$	$2x = 4$	$4y = -4$
$y = 1$	$x = 2$	$y = -1$

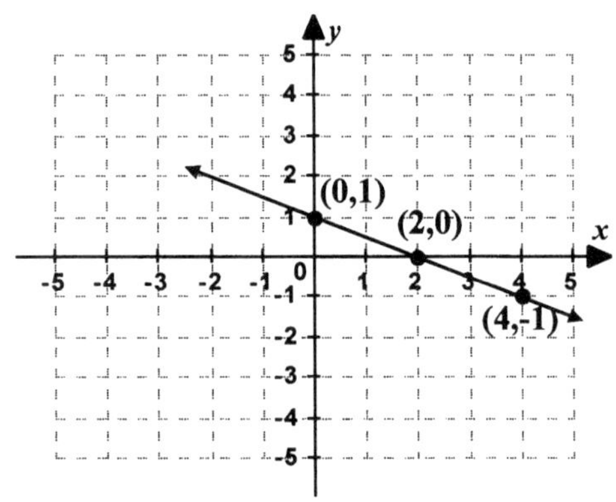

25. Choose any 3 values for x or y:

Let $x=0$: Let $y=0$: Let $x=1$:

$$\begin{array}{lll} y=2x+3 & y=2x+3 & y=2x+3 \\ y=2\cdot 0+3 & 0=2x+3 & y=2\cdot 1+3 \\ y=3 & -3=2x & y=5 \\ & -\frac{3}{2}=x & \end{array}$$

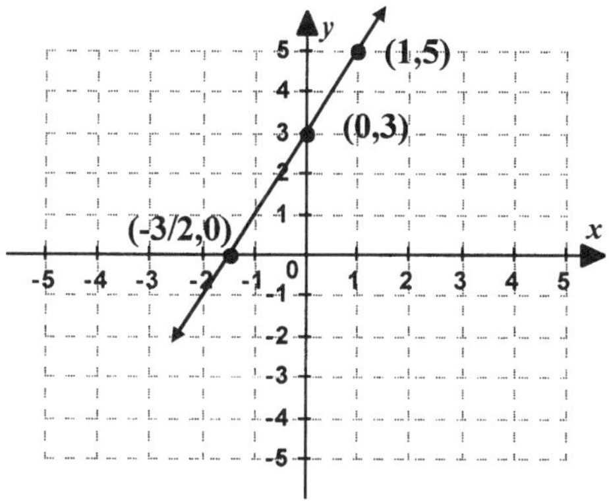

27. Choose any 3 values for x or y:

Let $x=0$: Let $x=1$: Let $x=2$:

$$\begin{array}{lll} y=2x & y=2x & y=2x \\ y=2\cdot 0 & y=2\cdot 1 & y=2\cdot 2 \\ y=0 & y=2 & y=4 \end{array}$$

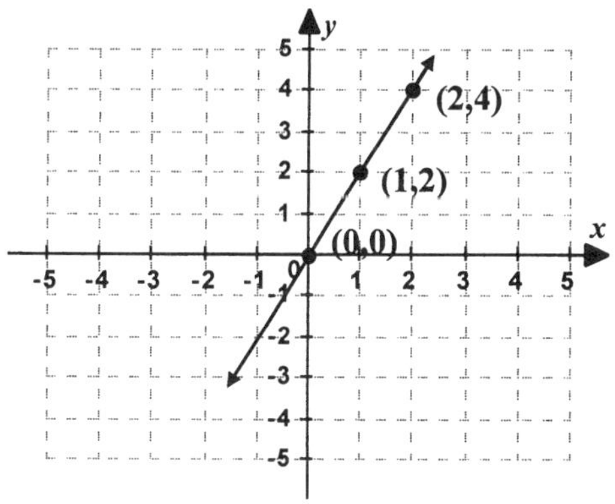

29. Choose any 3 values for x or y:

Let $x=0$: Let $x=3$: Let $x=-3$:

$$\begin{array}{lll} y=\frac{1}{3}x & y=\frac{1}{3}x & y=\frac{1}{3}x \\ y=\frac{1}{3}\cdot 0 & y=\frac{1}{3}\cdot\cdot 3 & y=\frac{1}{3}(-3) \\ y=0 & y=1 & y=-1 \end{array}$$

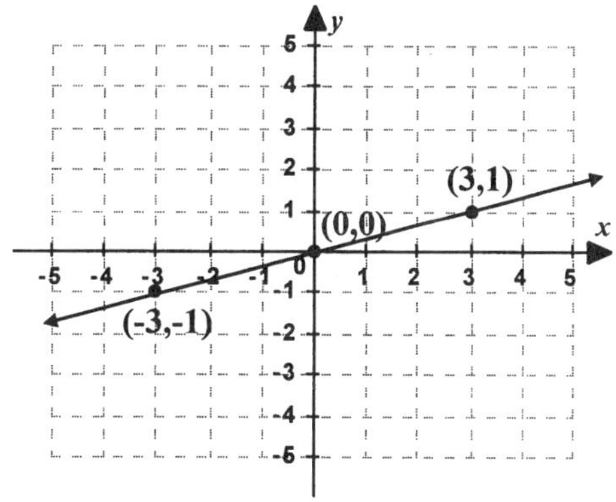

31. The equation $x=4$ is equivalent to the equation $x+0\cdot y=4$. Solutions are any ordered pairs that have x-coordinate of 4; y can be any number. Some solutions are: $(4,0),(4,2),(4,-2)$

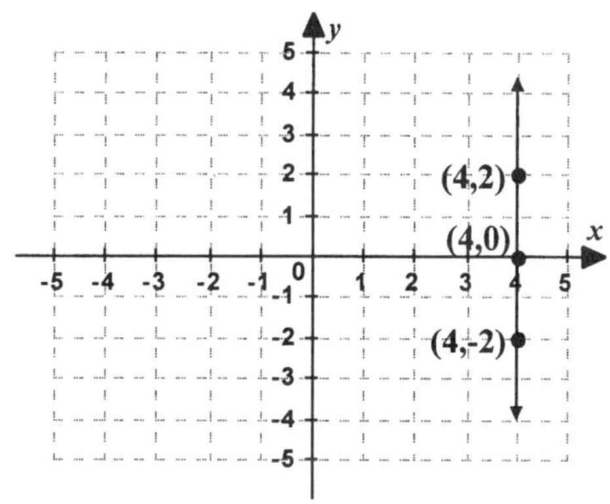

33. The equation $y=-2$ is equivalent to the equation $0x+y=-2$. Solutions are any ordered pairs that have y-coordinate of -2; x can be any number. Some solutions are: $(0,-2),(3,-2),(-3,-2)$

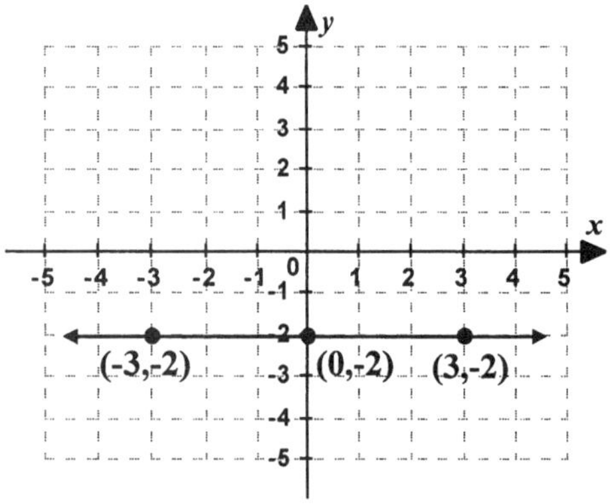

35. a. Let $x=0$:

$3x-2y=6$
$3\cdot 0-2y=6$
$-2y=6$
$y=-3$

This is not the equation because $(0,-3)$ is not on the graph.

b. Let $x=0$: Let $y=0$: Let $x=6$:

$2x-3y=6$ $2x-3y=6$ $2x-3y=6$
$2\cdot 0-3y=6$ $2x-3\cdot 0=6$ $2\cdot 6-3y=6$
$-3y=6$ $2x=6$ $-3y=-6$
$y=-2$ $x=3$ $y=2$

The points $(0,-2),(-2,0),(6,2)$ satisfy this equation and lie on the graph, so $2x-3y=6$ is the graph of the equation.

c. Let $x=0$:

$2x+3y=6$
$2\cdot 0+3y=6$
$3y=6$
$y=2$

This is not the equation because $(0,2)$ is not on the graph.

Applying the Concepts

37. Choose any 3 values for α or θ between 0° and 90°:

Let $\alpha=0°$: Let $\theta=0°$: Let $\alpha=40°$:

$\alpha+\theta=90°$ $\alpha+\theta=90°$ $\alpha+\theta=90°$
$0°+\theta=90°$ $\alpha+90°=90°$ $40°+\theta=90°$
$\theta=90°$ $\alpha=90°$ $\theta=50°$

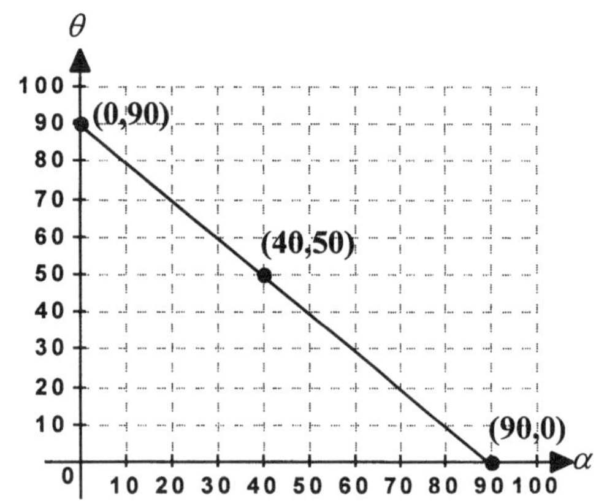

39. Choose any 3 values for g between 0 and 20 and find the corresponding y values:

Let $g=0$: Let $g=10$: Let $g=20$:

$y=7+2g$ $y=7+2g$ $y=7+2g$
$y=7+2\cdot 0$ $y=7+2\cdot 10$ $y=7+2\cdot 20$
$y=7$ $y=27$ $y=47$

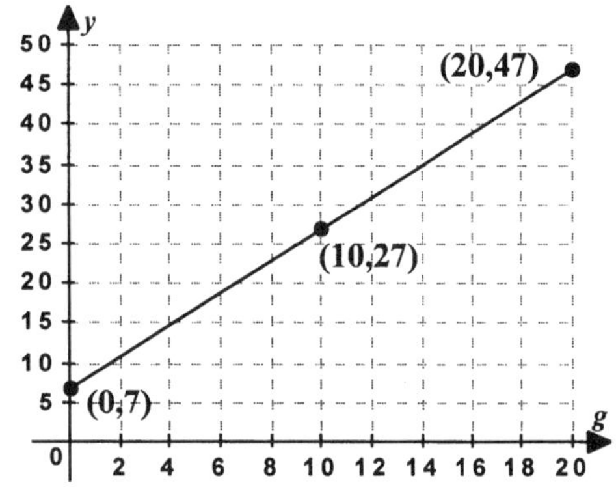

Note: Since g represents gallons of water delivered each month, g must be greater than or equal to 0 gal. (No negative values)

Review Problems

41. $$\frac{3}{5}\left(2\frac{1}{5}-1\frac{1}{10}\right)=\frac{3}{5}\left(\frac{11}{5}-\frac{11}{10}\right)$$
$$=\frac{3}{5}\left(\frac{22}{10}-\frac{11}{10}\right)$$
$$=\frac{3}{5}\left(\frac{11}{10}\right)$$
$$=\frac{33}{50}$$

43. $$\left(\frac{3}{5}+\frac{1}{3}\right)\left(\frac{3}{5}-\frac{1}{3}\right)=\left(\frac{9}{15}+\frac{5}{15}\right)\left(\frac{9}{15}-\frac{5}{15}\right)$$
$$=\left(\frac{14}{15}\right)\left(\frac{4}{15}\right)$$
$$=\frac{56}{225}$$

45. $$\left(\frac{7}{15}-\frac{11}{30}\right)^2=\left(\frac{14}{30}-\frac{11}{30}\right)^2$$
$$=\left(\frac{3}{30}\right)^2$$
$$=\left(\frac{1}{10}\right)^2$$
$$=\left(\frac{1}{10}\right)\left(\frac{1}{10}\right)$$
$$=\frac{1}{100}$$

Chapter 4 Review

1. $10x+7x=(10+7)x=17x$

3. $$\begin{aligned}2a+9a+3-6&=(2+9)a+3-6\\&=11a-3\end{aligned}$$

5. $$\begin{aligned}6x-x+4&=(6-1)x+4\\&=5x+4\end{aligned}$$

7. $$\begin{aligned}2a-6+8a+2&=2a+8a-6+2\\&=(2+8)a-6+2\\&=10a-4\end{aligned}$$

9. When $x=4$:
$$\begin{aligned}10x+2&=10\cdot4+2\\&=40+2\\&=42\end{aligned}$$

11. When $x=4$:
$$\begin{aligned}-2x+9&=-2\cdot4+9\\&=-8+9\\&=1\end{aligned}$$

13. When $x=-3$:
$$\begin{aligned}5x-2&=-17\\5(-3)-2&=-17\\-15-2&=-17\\-17&=-17 \text{ true}\end{aligned}$$
$x=-3$ is a solution.

15. $$\begin{aligned}x-5&=4\\x-5+5&=4+5\\x&=9\end{aligned}$$

17. $$\begin{aligned}2x+1&=7\\2x+1+(-1)&=7+(-1)\\2x&=6\\\frac{2x}{2}&=\frac{6}{2}\\x&=3\end{aligned}$$

19. $$\begin{aligned}2x+4&=3x-5\\2x+4+(-2x)&=3x-5+(-2x)\\4&=x-5\\4+5&=x-5+5\\9&=x\end{aligned}$$

21. $$\begin{aligned}3(x-2)&=9\\3x-6&=9\\3x&=15\\x&=5\end{aligned}$$

23. $$\begin{aligned}3(2x+1)-4&=-7\\6x+3-4&=-7\\6x-1&=-7\\6x&=-6\\x&=-1\end{aligned}$$

25. $$5x+\frac{3}{8}=-\frac{1}{4} \qquad \text{LCD}=8$$
$$\begin{aligned}8\left(5x+\frac{3}{8}\right)&=8\left(-\frac{1}{4}\right)\\8(5x)+8\left(\frac{3}{8}\right)&=8\left(-\frac{1}{4}\right)\\40x+3&=-2\\40x&=-5\\x&=-\frac{5}{40}\\x&=-\frac{1}{8}\end{aligned}$$

27. Let x = number asked for

The sum of a number and -3 is -5

$$\underbrace{x+(-3)}_{\text{The sum of a number and } -3} \;=\; -5$$

$$\begin{aligned}x+(-3)&=-5\\x+(-3)+3&=-5+3\\x&=-2\end{aligned}$$
The number is –2.

29. Let x = number asked for

$\underbrace{\text{3 times}}$ $\underbrace{\text{the sum of a number and 2}}$ is -6.

$$\downarrow \qquad \downarrow \qquad \downarrow \downarrow$$

$$3\times \qquad (x+2) \qquad = -6$$

$$3(x+2) = -6$$
$$3x+6 = -6$$
$$3x = -12$$
$$x = -4$$

The number is –4.

31. Let x = width (in m)
Then $2x$ = length (in m)

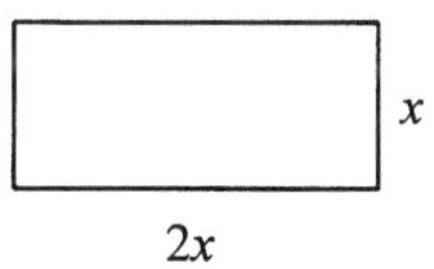

$$P = \text{sum of all sides}$$
$$42 = x+x+2x+2x$$
$$42 = 6x$$
$$7 = x$$

The width is $x = 7\text{ m}$.
The length is $2x = 2\cdot 7 = 14\text{ m}$.

33. Let $x = -2$:

$$3x+2y = 6$$
$$3(-2)+2y - 6$$
$$-6+2y = 6$$
$$2y = 12$$
$$y = 6$$

35. Let $x = 0$:

$$3x+2y = 6$$
$$3(0)+2y = 6$$
$$0+2y = 6$$
$$2y = 6$$
$$y = 3$$

37. Let $y = 3$:

$$3x+2y = 6$$
$$3x+2\cdot 3 = 6$$
$$3x+6 = 6$$
$$3x = 0$$
$$x = 0$$

39. Let $y = 0$:

$$3x+2y = 6$$
$$3x+2\cdot 0 = 6$$
$$3x+0 = 6$$
$$3x = 6$$
$$x = 2$$

41-47.

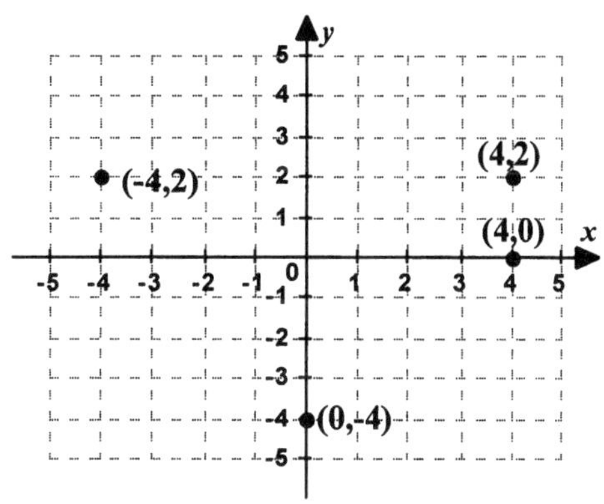

49. Choose any 3 values for x or y:

Let $x=0$: Let $y=0$: Let $x=1$:

$x+y=2 \quad x+y=2 \quad x+y=2$

$0+y=2 \quad x+0=2 \quad 1+y=2$

$y=2 \quad x=2 \quad y=1$

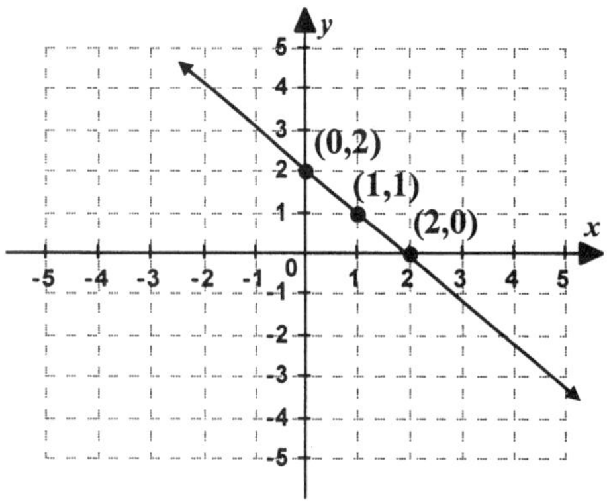

51. Choose any 3 values for x or y:

Let $x=0$: Let $y=0$: Let $x=4$:

$3x+2y=6 \quad 3x+2y=6 \quad 3x+2y=6$

$3\cdot 0+2y=6 \quad 3x+2\cdot 0=6 \quad 3\cdot 4+2y=6$

$0+2y=6 \quad 3x+0=6 \quad 12+2y=6$

$2y=6 \quad 3x=6 \quad 2y=-6$

$y=3 \quad x=2 \quad y=-3$

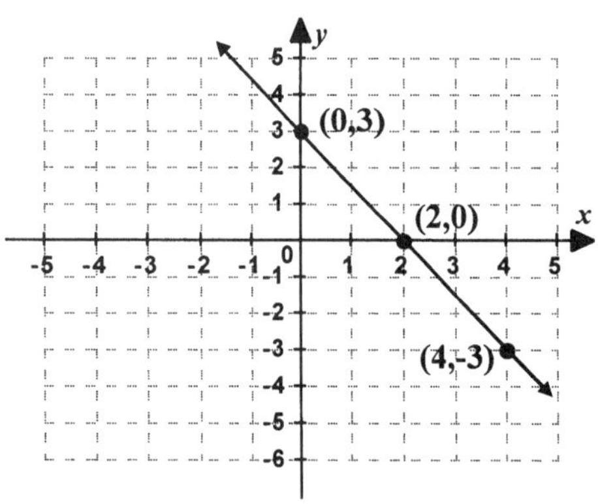

53. Choose any 3 values for x or y:

Let $x=0$: Let $x=2$: Let $x=4$:

$y=\frac{1}{2}x \quad y=\frac{1}{2}x \quad y=\frac{1}{2}x$

$y=\frac{1}{2}\cdot 0 \quad y=\frac{1}{2}\cdot 2 \quad y=\frac{1}{2}\cdot 4$

$y=0 \quad y=1 \quad y=2$

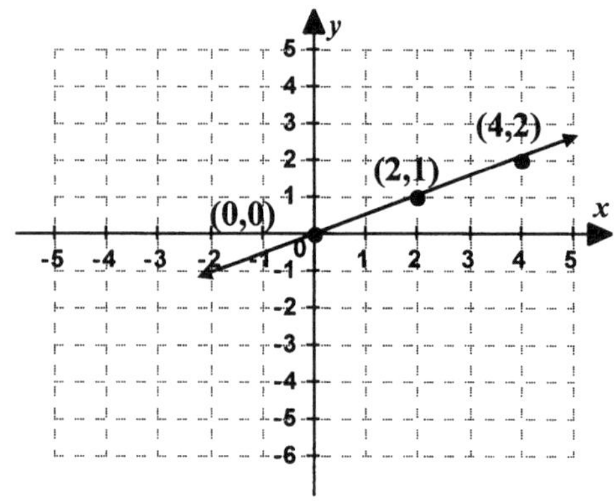

55. Choose any 3 values for x or y:

Let $x=0$: Let $y=0$: Let $x=1$:

$y=2x-3 \quad y=2x-3 \quad y=2x-3$

$y=2\cdot 0-3 \quad 0=2x-3 \quad y=2\cdot 1-3$

$y=0-3 \quad 3=2x \quad y=2-3$

$y=-3 \quad \frac{3}{2}=x \quad y=-1$

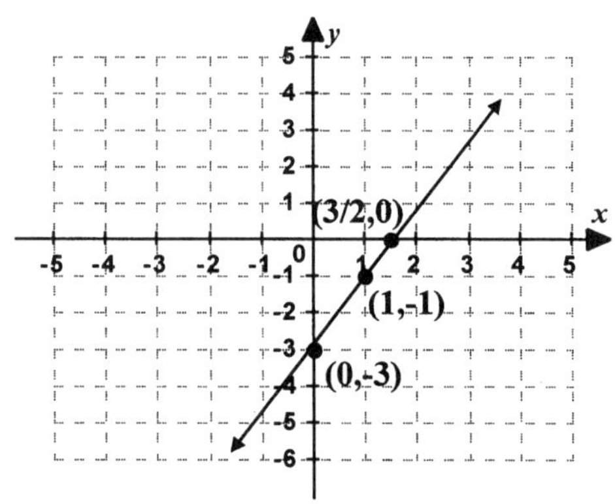

57. The equation $x=-2$ is equivalent to $x+0y=-2$. Solutions are any ordered pairs with $x=-2$; y can be any value. Some solutions are $(-2,0),(-2,3),(-2,-3)$

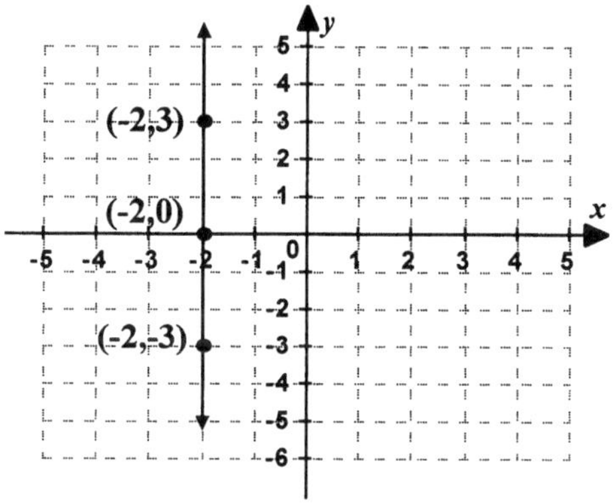

Chapters 1-4 Cumulative Review

1.
$$\begin{array}{r} \overset{1}{7},\overset{1}{5}20 \\ 599 \\ +\ 8,640 \\ \hline 16,759 \end{array}$$

3.
$$\begin{array}{r} 12 \\ 13\overline{)156} \\ 13\downarrow \\ \hline 26 \\ \underline{26} \\ 0 \end{array}$$

5.
$$\begin{array}{r} 490\ R2 \\ 64\overline{)31362} \\ 256\downarrow \\ \hline 576 \\ 576\downarrow \\ \hline 02 \\ 0 \\ \hline 2 \end{array}$$
$$\frac{31,362}{64} = 490\frac{2}{64} = 490\frac{1}{32}$$

7. $12 + 81 \div 3^2 = 12 + 81 \div 9$
$= 12 + 9$
$= 21$

9. $-25 + (-13) = -39$

11. $\frac{-39}{-3} = 13$

13. $\frac{3 \cdot 2^3 - 3 \cdot 4^2}{-2^3} = \frac{3 \cdot 8 - 3 \cdot 16}{-8}$
$= \frac{24 - 48}{-8}$
$= \frac{-24}{-8}$
$= 3$

15. $\frac{-6(2^3) - 4^2}{2(2)^2} = \frac{-6 \cdot 8 - 16}{2 \cdot 4}$
$= \frac{-48 - 16}{8}$
$= \frac{-64}{8}$
$= -8$

17. $17 \div \left(\frac{1}{3}\right)^2 = 17 \div \left(\frac{1}{3} \cdot \frac{1}{3}\right)$
$= 17 \div \frac{1}{9}$
$= 17 \cdot \frac{9}{1}$
$= 153$

19. $\left(16 \div 1\frac{1}{4}\right) \div 2 = \left(\frac{16}{1} \div \frac{5}{4}\right) \div \frac{2}{1}$
$= \frac{16}{1} \cdot \frac{4}{5} \cdot \frac{1}{2}$
$= \frac{16 \cdot 2 \cdot \cancel{2} \cdot 1}{1 \cdot 5 \cdot \cancel{2}}$
$= \frac{32}{5}$
$= 6\frac{2}{5}$

21. $3(z + 2) - 2(3z - 1)$
$= 3(z + 2) + (-2)(3z - 1)$
$= 3z + 6 + (-6z) + 2$
$= -3z + 8$

23. $13+\frac{3}{14}\div\frac{5}{42}=13+\frac{3}{14}\cdot\frac{42}{5}$
$$=13+\frac{3\cdot\cancel{14}\cdot 3}{\cancel{14}\cdot 5}$$
$$=13+\frac{9}{5}$$
$$=13\frac{9}{5}$$
$$=14\frac{4}{5}$$

25. $5a-3a+6a=-8+(-8)$
$$8a=-16$$
$$a=-2$$

27.

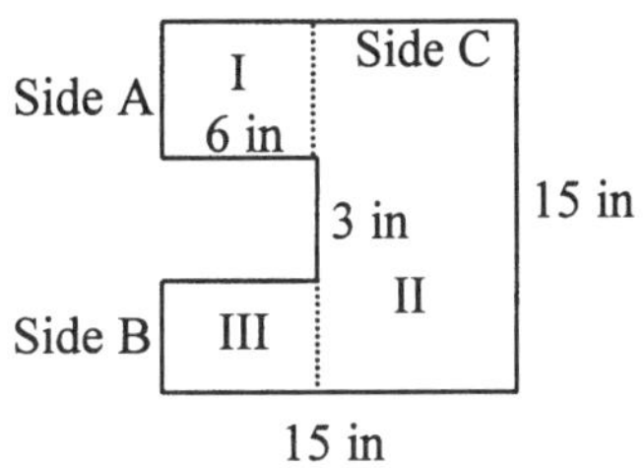

Find the lengths of the missing sides:
Side A + Side B =15 in – 3 in = 12 in.
Because no other information is given, we will assume Side A and Side B are equal in length, and both are 6 in.
Side C = 15 in – 6 in = 9 in

$P=$ sum of all sides
$$=15+15+15+6+6+3+6+6$$
$$=72\text{ in}$$

To find the area, separate the figure into 3 rectangular regions:

$A=$ Area I + Area II + Area III
$$=lw+lw+lw$$
$$=6\cdot 6+9\cdot 15+6\cdot 6$$
$$=36+135+36$$
$$=207\text{ in}^2$$

29. $-15-(-62)=-15+62=47$

31.

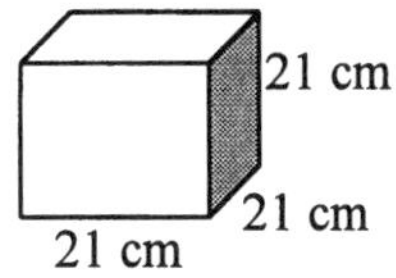

The volume of the cube is:
$$V=s^3$$
$$=(21\text{ cm})^3$$
$$=(21\text{ cm})(21\text{ cm})(21\text{ cm})$$
$$=9,261\text{ cm}^3$$

33. $\underbrace{2/3\text{ of}}_{\frac{2}{3}\times}\quad\underbrace{\text{the product of 7 and 9}}_{(7\times 9)}$

$$\frac{2}{3}(7\cdot 9)=\frac{2}{3}\cdot\frac{63}{1}=\frac{2\cdot\cancel{3}\cdot 21}{\cancel{3}\cdot 1}=42$$

35.

Let $x=10$:	Let $y=2$:	Let $x=4$:
$2x-2y=12$	$2x-2y=12$	$2x-2y=12$
$2\cdot 10-2y=12$	$2x-2\cdot 2=12$	$2\cdot 4-2y=12$
$20-2y=12$	$2x-4=12$	$8-2y=12$
$-2y=-8$	$2x=16$	$-2y=4$
$y=4$	$x=8$	$y=-2$
$(10,4)$	$(8,2)$	$(4,-2)$

37. $18\frac{1}{3}\cdot 12=\frac{55}{3}\cdot\frac{12}{1}=\frac{55\cdot\cancel{3}\cdot 4}{\cancel{3}\cdot 1}=220\text{ in.}$

39. $V=lwh$
$$=(36\text{in})(24\text{in})(12\text{in})$$
$$=10{,}368\text{ in}^3$$

Chapter 4 Test

1. $9x-3x+7-12=(9-3)x+7-12$
$=6x-5$

3. When $x=3$:
$3x-12=3\cdot3-12$
$=9-12$
$=-3$

5. When $x=-1$:
$4x-3=-7$
$4(-1)-3=-7$
$-4-3=-7$
$-7=-7$ true
$x=-1$ is a solution.

7. $\frac{2}{3}y=18$ LCD = 3
$3\left(\frac{2}{3}y\right)=3(18)$
$2y=54$
$y=27$

9. $2(x-5)=-8$
$2x-10=-8$
$2x-10+10=-8+10$
$2x=2$
$x=1$

11. $6(3x-2)-8=4x-6$
$18x-12-8=4x-6$
$18x-20=4x-6$
$18x-20+(-4x)=4x-6+(-4x)$
$14x-20=-6$
$14x-20+20=-6+20$
$14x=14$
$x=1$

13. $r=\frac{d}{t}=\frac{3{,}100 \text{ mi}}{140 \text{ hr}}\approx 22.14$ or 22 mi/hr

15. Let x = Susan's age.
Then $x-5$ is Karen's age.

	Age Today	Age 3 years ago
Susan	x	$x-3$
Karen	$x-5$	$(x-5)-3=x-8$
Sum		$(x-3)+(x-8)$

The sum of their ages 3 years ago was 11
$(x-3)+(x-8) = 11$

$(x-3)+(x-8)=11$
$x-3+x-8=11$
$2x-11=11$
$2x=22$
$x=11$
Susan's age is $x=11$ years.
Karen's age is $x-5=11-5=6$ years.

17.

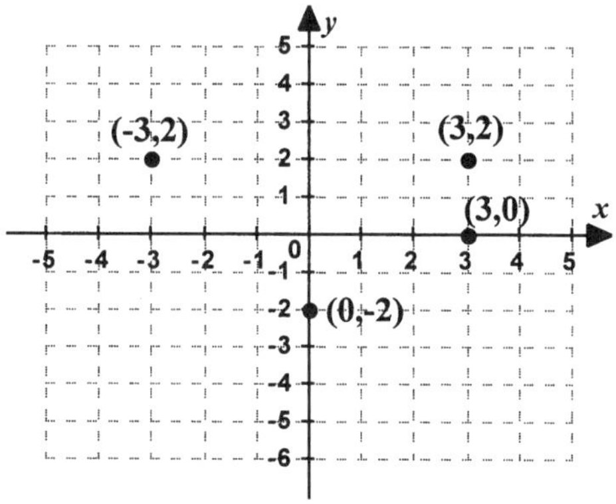

19. Choose any 3 values for x or y:

Let $x = 0$:	Let $y = 0$:	Let $x = 2$:
$y = 2x - 1$	$y = 2x - 1$	$y = 2x - 1$
$y = 2 \cdot 0 - 1$	$0 = 2x - 1$	$y = 2 \cdot 2 - 1$
$y = -1$	$1 = 2x$	$y = 4 - 1$
	$\frac{1}{2} = x$	$y = 3$

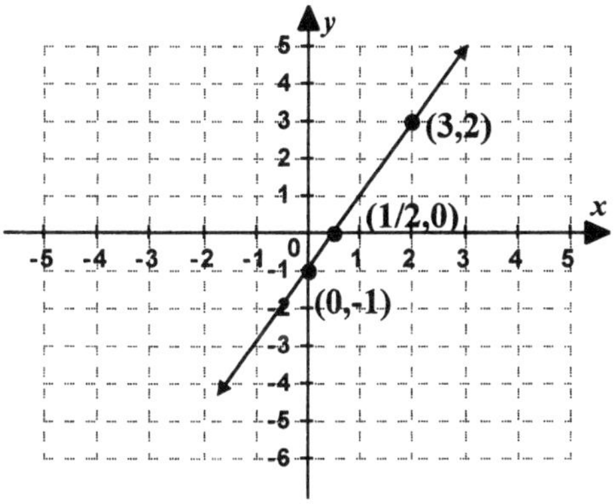

Chapter 5 Preview

1.

$$\begin{array}{r} 25{,}430 \\ 2{,}897 \\ +379{,}600 \\ \hline 407{,}927 \end{array}$$

3.

$$\begin{array}{r} 2{,}000 \\ -1{,}564 \\ \hline 436 \end{array}$$

5.

$$\begin{array}{r} 305 \\ \times 436 \\ \hline 1830 \\ 9150 \\ \underline{122000} \\ 132{,}980 \end{array}$$

7.

$$\begin{aligned} 480+12(32)^2 &= 480+12(1{,}024) \\ &= 480+12{,}288 \\ &= 12{,}768 \end{aligned}$$

9.

$$\begin{array}{r} 1{,}848 \\ 27\overline{)49{,}896} \\ \underline{27\downarrow} \\ 228 \\ \underline{216\downarrow} \\ 129 \\ \underline{108\downarrow} \\ 216 \\ \underline{216} \\ 0 \end{array}$$

11. $5^2+7^2=25+49=74$

13. $10x-15x=(10-15)x=-5x$

15. $\dfrac{38}{100}=\dfrac{\cancel{2}\cdot 19}{\cancel{2}\cdot 50}=\dfrac{19}{50}$

17.

$$\begin{aligned} 5a-420 &= -3a+980 \\ 5a+3a-420 &= -3a+980+3a \\ 8a-420 &= 980 \\ 8a-420+420 &= 980+420 \\ 8a &= 1{,}400 \\ a &= \frac{1{,}400}{8} \\ a &= 175 \end{aligned}$$

19. $48=2\cdot 2\cdot 2\cdot 2\cdot 3=2^4\cdot 3$

Chapter 5 Pretest

1. 4.013 in words:
four and thirteen thousandths

3. Thirty-four hundredths is 0.34

5. Add trailing 0's so all numbers have an equal number of digits. Then order:
0.04
0.40
0.41
0.45
0.50
0.51

7.
$$\begin{array}{r} 7.36 \\ +\,8.05 \\ \hline 15.41 \end{array}$$

9.
$$\begin{array}{r} 3.6 \\ \times\,2.7 \\ \hline 252 \\ \underline{720} \\ 9.72 \end{array}$$

11.
$$\begin{array}{r} 4.12 \\ 32\overline{)131.84} \\ \underline{128\downarrow} \\ 38 \\ \underline{32\downarrow} \\ 64 \\ \underline{64} \\ 0 \end{array}$$

13.
$$\begin{array}{r} 0.42 \\ 50\overline{)21.00} \\ \underline{200\downarrow} \\ 100 \\ \underline{100} \\ 0 \end{array}$$

15.
$$\begin{array}{r} \$20.00 \\ -16.93 \\ \hline \$3.07 \end{array}$$

17. $\sqrt{36}+2\sqrt{100}=6+2(10)=6+20=26$

19.
$$\begin{aligned} c &= \sqrt{a^2+b^2} \\ &= \sqrt{9^2+12^2} \\ &= \sqrt{81+144} \\ &= \sqrt{225} \\ &= 15 \end{aligned}$$

21.
$$\begin{aligned} \sqrt{18x^3y^2} &= \sqrt{2\cdot 3\cdot 3\cdot x\cdot x\cdot x\cdot y\cdot y} \\ &= 3\cdot x\cdot y\sqrt{2\cdot x} \\ &= 3xy\sqrt{2x} \end{aligned}$$

23.
$$\begin{aligned} 3\sqrt{12}+5\sqrt{27} &= 3\sqrt{4\cdot 3}+5\sqrt{9\cdot 3} \\ &= 3\cdot 2\sqrt{3}+5\cdot 3\sqrt{3} \\ &= 6\sqrt{3}+15\sqrt{3} \\ &= 21\sqrt{3} \end{aligned}$$

Section 5.1

1. 0.3 in words:
three tenths

3. 0.015 in words:
fifteen thousandths

5. 3.4 in words:
three and four tenths

7. 52.7 in words:
fifty-two and seven tenths

9. $405.36 = 405\frac{36}{100}$

11. $9.009 = 9\frac{9}{1,000}$

13. $1.234 = 1\frac{234}{1,000}$

15. $0.00305 = \frac{305}{100,000}$

17. 458.327
↑
tens

19. 29.52
↑
tenths

21. 0.00375
↑
hundred thousandths

23. 275.01
↑
ones

25. 539.76
↑
hundreds

27. Fifty-five hundredths is 0.55

29. Six and nine tenths is 6.9

31. Eleven and eleven hundredths is 11.11

33. One hundred and two hundredths is 100.02

35. Three thousand and three thousandths is 3,000.003

37. Look at the digit to the right of the one you are rounding to:
Whole number 47.**5**479
$5 \geq 5$ so add 1 to the ones place and delete all digits to the right 48

Tenth 47.5**4**79
$4 < 5$ so leave the tenths place as is and delete all digits to the right 47.5

Hundredth 47.54**7**9
$7 \geq 5$ so add 1 to the hundredths place and delete all digits to the right 47.55

Thousandth 47.547**9**
$9 \geq 5$ so add 1 to the thousandths place and delete all digits to the right 47.548

39. Look at the digit to the right of the one you are rounding to:
Whole number 0.**8**175
8 ≥ 5 so add 1 to the ones place and delete all digits to the right 1

Tenth 0.8**1**75
1 < 5 so leave the tenths place as is and delete all digits to the right 0.8

Hundredth 0.81**7**5
7 ≥ 5 so add 1 to the hundredths place and delete all digits to the right 0.82

Thousandth 0.817**5**
5 ≥ 5 so add 1 to the thousandths place and delete all digits to the right 0.818

41. Look at the digit to the right of the one you are rounding to:
Whole number 0.**1**562
1 < 5 so leave the ones place as is and delete all digits to the right 0

Tenth 0.1**5**62
5 ≥ 5 so add 1 to the tenths place and delete all digits to the right 0.2

Hundredth 0.15**6**2
6 ≥ 5 so add 1 to the hundredths place and delete all digits to the right 0.16

Thousandth 0.156**2**
2 < 5 so leave the thousandths place as is and delete all digits to the right 0.156

43. Look at the digit to the right of the one you are rounding to:
Whole number 2,789.**3**241
3 < 5 so leave the ones place as is and delete all digits to the right 2,789

Tenth 2,789.3**2**41
2 < 5 so leave the tenths place as is and delete all digits to the right 2,789.3

Hundredth 2,789.32**4**1
4 < 5 so leave the hundredths place as is and delete all digits to the right 2,789.32

Thousandth 2,789.324**1**
1 < 5 so leave the thousandths place as is and delete all digits to the right 2,789.324

45. Look at the digit to the right of the one you are rounding to:
Whole number 99.**9**999
9 ≥ 5 so add 1 to the ones place (and carry) and delete all digits to the right 100

Tenth 99.9**9**99
9 ≥ 5 so add 1 to the tenths place (and carry) and delete all digits to the right
100.0

Hundredth 99.99**9**9
9 ≥ 5 so add 1 to the hundredths place (and carry) and delete all digits to the right
100.00

Thousandth 99.999**9**
9 ≥ 5 so add 1 to the thousandths place (and carry) and delete all digits to the right
100.000

Applying the Concepts

47. 3.11 in words:
Three and eleven hundredths
2.5 in words:
Two and five tenths

49. Look at the digit to the right of the one you are rounding to:
Hundredth 186,282.39$\underline{7}$6
$7 \geq 5$ so add 1 to the hundredths place (and carry) and delete all digits to the right
186,282.40

51. 0.15 in words:
Fifteen hundredths

53. For each date, read the corresponding price at the top of the bar as measured by the vertical axis.

Date	Price (dollars)
4/5/2004	2.126
4/12/04	2.157
4/19/04	2.148
4/26/04	2.124

55. To compare the numbers, add trailing zeros where needed so all numbers have the same amount of digits. Then arrange from smallest to largest by reading from left to right:

Original Set	Ordered Set
0.020	0.002
0.050	0.005
0.025	0.020
0.052	0.025
0.005	0.050
0.002	0.052

57. Look at the digit to the right of the tens place:
7.4$\underline{5}$1 $5 \geq 5$ so add 1 to the tens place and delete all digits to the right 7.5
7.4$\underline{4}$9 $4 < 5$ so leave the tens place as is and delete all digits to the right 7.4
7.5$\underline{4}$ $4 < 5$ so leave the tens place as is and delete all digits to the right 7.5
7.5$\underline{6}$ $6 \geq 5$ so add 1 to the tens place and delete all digits to the right 7.6

The answers are 7.451 and 7.54

59. $0.25 = \frac{25}{100} = \frac{1}{4}$

61. $0.35 = \frac{35}{100} = \frac{7}{20}$

63. $0.125 = \frac{125}{1{,}000} = \frac{1}{8}$

65. $0.625 = \frac{625}{1{,}000} = \frac{5}{8}$

67. $0.875 = \frac{875}{1{,}000} = \frac{7}{8}$

Estimating

69. Add trailing zeros so that both numbers have an equal number of digits.
9.90
9.99
By reading the digits from left to right, the larger number is 9.99, which is closer to 10.

71. Add trailing zeros so that both numbers have an equal number of digits.
10.50
10.05
By reading the digits from left to right, the smaller number is 10.05, which is closer to 10.

73. Add trailing zeros so that both numbers have an equal number of digits.
0.50
0.05
By reading the digits from left to right, the smaller number is 0.05, which is closer to 0.

75. Add trailing zeros so that both numbers have an equal number of digits.
0.01
0.02
By reading the digits from left to right, the smaller number is 0.01, which is closer to 0.

Review Problems

77.

$$\begin{array}{rcrcr} 4\dfrac{3}{10} & = & 4\dfrac{3\cdot 10}{10\cdot 10} & = & 4\dfrac{30}{100} \\ +\ 2\dfrac{1}{100} & = & +\ 2\dfrac{1}{100} & = & +\ 2\dfrac{1}{100} \\ \hline & & & & 6\dfrac{31}{100} \end{array}$$

79.

$$\begin{array}{rcrcr} 8\dfrac{5}{10} & = & 8\dfrac{5\cdot 10}{10\cdot 10} & = & 8\dfrac{50}{100} \\ -\ 2\dfrac{4}{100} & = & -\ 2\dfrac{4}{100} & = & -\ 2\dfrac{4}{100} \\ \hline & & & & 6\dfrac{46}{100} = 6\dfrac{23}{50} \end{array}$$

81.

$$\begin{array}{rcrcr} 5\dfrac{1}{10} & = & 5\dfrac{1\cdot 100}{10\cdot 100} & = & 5\dfrac{100}{1000} \\ 6\dfrac{2}{100} & = & 6\dfrac{2\cdot 10}{100\cdot 10} & = & 6\dfrac{20}{1000} \\ +7\dfrac{3}{1000} & = & +7\dfrac{3}{1000} & = & +7\dfrac{3}{1000} \\ \hline & & & & 18\dfrac{123}{1{,}000} \end{array}$$

83. $2+\dfrac{3}{x}=\dfrac{2\cdot x}{1\cdot x}+\dfrac{3}{x}=\dfrac{2x+3}{x}$

85. $x-\dfrac{3}{4}=\dfrac{x\cdot 4}{1\cdot 4}-\dfrac{3}{4}=\dfrac{4x-3}{4}$

Improving Your Quantitative Literacy

87. a. Reading the graph from left to right, find where the costs are falling: 1990, 1991, 2000, 2002, 2003

b. Reading the graph from left to right, find where the line is the steepest over two years: 1998 to 2000

c. Reading the price graph from left to right, find the steepest drop over a 1 year period: 1985

d. Answers will vary.

Section 5.2

1.

$$\begin{array}{r} 2.91 \\ +\ \underline{3.28} \\ 6.19 \end{array}$$

3.

$$\begin{array}{r} 0.04 \\ 0.31 \\ +\ 0.78 \\ \hline 1.13 \end{array}$$

5. $3.89+(-2.4)$ *subtract absolute values*

$$\begin{array}{r} 3.89 \\ -\ 2.40 \\ \hline 1.49 \end{array}$$

The answer has the sign of the larger absolute value: 1.49

7. Note: $1.81 = 1.810$ and $2.7 = 2.700$

$$\begin{array}{r} 4.532 \\ 1.810 \\ +\ 2.700 \\ \hline 9.042 \end{array}$$

9. $0.081+(-5)+2.94$ *subtract absolute values*

$=-4.919+2.94$ *subtract absolute values*

$=-1.979$

11. Note: $6.78 = 6.7800$

$0.004 = 0.0040$

$$\begin{array}{r} 5.0003 \\ 6.7800 \\ +\ 0.0040 \\ \hline 11.7843 \end{array}$$

13.

$$\begin{array}{r} 7.123 \\ 8.120 \\ +\ 9.100 \\ \hline 24.343 \end{array}$$

15.

$$\begin{array}{r} 9.001 \\ 8.010 \\ +\ 7.100 \\ \hline 24.111 \end{array}$$

17.

$$\begin{array}{r} 89.7854 \\ 3.4000 \\ 65.3500 \\ +\ 100.0060 \\ \hline 258.5414 \end{array}$$

19.

$$\begin{array}{r} 543.21 \\ +\ 123.45 \\ \hline 666.66 \end{array}$$

21.

$$\begin{array}{r} 99.34 \\ -\ 88.23 \\ \hline 11.11 \end{array}$$

23. $2.4-5.97$ *subtract absolute values*

$$\begin{array}{r} 5.97 \\ -\ 2.40 \\ \hline 3.57 \end{array}$$

The answer has the sign of the larger absolute value: -3.57

25.

$$\begin{array}{r} 6.30 \\ -\ 2.08 \\ \hline 4.22 \end{array}$$

27. $28.96-149.37$ *subtract absolute values*

$$\begin{array}{r} 149.37 \\ -\ 28.96 \\ \hline 120.41 \end{array}$$

The answer has the sign of the larger absolute value: -120.41

29. $\begin{array}{r} 45.000 \\ -\ 0.067 \\ \hline 44.933 \end{array}$

31. $-8-0.327=-8+(-0.327)$

add the absolute values

$\begin{array}{r} 8.000 \\ +\ 0.327 \\ \hline 8.327 \end{array}$

The answer has the common sign: -8.327

33. $\begin{array}{r} 765.432 \\ -\ 234.567 \\ \hline 530.865 \end{array}$

35. $\begin{array}{r} 34.07 \\ -\ 6.18 \\ \hline 27.89 \end{array}$

37. $\begin{array}{r} 40.04 \\ -\ 4.40 \\ \hline 35.64 \end{array}$

39. $\begin{array}{r} 768.436 \\ -\ 356.998 \\ \hline 411.438 \end{array}$

41. $(7.8-4.3)+2.5$
$=3.5+2.5$
$=6$

43. $7.8-(4.3+2.5)$
$=7.8-6.8$
$=1$

45. $(9.7-5.2)-1.4$
$=4.5-1.4$
$=3.1$

47. $9.7-(5.2-1.4)$
$=9.7-3.8$
$=5.9$

49. $(8.2+0.072)-5$
$=8.272-5$
$=3.272$

51. Let x = the number
$0.035+x=4.036$
$x=4.036-0.035$
$x=4.001$

Applying the Concepts

53. Find the total of the amounts spent at each store:

$\begin{array}{r} \$25.37 \\ 39.41 \\ +\ 52.04 \\ \hline \$116.82 \end{array}$

55. Find the sum of the highest prices for each album:

$\begin{array}{r} \$20.79 \\ 13.99 \\ 27.28 \\ 13.95 \\ +\ 18.81 \\ \hline \$94.82 \end{array}$

57. Total deductions:

$\begin{array}{r} \$311.93 \\ 158.21 \\ +\ 64.72 \\ \hline \$534.86 \end{array}$

Gross Pay:	\$2,105.96
−Total Deductions:	−534.86
Take-home Pay	\$1,571.10

59. $P = 2l + 2w$

$= 2(1.41 \text{ in}) + 2(0.84 \text{ in})$

$= 2.82 \text{ in} + 1.68 \text{ in}$

$= 4.5 \text{ in}$

61. $10.00 − 4.57 = $5.43

They receive $5.43 in change.

63. a.

2002 subscribers:	66.9 million
− 1990 subscribers:	− 50.5 million
Difference:	16.4 million

b.

1990 subscribers:	62.3 million
− 2002 subscribers:	− 55.2 million
Difference:	7.1 million

c. There was a greater increase in cable subscribers with 16.4 million.

65. a. From 2006 on the horizontal axis, draw a vertical line up to the graph. It is slightly more than 15 million.

b. From 2005 on the horizontal axis, draw a vertical line up to the graph. It is slightly less than 10 million.

c. Answers will vary:

2006 sales:	15.5
− 2005 sales:	−9.5
Increase:	6.0 million

67. Above each car, draw a bar with height corresponding to the time as measured by the vertical axis.

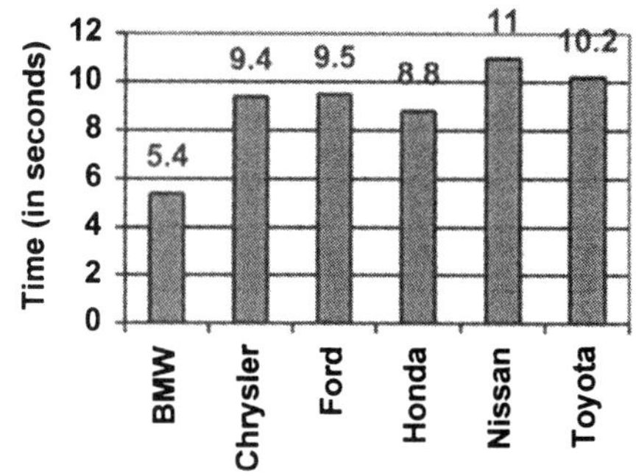

Slowest car:	11.0
− Fastest car:	− 5.4
Time difference:	5.6 seconds

Estimating

69. $0.005 + 0.090 = 0.095$ rounds to 0.1

Review Problems

71. $\frac{1}{10} \cdot \frac{3}{10} = \frac{1 \cdot 3}{10 \cdot 10} = \frac{3}{100}$

73. $\frac{3}{100} \cdot \frac{17}{100} = \frac{3 \cdot 17}{100 \cdot 100} = \frac{51}{10{,}000}$

75. $-5\left(-\frac{3}{10}\right) = \frac{5}{1} \cdot \frac{3}{10} = \frac{5 \cdot 3}{1 \cdot 10} = \frac{15}{10} = \frac{3}{2} = 1\frac{1}{2}$

77.

$$\begin{array}{r} 56 \\ \times 25 \\ \hline 280 \\ 1{,}120 \\ \hline 1{,}400 \end{array}$$

Extending the Concepts

79. $P = 2l + 2w \qquad let\ P = 9.5, l = 2.75$

$9.5 = 2(2.75) + 2w$

$9.5 = 5.5 + 2w$

$9.5 + (-5.5) = 5.5 + 2w + (-5.5)$

$4 = 2w$

$2 = w$

The width is 2 inches.

81.

Amount given cashier:	\$20.01
– Total bill:	– 16.76
Change:	\$3.25

You receive \$3.25 in change, probably as three \$1 bills and a quarter.

83. $2.5, 2.75, 3, \ldots$

The first number in the sequence is 2.5. Each number thereafter comes from adding 0.25 to the previous number, meaning the sequence is *arithmetic.* The next number in the sequence is $3 + 0.25 = 3.25$

Improving Your Quantitative Literacy

85. a.

2006 users:	163.7 million
– 2003 users:	– 147.6 million
Increase:	16.1 million

b. The difference between sequential years:

Years	Increase (in millions)
2003 to 2004	$154.2 - 147.5 = 6.7$
2004 to 2005	$159.7 - 154.2 = 5.5$
2005 to 2006	$163.7 - 159.7 = 4.0$
2006 to 2007	$167.5 - 163.7 = 3.8$

The largest increase was 2003 to 2004.

c. Answers will vary.
An estimated increase is 3.6 million, making the number of users in 2008:
$167.5 + 3.6 = 171.1$ million.

Section 5.3

1.
$$\begin{array}{rl} 0.7 & \textit{1 decimal place} \\ \times\ 0.4 & \textit{1 decimal place} \\ \hline 0.28 & \textit{2 decimal places} \end{array}$$

3.
$$\begin{array}{rl} 0.07 & \textit{2 decimal places} \\ \times\ 0.4 & \textit{1 decimal place} \\ \hline 0.028 & \textit{3 decimal places} \end{array}$$

5.
$$\begin{array}{rl} 0.03 & \textit{2 decimal places} \\ \times\ 0.09 & \textit{2 decimal places} \\ \hline 0.0027 & \textit{4 decimal places} \end{array}$$

7.
$$\begin{array}{rl} 2.6 & \textit{1 decimal place} \\ \times\ 0.3 & \textit{1 decimal place} \\ \hline 0.78 & \textit{2 decimal places} \end{array}$$

$-2.6(0.3) = -0.78$

9.
$$\begin{array}{rl} 0.9 & \textit{1 decimal place} \\ \times\ 0.88 & \textit{2 decimal places} \\ \hline 72 & \\ \underline{720} & \\ 0.792 & \textit{3 decimal places} \end{array}$$

11.
$$\begin{array}{rl} 3.12 & \textit{2 decimal places} \\ \times\ 0.005 & \textit{3 decimal places} \\ \hline 0.01560 & \textit{5 decimal places} \end{array}$$

The trailing zero can be dropped: 0.0156

13.
$$\begin{array}{rl} 4.003 & \textit{3 decimal places} \\ \times\ 6.07 & \textit{2 decimal places} \\ \hline 28021 & \\ \underline{2401800} & \\ 24.29821 & \textit{5 decimal places} \end{array}$$

15.
$$\begin{array}{rl} 0.006 & \textit{3 decimal places} \\ \times\ 5 & \textit{0 decimal places} \\ \hline 0.030 & \textit{3 decimal places} \end{array}$$

Drop the trailing zero: $-5(0.006) = -0.03$

17.
$$\begin{array}{rl} 75.14 & \textit{2 decimal places} \\ \times\ 2.5 & \textit{1 decimal place} \\ \hline 37570 & \\ \underline{150280} & \\ 187.850 & \textit{3 decimal places} \end{array}$$

The trailing zero can be dropped: 187.85

19.
$$\begin{array}{rl} 0.1 & \textit{1 decimal place} \\ \times\ 0.02 & \textit{2 decimal places} \\ \hline 0.002 & \textit{3 decimal places} \end{array}$$

21.
$$\begin{array}{rl} 2.796 & \textit{3 decimal places} \\ \times\ 10 & \textit{0 decimal places} \\ \hline 27.960 & \textit{3 decimal places} \end{array}$$

The trailing zero can be dropped, and $2.796(-10) = -27.96$

(Shortcut for multiplying by 1$\underline{0}$ is to move decimal point 1 place to the right)

23.
$$\begin{array}{rl} 0.0043 & \textit{4 decimal places} \\ \times\ 100 & \textit{0 decimal places} \\ \hline 0.4300 & \textit{4 decimal places} \end{array}$$

The trailing zero can be dropped: 0.43
(Shortcut for multiplying by 1$\underline{00}$ is to move decimal point 2 places to the right)

25.
$$\begin{array}{rl} 49.94 & \textit{2 decimal places} \\ \times 1{,}000 & \textit{0 decimal places} \\ \hline 49{,}940.00 & \textit{2 decimal places} \end{array}$$

The trailing zeros can be dropped: 49,940
(Shortcut for multiplying by 1$\underline{000}$ is to move decimal point 3 places to the right)

27.

$$\begin{array}{rl} 987.654 & \textit{3 decimal places} \\ \underline{\times 10{,}000} & \textit{0 decimal places} \\ 9{,}876{,}540.000 & \textit{3 decimal places} \end{array}$$

The trailing zeros can be dropped:

9,876,540

(Shortcut for multiplying by 10,000 is to move decimal point 4 places to the right)

29. $2.1(3.5-2.6)=2.1(0.9)$
$=1.89$

31. $-0.05(0.02+0.03)=-0.05(0.05)$
$=-0.0025$

33. $2.02(0.03-2.5)=2.02(-2.47)$
$=-4.9894$

35. $(2.1+0.03)(3.4+0.05)=(2.13)(3.45)$
$=7.3485$

37. $(2.1-0.1)(2.1+0.1)=(2.0)(2.2)$
$=4.4$

39. $3.08-0.2(5+0.03)=3.08-0.2(5.03)$
$=3.08-1.006$
$=2.074$

41. $4.23-5(0.04+0.09)=4.23-5(0.13)$
$=4.23-0.65$
$=3.58$

43. $2.5+10(4.3)^2=2.5+10(18.49)$
$=2.5+184.9$
$=187.4$

45. $100(1+0.08)^2=100(1.08)^2$
$=100(1.1664)$
$=116.64$

47. $(1.5)^2+(2.5)^2+(3.5)^2$
$=(2.25)+(6.25)+(12.25)$
$=20.75$

Applying the Concepts

49. $\underbrace{\text{Product of 6 and}}\ \underbrace{\text{the sum of 0.001 and 0.02}}$

$6\times \qquad (0.001+0.02)$

$6(0.001+0.02)=6(0.021)$
$=0.126$

51. For example:

$$\begin{array}{ccc} 1.2 & 0.12 & 0.012 \\ \underline{\times 100} & \underline{\times 100} & \underline{\times 100} \\ 120.0=120 & 12.00=12 & 1.200=1.2 \end{array}$$

When multiplying by 100, the decimal point moves 2 places to the right.

53.

Cost per \$1,000:	\$9.66
Number of \$1,000's:	× 143
Total payment:	2898
	38640
	96600
	1,381.38

The monthly payment is \$1,381.38

55. Cost of the first minute + Cost of the next 19 minutes = Total Cost

$\$0.45 + \$0.35(19)$

$0.45+0.35(19)=0.45+6.65$
$=7.10$

The total cost is \$7.10

57. Cost for 2 days + Cost for 120 miles

$\$15(2) + \$0.12(120)$

$15(2)+0.12(120)=30+14.4$
$=44.0$

The total cost is \$44.40

59. Pay for 1st 36 hours + Pay for next 9 hours

$$\$5.92(36) \quad + \quad \$8.88(9)$$

$$\begin{aligned} 5.92(36)+8.88(9) &= 213.12+79.92 \\ &= 293.04 \end{aligned}$$

His total pay is \$293.04

61. $$\begin{aligned} A &= lw \\ &= (67 \text{ mm})(127 \text{ mm}) \\ &= 8{,}509 \text{ mm}^2 \end{aligned}$$

63. $$\begin{aligned} A &= lw \\ &= (1.41 \text{ in})(0.84 \text{ in}) \\ &= 1.1844 \text{ in}^2 \\ &\approx 1.18 \text{ in}^2 \end{aligned}$$

65. $$\begin{aligned} C &= 2\pi r \\ &= 2(3.14)(4 \text{ in}) \\ &= 25.12 \text{ in} \end{aligned} \qquad \begin{aligned} A &= \pi r^2 \\ &= (3.14)(4 \text{ in})^2 \\ &= (3.14)(16 \text{ in}^2) \\ &= 50.24 \text{ in}^2 \end{aligned}$$

67. $$\begin{aligned} C &= 2\pi r \\ &= 2(3.14)(3{,}900 \text{ mi}) \\ &= 24{,}492 \text{ mi} \end{aligned}$$

69. Find the circumference:

$$\begin{aligned} C &= 2\pi r \\ &= 2(3.14)(26.75 \text{ in}) \\ &= 167.99 \text{ in} \\ &\approx 168 \text{ in} \end{aligned}$$

71. $$\begin{aligned} V &= \pi r^2 h \\ &= 3.14(2 \text{ ft})^2(8 \text{ ft}) \\ &= 3.14(4 \text{ ft}^2)(8 \text{ ft}) \\ &= 100.48 \text{ ft}^3 \end{aligned}$$

73. $$\begin{aligned} V &= \pi r^2 h \\ &= 3.14(2 \text{ ft})^2(4 \text{ ft}) \\ &= 3.14(4 \text{ ft}^2)(4 \text{ ft}) \\ &= 50.24 \text{ ft}^3 \end{aligned}$$

75. Note the radius is 1/2 of the diameter:

$$r = \frac{1}{2}d = \frac{1}{2}(26.5 \text{ mm}) = 13.25 \text{ mm}$$

a. $$\begin{aligned} C &= 2\pi r \\ &= 2(3.14)(13.25 \text{ mm}) \\ &= 83.21 \text{ mm} \end{aligned}$$

b. $$\begin{aligned} A &= \pi r^2 \\ &= (3.14)(13.25 \text{ mm})^2 \\ &= (3.14)(175.5625 \text{ mm}^2) \\ &= 551.26625 \text{ mm}^2 \\ &\approx 551.27 \text{ mm}^2 \end{aligned}$$

c. $$\begin{aligned} V &= \pi r^2 h \\ &= 3.14(13.25 \text{ mm})^2(2.00 \text{ mm}) \\ &= 3.14(175.5625 \text{ mm}^2)(2.00 \text{ mm}) \\ &= 1{,}102.5325 \text{ mm}^3 \\ &\approx 1{,}102.53 \text{ mm}^3 \end{aligned}$$

Review Problems

77.

$$\begin{array}{r} 1{,}879 \\ 2\overline{)3{,}758} \\ \underline{2\downarrow} \\ 17 \\ \underline{16\downarrow} \\ 15 \\ \underline{14\downarrow} \\ 18 \\ \underline{18} \\ 0 \end{array}$$

79.

```
       1,516 R4
33)50,032
   33↓
   170
   165↓
     53
     33↓
    202
    198
      4
```

81. $\dfrac{a^2b^3}{c^4} \div \dfrac{a^2b}{c^7}$

$= \dfrac{a^2b^3}{c^4} \cdot \dfrac{c^7}{a^2b}$

$= \dfrac{\cancel{a} \cdot \cancel{a} \cdot \cancel{b} \cdot b \cdot b \cdot \cancel{c} \cdot \cancel{c} \cdot \cancel{c} \cdot \cancel{c} \cdot c \cdot c \cdot c}{\cancel{c} \cdot \cancel{c} \cdot \cancel{c} \cdot \cancel{c} \cdot \cancel{a} \cdot \cancel{a} \cdot \cancel{b}}$

$= b^2c^3$

83. $\dfrac{10x^5}{21y^3} \div \dfrac{5x^3}{42y^5}$

$= \dfrac{10x^5}{21y^3} \cdot \dfrac{42y^5}{5x^3}$

$= \dfrac{420x^5y^5}{105y^3x^3}$

$= \dfrac{\cancel{105} \cdot 4 \cdot \cancel{x} \cdot \cancel{x} \cdot \cancel{x} \cdot x \cdot x \cdot \cancel{y} \cdot \cancel{y} \cdot \cancel{y} \cdot y \cdot y}{\cancel{105} \cdot \cancel{y} \cdot \cancel{y} \cdot \cancel{y} \cdot \cancel{x} \cdot \cancel{x} \cdot \cancel{x}}$

$= 4x^2y^2$

Improving your Quantitative Literacy

85. Volume of the center tank:

$V = \pi r^2 h$

$= 3.14(6\text{ ft})^2(16\text{ ft})$

$= 3.14(36\text{ ft}^2)(16\text{ ft})$

$= 1,808.64\text{ ft}^3$

Volume of the containment tank:

$V = \pi r^2 h$

$= 3.14(8\text{ ft})^2(4\text{ ft})$

$= 3.14(64\text{ ft}^2)(4\text{ ft})$

$= 803.84\text{ ft}^3$

No. The containment tank holds less.

Section 5.4

1.
```
     19.7
 20)394.0
    20↓
    194
    180↓
     140
     140
       0
```

3.
```
      6.2
 40)248.0
    240↓
      80
      80
       0
```

5.
```
    5.2
 5)26.0
   25↓
    10
    10
     0
```

7.
```
     11.04
 25)276.00
    25↓
     26
     25↓↓
      100
      100
        0
```

9.
```
   4.8
 6)28.8
   24↓
    48
    48
     0
```

11.
```
   9.7
 8)77.6
   72↓
    56
    56
     0
```

13.
```
     2.63
 35)92.05
    70↓
    220
    210↓
     105
     105
       0
```

15.
```
      4.24
 45)190.80
    180↓
     108
      90↓
      180
      180
        0
```

17. $\begin{array}{r} 2.55 \\ 34\overline{)86.70} \end{array}$

$\begin{array}{r} \underline{68\downarrow} \\ 187 \\ \underline{170\downarrow} \\ 170 \\ \underline{170} \\ 0 \end{array}$

19. $\begin{array}{r} 1.35 \\ 22\overline{)29.70} \end{array}$

$\begin{array}{r} \underline{22\downarrow} \\ 77 \\ \underline{66\downarrow} \\ 110 \\ \underline{110} \\ 0 \end{array}$

21. $4.5\overline{)29.25} \rightarrow \begin{array}{r} 6.5 \\ 45\overline{)292.5} \end{array}$

$\begin{array}{r} \underline{270\downarrow} \\ 225 \\ \underline{225} \\ 0 \end{array}$

23. $0.11\overline{)1.089} \rightarrow \begin{array}{r} 9.9 \\ 11\overline{)108.9} \end{array}$

$\begin{array}{r} \underline{99\downarrow} \\ 99 \\ \underline{99} \\ 0 \end{array}$

25. $2.3\overline{)0.115} \rightarrow \begin{array}{r} 0.05 \\ 23\overline{)1.15} \end{array}$

$\begin{array}{r} \underline{115} \\ 0 \end{array}$

27. $0.012\overline{)1.068} \rightarrow \begin{array}{r} 89 \\ 12\overline{)1,068} \end{array}$

$\begin{array}{r} \underline{96\downarrow} \\ 108 \\ \underline{108} \\ 0 \end{array}$

29. $1.1\overline{)2.42} \rightarrow \begin{array}{r} 2.2 \\ 11\overline{)24.2} \end{array}$

$\begin{array}{r} \underline{22\downarrow} \\ 22 \\ \underline{22} \\ 0 \end{array}$

31. $\begin{array}{r} 1.346 \\ 26\overline{)35.000} \end{array}$

$\begin{array}{r} \underline{26\downarrow} \\ 90 \\ \underline{78\downarrow} \\ 120 \\ \underline{104\downarrow} \\ 160 \\ \underline{156} \\ 4 \end{array}$

Rounding to the nearest hundredth: 1.35

33. $3.3\overline{)56} \rightarrow 33\overline{)560.000}$ (quotient: 16.969)

$$
\begin{array}{r}
16.969 \\
33\overline{)560.000} \\
\underline{33\downarrow} \\
230 \\
\underline{198\downarrow} \\
320 \\
\underline{297\downarrow} \\
230 \\
\underline{198} \\
320 \\
\underline{297} \\
23
\end{array}
$$

Rounding to the nearest hundredth: 16.97

35. $0.5\overline{)0.1234} \rightarrow 5\overline{)1.234}$ (quotient: 0.246)

$$
\begin{array}{r}
0.246 \\
5\overline{)1.234} \\
\underline{10\downarrow} \\
23 \\
\underline{20\downarrow} \\
34 \\
\underline{32} \\
2
\end{array}
$$

Rounding to the nearest hundredth: 0.25

37.

$$
\begin{array}{r}
2.714 \\
7\overline{)19.000} \\
\underline{14\downarrow} \\
50 \\
\underline{49\downarrow} \\
10 \\
\underline{7\downarrow} \\
30 \\
\underline{28} \\
2
\end{array}
$$

Rounding the to nearest hundredth: 2.71

39. $0.059\overline{)0.69} \rightarrow 59\overline{)690.000}$ (quotient: 11.694)

$$
\begin{array}{r}
11.694 \\
59\overline{)690.000} \\
\underline{59\downarrow} \\
100 \\
\underline{59\downarrow} \\
410 \\
\underline{354\downarrow} \\
560 \\
\underline{531\downarrow} \\
290 \\
\underline{236} \\
54
\end{array}
$$

Rounding the to nearest hundredth: 11.69

41. $0.3\overline{)6.99} \rightarrow 3\overline{)69.9}$

False.

43. $0.06\overline{)42} \rightarrow 0.6\overline{)420} \rightarrow 6\overline{)4,200}$

True.

Applying the Concepts

45. $12 \cancel{\text{km}}\left(\dfrac{1 \text{ mi}}{1.61 \cancel{\text{km}}}\right) = \dfrac{12}{1.61} \text{ mi}$

$1.61\overline{)12.0} \rightarrow 161\overline{)1200.00}$ (quotient: 7.45)

$$
\begin{array}{r}
7.45 \\
161\overline{)1200.00} \\
\underline{1127\downarrow} \\
730 \\
\underline{644\downarrow} \\
860 \\
\underline{805} \\
55
\end{array}
$$

1.61 km is approximately 7.5 miles.

47. For all players:

$$\frac{\text{Total earnings}}{\text{Tournaments}} = \text{Average earnings}$$

Web: $\dfrac{\$1{,}591{,}959}{25} = \$63{,}678$

Inkster: $\dfrac{\$1{,}337{,}256}{24} = \$55{,}719$

Pak: $\dfrac{\$959{,}926}{27} = \$35{,}553$

Sorenstam: $\dfrac{\$863{,}816}{22} = \$39{,}264$

Kane: $\dfrac{\$757{,}844}{30} = \$25{,}261$

49. $\dfrac{\text{Total wages}}{\text{hours worked}} = \text{wage per hour}$

$$\frac{\$33.90}{6} = \$5.65$$

She earns \$5.65 per hour.

51. $\dfrac{\text{Total miles}}{\text{gallons used}} = \text{miles per 1 gallon}$

$$\frac{336 \text{ miles}}{15 \text{ gallons}} = 22.4 \text{ miles per gallon}$$

53. First, find the amount of her total wages that came from overtime:

Total wage – regular wage = overtime wage

$$\$294.93 \quad - \quad 36(6.78) \quad = 294.93 - 244.08$$
$$= 50.85$$

$$\frac{\text{Overtime wages}}{\text{Overtime wage per hour}} = \text{Overtime hours}$$

$$\frac{\$50.85}{\$10.17 \text{ per hour}} = 5 \text{ hours}$$

She worked 5 hours overtime.

55. First find the amount of the call that comes from additional minutes:

Total cost of call:	\$2.33
– cost of 1st minute:	– 0.41
Total cost of additional minutes:	\$1.92

To find the number of additional minutes:

$$\frac{\text{Total cost of additional minutes}}{\text{cost per minute}} = \frac{\$1.92}{\$0.32}$$
$$= 6$$

The call was 7 minutes long (the 1st minute plus 6 additional minutes).

57. Complete the table as follows:

Class	Units		Value	Grade Points
Math	3	A	4	$3 \times 4 = 12$
Health	2	B	3	$2 \times 3 = 6$
History	3	B	3	$3 \times 3 = 9$
English	3	C	2	$3 \times 2 = 6$
Chemistry	4	C	2	$4 \times 2 = 8$
Total	15			41

$$\frac{\text{Total grade points}}{\text{Total units}} = \text{GPA}$$

$$\frac{41}{15} = 2.73$$

His GPA is 2.73

59. Complete the table as follows:

Class	Units		Value	Grade Points
Math	3	A	4	$3 \times 4 = 12$
Health	2	C	2	$2 \times 2 = 4$
History	3	B	3	$3 \times 3 = 9$
English	3	C	2	$3 \times 2 = 6$
Chemistry	4	C	2	$4 \times 2 = 8$
Total	15			39

$$\frac{\text{Total grade points}}{\text{Total units}} = \text{GPA}$$

$$\frac{39}{15} = 2.6$$

His GPA would have dropped from 2.73 to 2.6, a difference of 0.13

Calculator Exercises

61. $7 \div 9 = 0.77778$

63. $243 \div 0.791 = 307.20607$

65. $0.0503 \div 0.0709 = 0.70945$

Review Problems

67. $\frac{75}{100} = \frac{\cancel{25} \cdot 3}{\cancel{25} \cdot 4} = \frac{3}{4}$

69. $\frac{12x}{18xy} = \frac{\cancel{6} \cdot 2 \cdot \cancel{x}}{\cancel{6} \cdot 3 \cdot \cancel{x} \cdot y} = \frac{2}{3y}$

71. $\frac{75x^2y^3}{100xy^2} = \frac{\cancel{25} \cdot 3 \cdot \cancel{x} \cdot x \cdot \cancel{y} \cdot \cancel{y} \cdot y}{\cancel{25} \cdot 4 \cdot \cancel{x} \cdot \cancel{y} \cdot \cancel{y}} = \frac{3xy}{4}$

73. $\frac{3}{5} = \frac{3 \cdot 2}{5 \cdot 2} = \frac{6}{10}$

75. $\frac{3}{5} = \frac{3 \cdot 20}{5 \cdot 20} = \frac{60}{100}$

77. $\frac{4}{5} = \frac{4 \cdot 3x}{5 \cdot 3x} = \frac{12x}{15x}$

79. $\frac{4}{x} = \frac{4 \cdot 15}{x \cdot 15} = \frac{60}{15x}$

81. $\frac{6}{5x} = \frac{6 \cdot 3}{5x \cdot 3} = \frac{18}{15x}$

Improving your Quantitative Literacy

83. a. No. Assume the original ages were given to the nearest year (a whole number). The median could either be a whole number, or possibly a decimal with decimal part equal to 0.5.

For example:
Ages: 18, 22, 27, 30, 33
Median: (middle value) 27

Ages: 18, 22, 27, 30, 33, 35
Median: (mean of the middle 2 values:)

$$\text{Median} = \frac{27+30}{2} = \frac{57}{2} = 28.5$$

b. The ages in the chart must be means.

Section 5.5

1. $\frac{1}{8}$ of the circle is shaded:

$$\begin{array}{r} 0.125 \\ 8\overline{)1.000} \\ \underline{8\downarrow} \\ 20 \\ \underline{16\downarrow} \\ 40 \\ \underline{40} \\ 0 \end{array} \qquad \frac{1}{8} = 0.125$$

3. $\frac{5}{8}$ of the circle is shaded:

$$\begin{array}{r} 0.625 \\ 8\overline{)5.000} \\ \underline{48\downarrow} \\ 20 \\ \underline{16\downarrow} \\ 40 \\ \underline{40} \\ 0 \end{array} \qquad \frac{5}{8} = 0.625$$

5.

$$\begin{array}{r} 0.25 \\ 4\overline{)1.00} \\ \underline{8\downarrow} \\ 20 \\ \underline{20} \\ 0 \end{array} \quad \begin{array}{r} 0.5 \\ 4\overline{)2.0} \\ \underline{20} \\ 0 \end{array} \quad \begin{array}{r} 0.75 \\ 4\overline{)3.00} \\ \underline{28\downarrow} \\ 20 \\ \underline{20} \\ 0 \end{array} \quad \begin{array}{r} 1 \\ 4\overline{)4} \\ \underline{4} \\ 0 \end{array}$$

Fraction	$\frac{1}{4}$	$\frac{2}{4}$	$\frac{3}{4}$	$\frac{4}{4}$
Decimal	0.25	0.5	0.75	1

7.

$$\begin{array}{r} 0.166 \\ 6\overline{)\ 1.00} \\ \underline{6\downarrow} \\ 40 \\ \underline{36} \\ 4 \end{array} \quad \begin{array}{r} 0.33 \\ 6\overline{)2.00} \\ \underline{18\downarrow} \\ 20 \\ \underline{18} \\ 2 \end{array} \quad \begin{array}{r} 0.5 \\ 6\overline{)3.0} \\ \underline{30} \\ 0 \end{array}$$

$$\begin{array}{r} 0.66 \\ 6\overline{)4.00} \\ \underline{36\downarrow} \\ 40 \\ \underline{36} \\ 4 \end{array} \quad \begin{array}{r} 0.833 \\ 6\overline{)5.000} \\ \underline{48\downarrow} \\ 20 \\ \underline{18\downarrow} \\ 20 \\ \underline{18} \\ 2 \end{array} \quad \begin{array}{r} 1 \\ 6\overline{)6} \\ \underline{6} \\ 0 \end{array}$$

Fraction	$\frac{1}{6}$	$\frac{2}{6}$	$\frac{3}{6}$	$\frac{4}{6}$	$\frac{5}{6}$	$\frac{6}{6}$
Decimal	$0.1\overline{6}$	$0.\overline{3}$	0.5	$0.\overline{6}$	$0.8\overline{3}$	1

9.

$$\begin{array}{r} 0.48 \\ 25\overline{)12.00} \\ \underline{100\downarrow} \\ 200 \\ \underline{200} \\ 0 \end{array} \qquad \frac{12}{25} = 0.48$$

11.

$$\begin{array}{r} 0.4375 \\ 32\overline{)14.0000} \\ \underline{128\downarrow} \\ 120 \\ \underline{96\downarrow} \\ 240 \\ \underline{224\downarrow} \\ 160 \\ \underline{160} \\ 0 \end{array} \qquad \frac{14}{32} = 0.4375$$

13. $$\begin{array}{r} 0.923 \\ 13\overline{)12.000} \\ \underline{117\downarrow} \\ 30 \\ \underline{26\downarrow} \\ 40 \\ \underline{39} \\ 1 \end{array}$$ $\frac{12}{13} \approx 0.92$

15. $$\begin{array}{r} 0.272 \\ 11\overline{)3.000} \\ \underline{22\downarrow} \\ 80 \\ \underline{77\downarrow} \\ 30 \\ \underline{22} \\ 8 \end{array}$$ $\frac{3}{11} \approx 0.27$

17. $$\begin{array}{r} 0.086 \\ 23\overline{)2.000} \\ \underline{184\downarrow} \\ 160 \\ \underline{138} \\ 22 \end{array}$$ $\frac{3}{23} \approx 0.09$

19. $$\begin{array}{r} 0.279 \\ 43\overline{)12.000} \\ \underline{86\downarrow} \\ 340 \\ \underline{301\downarrow} \\ 390 \\ \underline{387} \\ 3 \end{array}$$ $\frac{12}{43} \approx 0.28$

21. The table is as follows:

Decimal	Fraction
0.125	$0.125 = \frac{125}{1,000} = \frac{1}{8}$
0.250	$0.250 = \frac{250}{1,000} = \frac{1}{4}$
0.375	$0.375 = \frac{375}{1,000} = \frac{3}{8}$
0.500	$0.500 = \frac{500}{1,000} = \frac{1}{2}$
0.625	$0.625 = \frac{625}{1,000} = \frac{5}{8}$
0.750	$0.750 = \frac{750}{1,000} = \frac{3}{4}$
0.875	$0.875 = \frac{875}{1,000} = \frac{7}{8}$

23. $0.15 = \frac{15}{100} = \frac{3}{20}$

25. $0.08 = \frac{8}{100} = \frac{2}{25}$

27. $0.375 = \frac{375}{1,000} = \frac{3}{8}$

29. $5.6 = 5\frac{6}{10} = 5\frac{3}{5}$

31. $5.06 = 5\frac{6}{100} = 5\frac{3}{50}$

33. $1.22 = 1\frac{22}{100} = 1\frac{11}{50}$

35. $$\begin{aligned} \frac{1}{2}(2.3+2.5) &= \frac{1}{2}(4.8) \\ &= 2.4 \end{aligned}$$

37. $\frac{3}{8}(4.7) = \frac{3(4.7)}{8}$
$= \frac{14.1}{8}$
$= 1.7625$

39. $3.4 - \frac{1}{2}(0.76) = 3.4 - 0.38$
$= 3.02$

41. $\frac{2}{5}(0.3) + \frac{3}{5}(0.3) = 0.4(0.3) + 0.6(0.3)$
$= 0.12 + 0.18$
$= 0.3$

43. $6\left(\frac{3}{5}\right)(0.02) = 6(0.6)(0.02)$
$= 0.072$

45. $\frac{5}{8} + 0.35\left(\frac{1}{2}\right) = \frac{5}{8} + 0.175$
$= 0.625 + 0.175$
$= 0.8$

47. $\left(\frac{1}{3}\right)^2 (5.4) + \left(\frac{1}{2}\right)^3 (3.2)$
$= \left(\frac{1}{9}\right)(5.4) + \left(\frac{1}{8}\right)(3.2)$
$= 0.6 + 0.4$
$= 1$

49. $(0.25)^2 + \left(\frac{1}{4}\right)^2 (3) = 0.0625 + \left(\frac{1}{16}\right)(3)$
$= 0.0625 + 0.1875$
$= 0.25$

Applying the Concepts

51.

Cost per pound	×	Number of pounds	=	Total Cost
\$2.59	×	$3\frac{1}{4}$	=	$2.59(3.25)$
			=	8.4175

Rounding to the nearest cent, the cost is \$8.42.

53. Markdown $= \frac{1}{3}(\$57.99) = \$19.33$

Sale price = Original price − Markdown
$= \$57.99 - \$19.33$
$= \$38.66$

55. Find the perimeter of the shaded region by summing the perimeters of the 3 small shaded triangles:
$3P = 3(1 \text{ in} + 1 \text{ in} + 1 \text{ in}) = 3(3 \text{ in}) = 9 \text{ in}$

57.

$$\begin{array}{r} 0.75 \\ 4\overline{)3.00} \\ \underline{28\downarrow} \\ 20 \\ \underline{20} \\ 0 \end{array} \qquad \begin{array}{r} 0.562 \\ 16\overline{)9.000} \\ \underline{80\downarrow} \\ 100 \\ \underline{96\downarrow} \\ 40 \\ \underline{32} \\ 8 \end{array} \qquad \begin{array}{r} 0.093 \\ 32\overline{)3.000} \\ \underline{288\downarrow} \\ 120 \\ \underline{96} \\ 24 \end{array}$$

$$\begin{array}{r} 0.218 \\ 32\overline{)7.000} \\ \underline{64\downarrow} \\ 60 \\ \underline{32\downarrow} \\ 280 \\ \underline{256} \\ 24 \end{array} \qquad \begin{array}{r} 0.062 \\ 16\overline{)1.000} \\ \underline{96\downarrow} \\ 40 \\ \underline{32} \\ 8 \end{array}$$

Date	Gain/Loss($)	Decimal
March 6	$\frac{3}{4}$	0.75
March 7	$\frac{9}{16}$	0.56
March 8	$\frac{3}{32}$	0.09
March 9	$\frac{7}{32}$	0.22
March 10	$\frac{1}{16}$	0.06

59. Calories in ground beef $= 50.75\left(4\frac{1}{2}\right)$

$$= 50.75(4.5)$$
$$= 228.375$$

Calories in halibut $= 27.5\left(4\frac{1}{2}\right)$

$$= 27.5(4.5)$$
$$= 123.75$$

Difference in calories $= 228.375 - 123.75$

$$= 104.625 \text{ cal}$$

61.
$$\begin{aligned} V &= \frac{4}{3}\pi r^3 \\ &= \frac{4}{3}(3.14)(2\text{ mi})^3 \\ &= \frac{4}{3}(3.14)(8\text{ mi}^3) \\ &= \frac{4}{3}(25.12)\text{mi}^3 \\ &= 33.49\overline{3}\text{ mi}^3 \end{aligned}$$

63.
$$\begin{aligned} V &= \frac{4}{3}\pi r^3 \\ &= \frac{4}{3}(3.14)(3.9\text{ in})^3 \\ &= \frac{4}{3}(3.14)(59.319\text{in}^3) \\ &= \frac{4}{3}(186.26166\text{ in}^3) \\ &= 248.34888\text{ in}^3 \end{aligned}$$

The volume is 248.35 in^3.

65. 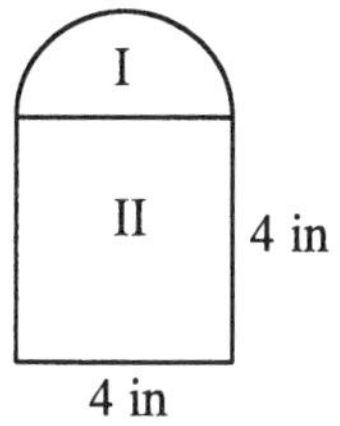

Separate into 2 regions. Note the radius of the circle is $\frac{1}{2}(\text{diameter}) = \frac{1}{2}(4 \text{ in}) = 2 \text{ in}$.

Also, the area of *half* a circle is $\frac{1}{2}(\pi r^2)$.

$$\begin{aligned}
\text{Area} &= \text{Area I} + \text{Area II} \\
&= \frac{1}{2}\pi r^2 + lw \\
&= \frac{1}{2}(3.14)(2)^2 + 4(4) \\
&= \frac{1}{2}(3.14)(4) + 16 \\
&= 6.28 + 16 \\
&= 22.28
\end{aligned}$$

The total area is 22.28 in^2

67. Separate into 2 solids (a cylinder and hemisphere). Note that the volume of *half* a sphere is $\frac{1}{2}\left(\frac{4}{3}\pi r^3\right)$

$$\begin{aligned}
V &= \text{Cylinder volume} + \text{hemisphere volume} \\
&= \pi r^2 h + \frac{1}{2}\left(\frac{4}{3}\pi r^3\right) \\
&= 3.14(3)^2(6) + \frac{1}{2}\left(\frac{4}{3}\right)(3.14)(3)^3 \\
&= 3.14(9)(6) + \frac{2}{3}(3.14)(27) \\
&= 169.56 + 56.52 \\
&= 226.08
\end{aligned}$$

The volume is 226.1 ft^3

Calculator Problems

69. $\frac{7}{9} = 0.\overline{7}$

71. $\frac{123}{999} = 0.\overline{123}$

73. Convert $\frac{1}{5}$ mi $= 0.2$ mi.

To calculate the costs after the first 0.2 miles, note that \$0.25 for $\frac{1}{5}$ of a mile is equivalent to \$1.25 for 1 mile (multiply both quantities by 5).

$$\underbrace{\$12.5 \text{ for } \text{1st 0.2 miles}}_{} + \underbrace{\$1.25 \text{ per mile for additional miles}}_{} = \text{Total Cost}$$

$$\$1.25 \quad + \$1.25(7.5 - 0.2)$$

$$\begin{aligned}
1.25 + 1.25(7.5 - 0.2) &= 1.25 + 1.25(7.3) \\
&= 1.25 + 9.125 \\
&= 10.375
\end{aligned}$$

The total cost is \$10.38

75. See discussion above in problem #73.

$$\underbrace{\$12.5 \text{ for } \text{1st 0.2 miles}}_{} + \underbrace{\$1.25 \text{ per mile for additional miles}}_{} = \text{Total Cost}$$

$$\$1.25 \quad + \$1.25(12.4 - 0.2)$$

$$\begin{aligned}
1.25 + 1.25(12.4 - 0.2) &= 1.25 + 1.25(12.2) \\
&= 1.25 + 15.25 \\
&= 16.50
\end{aligned}$$

The total cost is \$16.50. The meter is correct.

Review Problems

77.
$$x+3=-6$$
$$x+3+(-3)=-6+(-3)$$
$$x=-9$$

79.
$$\frac{1}{7}a=-7$$
$$7\left(\frac{1}{7}a\right)=7(-7)$$
$$a=-49$$

81.
$$5n+4=-26$$
$$5n+4+(-4)=-26+(-4)$$
$$5n=-30$$
$$\frac{5n}{5}=\frac{-30}{5}$$
$$n=-6$$

83.
$$5x+8=3x+2$$
$$5x+8+(-8)=3x+2+(-8)$$
$$5x=3x+(-6)$$
$$5x+(-3x)=3x+(-6)+(-3x)$$
$$2x=-6$$
$$x=-3$$

85.
$$2(x+3)=10$$
$$2x+6=10$$
$$2x=4$$
$$x=2$$

87.
$$3(y-4)+5=-4$$
$$3y-12+5=-4$$
$$3y-7=-4$$
$$3y=3$$
$$y=1$$

Improving your Quantitative Literacy

89. a. $\text{iTunes} < \frac{1}{3}(\text{Walmart})$

$\$9.99 < \frac{1}{3}(\$27.28)$

$\$9.99 < \$9.09$

false

b. $\text{Walmart} > 3(\text{iTunes})$

$\$27.28 > 3(\$9.99)$

$\$27.28 > \$29.97$

false

c. $\text{Walmart is} < 3(\text{iTunes})$

$\$27.28 < 3(\$9.99)$

$\$27.28 < \$29.97$

true

Section 5.6

1.
$$\begin{aligned} x+3.7 &= 2.2 \\ x+3.7+(-3.7) &= 2.2+(-3.7) \\ x &= -1.5 \end{aligned}$$

3.
$$\begin{aligned} x-0.45 &= 0.32 \\ x-0.45+0.45 &= 0.32+0.45 \\ x &= 0.77 \end{aligned}$$

5.
$$\begin{aligned} 8a &= 1.2 \\ \frac{8a}{8} &= \frac{1.2}{8} \\ a &= 0.15 \end{aligned}$$

7.
$$\begin{aligned} -4y &= 1.4 \\ \frac{-4y}{-4} &= \frac{1.4}{-4} \\ y &= -0.35 \end{aligned}$$

9.
$$\begin{aligned} 0.5n &= -0.4 \\ \frac{0.5n}{0.5} &= \frac{-0.4}{0.5} \\ n &= -0.8 \end{aligned}$$

11.
$$\begin{aligned} 4x-4.7 &= 3.5 \\ 4x-4.7+4.7 &= 3.5+4.7 \\ 4x &= 8.2 \\ \frac{4x}{4} &= \frac{8.2}{4} \\ x &= 2.05 \end{aligned}$$

13.
$$\begin{aligned} 0.02+5y &= -0.3 \\ 0.02+5y+(-0.02) &= -0.3+(-0.02) \\ 5y &= -0.32 \\ \frac{5y}{5} &= \frac{-0.32}{5} \\ y &= -0.064 \end{aligned}$$

15.
$$\begin{aligned} \frac{1}{3}x-2.99 &= 1.02 \\ \frac{1}{3}x-2.99+2.99 &= 1.02+2.99 \\ \frac{1}{3}x &= 4.01 \\ 3\left(\frac{1}{3}x\right) &= 3(4.01) \\ x &= 12.03 \end{aligned}$$

17.
$$\begin{aligned} 7n-0.32 &= 5n+0.56 \\ 7n-0.32+0.32 &= 5n+0.56+0.32 \\ 7n &= 5n+0.88 \\ 7n+(-5n) &= 5n+0.88+(-5n) \\ 2n &= 0.88 \\ \frac{2n}{2} &= \frac{0.88}{2} \\ n &= 0.44 \end{aligned}$$

19.
$$\begin{aligned} 3a+4.6 &= 7a+5.3 \\ 3a+4.6+(-4.6) &= 7a+5.3+(-4.6) \\ 3a &= 7a+0.7 \\ 3a+(-7a) &= 7a+0.7+(-7a) \\ -4a &= 0.7 \\ \frac{-4a}{-4} &= \frac{0.7}{-4} \\ a &= -0.175 \end{aligned}$$

21.
$$\begin{aligned} 0.5x+0.1(x+20) &= 3.2 \\ 0.5x+0.1x+2 &= 3.2 \\ 0.6x+2 &= 3.2 \\ 0.6x &= 1.2 \\ x &= 2 \end{aligned}$$

23.
$$\begin{aligned} 0.08x+0.09(x+2000) &= 690 \\ 0.08x+0.09x+180 &= 690 \\ 0.17x+180 &= 690 \\ 0.17x &= 510 \\ x &= 3{,}000 \end{aligned}$$

Applying the Concepts

25. Let x = miles driven

$$\frac{\$10}{\text{day}} \times \text{days} + \frac{\$0.16}{\text{mile}} \times \text{miles} = \text{Total cost}$$

$$\$10(1) + \$0.16x = \$23.92$$

$$10(1)+0.16x = 23.92$$
$$10+0.16x = 23.92$$
$$0.16x = 13.92$$
$$x = 87$$

The car was driven 87 miles.

27. Let x = miles driven

$$\frac{\$9}{\text{day}} \times \text{days} + \frac{\$0.15}{\text{mile}} \times \text{miles} = \text{Total cost}$$

$$\$9(2) + \$0.15x = \$40.05$$

$$9(2)+0.15x = 40.05$$
$$18+0.15x = 40.05$$
$$0.15x = 22.05$$
$$x = 147$$

The car was driven 147 miles.

29. Let x = number of nickels
Then $x + 10$ = number of dimes

$$\underbrace{\text{Amount of money in nickels}} + \underbrace{\text{Amount of money in dimes}} = \underbrace{\text{Total amount}}$$

$$\$0.05(x) + \$0.10(x+10) = \$2.20$$

$$0.05x+0.1(x+10) = 2.2$$
$$0.05x+0.1x+1 = 2.2$$
$$0.15x+1 = 2.2$$
$$0.15x = 1.2$$
$$x = 8$$

She has $x = 8$ nickels and
$x+10 = 8+10 = 18$ dimes.

31. Let x = number of dimes
Then $2x$ = number of quarters

$$\underbrace{\text{Amount of money in dimes}} + \underbrace{\text{Amount of money in quarters}} = \underbrace{\text{Total amount}}$$

$$\$0.10(x) + \$0.25(2x) = \$9.60$$

$$0.1x+0.25(2x) = 9.6$$
$$0.1x+0.5x = 9.6$$
$$0.6x = 9.6$$
$$x = 16$$

She has $x = 16$ dimes and
$2x = 2(16) = 32$ quarters.

33. Let x = total minutes of call
Then $x-1$ = number of additional minutes

$$\underbrace{\$.41 \text{ for 1st minute}} + \underbrace{\frac{\$0.32}{\text{minute}} \times \text{minutes}} = \underbrace{\text{Total}}$$

$$\$0.41 + \$0.32(x-1) = \$5.21$$

$$0.41+0.32(x-1) = 5.21$$
$$0.41+0.32x-0.32 = 5.21$$
$$0.32x+0.09 = 5.21$$
$$0.32x = 5.12$$
$$x = 16$$

The call was $x = 16$ minutes.

35. Let x = number of nickels
Then $x+3$ = number of dimes
and $x+5$ = number of quarters

$$\underbrace{\begin{matrix}\text{Amount}\\ \text{of money}\\ \text{in nickels}\end{matrix}}+\underbrace{\begin{matrix}\text{Amount}\\ \text{of money}\\ \text{in dimes}\end{matrix}}+\underbrace{\begin{matrix}\text{Amount}\\ \text{of money}\\ \text{in quarters}\end{matrix}}=\underbrace{\begin{matrix}\text{Total}\\ \text{value}\end{matrix}}$$

$$0.05x \quad + 0.10(x+3) + 0.25(x+5) = 4.35$$

$$0.05x + 0.1(x+3) + 0.25(x+5) = 4.35$$

$$0.05x + 0.1x + 0.3 + 0.25x + 1.25 = 4.35$$

$$0.4x + 1.55 = 4.35$$

$$0.4x = 2.8$$

$$x = 7$$

There are $x = 7$ nickels,
$x+3 = 7+3 = 10$ dimes, and
$x+5 = 7+5 = 12$ quarters.

Review Problems

37. $5^3 = 5 \cdot 5 \cdot 5 = 125$

39. $(-3)^2 = (-3)(-3) = 9$

41. $\left(\frac{1}{3}\right)^4 = \left(\frac{1}{3}\right)\left(\frac{1}{3}\right)\left(\frac{1}{3}\right)\left(\frac{1}{3}\right) = \frac{1}{81}$

43. $\left(-\frac{5}{6}\right)^2 = \left(-\frac{5}{6}\right)\left(-\frac{5}{6}\right) = \frac{25}{36}$

45. $(0.5)^2 = (0.5)(0.5) = 0.25$

47. $(-1.2)^2 = (-1.2)(-1.2) = 1.44$

Section 5.7

1. $\sqrt{64} = 8$, because $8^2 = 64$

3. $\sqrt{81} = 9$, because $9^2 = 81$

5. $\sqrt{36} = 6$, because $6^2 = 36$

7. $\sqrt{25} = 5$, because $5^2 = 25$

9. $3\sqrt{25} = 3 \cdot 5 = 15$

11. $6\sqrt{64} = 6 \cdot 8 = 48$

13. $15\sqrt{9} = 15 \cdot 3 = 45$

15. $16\sqrt{9} = 16 \cdot 3 = 48$

17. $$\begin{aligned} \sqrt{49} + \sqrt{64} &= 7 + 8 \\ &= 15 \end{aligned}$$

19. $$\begin{aligned} \sqrt{16} - \sqrt{9} &= 4 - 3 \\ &= 1 \end{aligned}$$

21. $$\begin{aligned} 3\sqrt{25} + 9\sqrt{49} &= 3 \cdot 5 + 9 \cdot 7 \\ &= 15 + 63 \\ &= 78 \end{aligned}$$

23. $$\begin{aligned} 15\sqrt{9} - 9\sqrt{16} &= 15 \cdot 3 - 9 \cdot 4 \\ &= 45 - 36 \\ &= 9 \end{aligned}$$

25. $\sqrt{\frac{16}{49}} = \frac{4}{7}$, because $\left(\frac{4}{7}\right)^2 = \frac{16}{49}$

27. $\sqrt{\frac{36}{64}} = \frac{6}{8} = \frac{3}{4}$, because $\left(\frac{6}{8}\right)^2 = \frac{36}{64}$

29. Work each side of the equation separately:
$$\begin{aligned} \sqrt{4} + \sqrt{9} &= \sqrt{4+9} \\ 2 + 3 &= \sqrt{13} \\ 5 &= \sqrt{13} \quad \text{false} \end{aligned}$$

31. Work each side of the equation separately:
$$\begin{aligned} \sqrt{25 \cdot 9} &= \sqrt{25} \cdot \sqrt{9} \\ \sqrt{225} &= 5 \cdot 3 \\ 15 &= 15 \quad \text{true} \end{aligned}$$

33. $$\begin{aligned} c &= \sqrt{a^2 + b^2} \\ &= \sqrt{6^2 + 8^2} \\ &= \sqrt{36 + 64} \\ &= \sqrt{100} \\ &= 10 \\ c &= 10 \text{ in.} \end{aligned}$$

35. $$\begin{aligned} c &= \sqrt{a^2 + b^2} \\ &= \sqrt{5^2 + 12^2} \\ &= \sqrt{25 + 144} \\ &= \sqrt{169} \\ &= 13 \\ c &= 13 \text{ ft.} \end{aligned}$$

37. $$\begin{aligned} c &= \sqrt{a^2 + b^2} \\ &= \sqrt{4^2 + 5^2} \\ &= \sqrt{16 + 25} \\ &= \sqrt{41} \\ &\approx 6.403124 \\ c &= 6.40 \text{ in.} \end{aligned}$$

39. $c = \sqrt{a^2 + b^2}$

$= \sqrt{9^2 + 15^2}$

$= \sqrt{81 + 225}$

$= \sqrt{306}$

≈ 17.492856

$c = 17.49$ m.

Applying the Concepts

41. Let c = length of wire (ft)

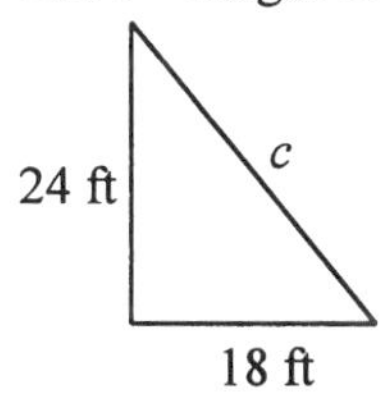

$c = \sqrt{a^2 + b^2}$

$= \sqrt{24^2 + 18^2}$

$= \sqrt{576 + 324}$

$= \sqrt{900}$

$= 30$

The wire is 30 feet long.

43. Make a sketch. Label all known and unknown quantities.
Let c = length of ladder (ft)

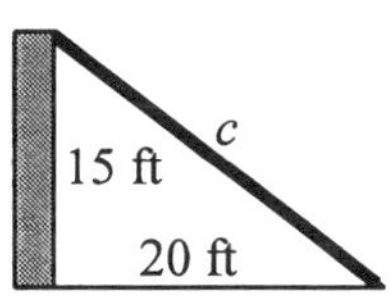

$c = \sqrt{a^2 + b^2}$

$= \sqrt{15^2 + 20^2}$

$= \sqrt{225 + 400}$

$= \sqrt{625}$

$= 25$

The ladder is 25 feet long.

Calculator Problems

45. $\sqrt{1.25} \approx 1.118034$

$= 1.1180$ (to nearest ten thousandth)

47. $\sqrt{125} \approx 11.180340$

$= 11.1803$ (to nearest ten thousandth)

49. $\sqrt{324} = 18$

$= 18.0000$ (to nearest ten thousandth)

51. $2\sqrt{3} \approx 2(1.732051)$

$= 3.464102$

$= 3.46$ (to nearest hundredth)

53. $5\sqrt{5} \approx 5(2.236068)$

$= 11.18034$

$= 11.18$ (to nearest hundredth)

55. $\dfrac{\sqrt{3}}{3} \approx \dfrac{1.732051}{3}$

≈ 0.577350

$= 0.58$ (to nearest hundredth)

57. $\sqrt{\dfrac{1}{3}} \approx \sqrt{0.333333}$

≈ 0.577340

$= 0.58$ (to nearest hundredth)

59. $\sqrt{12} + \sqrt{75} \approx 3.464102 + 8.660254$

$= 12.124356$

$= 12.124$ (to nearest thousandth)

61. $\sqrt{87} \approx 9.327379$

$= 9.327$ (to nearest thousandth)

63. $2\sqrt{3}+5\sqrt{3}\approx 2(1.732051)+5(1.732051)$
$=3.464102+8.660255$
$=12.124357$
$=12.124$ (to nearest thousandth)

65. $7\sqrt{3}\approx 7(1.732051)$
$=12.124357$
$=12.124$ (to nearest thousandth)

67. The table is as follows:

Height, h (ft)	Distance, d (miles)
10	$d=\sqrt{\frac{3\cdot 10}{2}}=\sqrt{15}\approx 4$
50	$d=\sqrt{\frac{3\cdot 50}{2}}=\sqrt{75}\approx 9$
90	$d=\sqrt{\frac{3\cdot 90}{2}}=\sqrt{135}\approx 12$
130	$d=\sqrt{\frac{3\cdot 130}{2}}=\sqrt{195}\approx 14$
170	$d=\sqrt{\frac{3\cdot 170}{2}}=\sqrt{255}\approx 16$
190	$d=\sqrt{\frac{3\cdot 190}{2}}=\sqrt{285}\approx 17$

Review Problems

69. $\frac{5}{7}\cdot\frac{14}{25}=\frac{\not{5}\cdot 2\cdot\not{7}}{\not{7}\cdot\not{5}\cdot 5}=\frac{2}{5}$

71.
$$\begin{array}{rcrcr} 4\frac{3}{10} & = & 4\frac{3\cdot 10}{10\cdot 10} & = & 4\frac{30}{100} \\ +\ 5\frac{2}{100} & = & +\ 5\frac{2}{100} & = & +\ 5\frac{2}{100} \\ \hline & & & & 9\frac{32}{100}=9\frac{8}{25} \end{array}$$

73. $3\frac{2}{10}\cdot 2\frac{5}{10}=\frac{32}{10}\cdot\frac{25}{10}=\frac{\not{2}\cdot\not{2}\cdot 8\cdot\not{5}\cdot\not{5}}{\not{2}\cdot\not{5}\cdot\not{2}\cdot\not{5}}=8$

75.
$$\begin{array}{rcr} 7\frac{1}{10} & = & 6\frac{11}{10} \\ -\ 4\frac{3}{10} & = & -\ 4\frac{3}{10} \\ \hline & & 2\frac{8}{10}=2\frac{4}{5} \end{array}$$

Extending the Concepts

77.

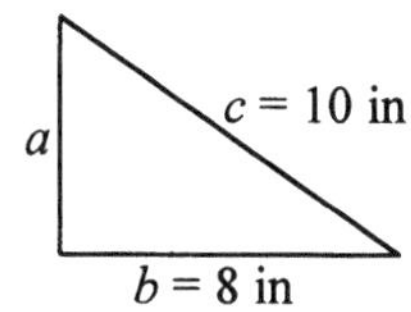

$a^2+b^2=c^2$
$a^2+8^2=10^2$
$a^2+64=100$
$a^2=36$
$a=6$
The other side is 6 inches long.

$A=\frac{1}{2}bh$
$=\frac{1}{2}(8\text{ in})(6\text{ in})$
$=24\text{ in}^2$

79.

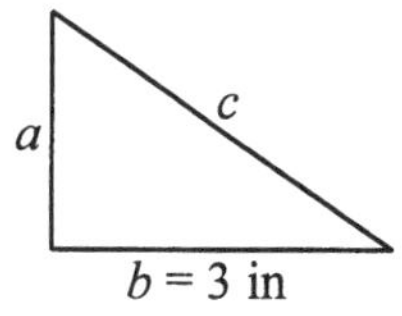

First find the height using the area formula:

$$A = \frac{1}{2}bh$$

$$6 = \frac{1}{2}(3)h$$

$$6 = \frac{3}{2}h$$

$$\frac{2}{3}(6) = \frac{2}{3}\left(\frac{3}{2}h\right)$$

$$h = 4 \text{ ft}$$

$$\begin{aligned} c &= \sqrt{a^2 + b^2} \\ &= \sqrt{4^2 + 3^2} \\ &= \sqrt{16 + 9} \\ &= \sqrt{25} \\ &= 5 \end{aligned}$$

The hypotenuse is 5 feet long.

Improving Your Quantitative Literacy

81. a. Answers will vary. Look at the trend:
From 1971 to 1981: 3.7 student decrease
From 1981 to 1991: 1.7 student decrease
From 1991 to 2001: 1.2 student decrease
If the trend continues, there may be a 1.0 student decrease in from 2001 to 2011:
$15.8 - 1.0 = 14.8$

b. Answers will vary. Since 1996 is half way between 1991 and 2001, you could use the mean as an estimate:

$$\text{Mean} = \frac{17 + 15.8}{2} = 16.4 \text{ students}$$

Section 5.8

1. $\sqrt{12} = \sqrt{2\cdot 2\cdot 3}$
$= \sqrt{2\cdot 2}\cdot\sqrt{3}$
$= 2\sqrt{3}$

3. $\sqrt{20} = \sqrt{2\cdot 2\cdot 5}$
$= \sqrt{2\cdot 2}\cdot\sqrt{5}$
$= 2\sqrt{5}$

5. $\sqrt{72} = \sqrt{2\cdot 2\cdot 2\cdot 3\cdot 3}$
$= \sqrt{2\cdot 2}\cdot\sqrt{3\cdot 3}\cdot\sqrt{2}$
$= 2\cdot 3\cdot\sqrt{2}$
$= 6\sqrt{2}$

7. $\sqrt{98} = \sqrt{7\cdot 7\cdot 2}$
$= \sqrt{7\cdot 7}\cdot\sqrt{2}$
$= 7\sqrt{2}$

9. $\sqrt{28} = \sqrt{2\cdot 2\cdot 7}$
$= \sqrt{2\cdot 2}\cdot\sqrt{7}$
$= 2\sqrt{7}$

11. $\sqrt{200} = \sqrt{2\cdot 2\cdot 2\cdot 5\cdot 5}$
$= \sqrt{2\cdot 2}\cdot\sqrt{5\cdot 5}\cdot\sqrt{2}$
$= 2\cdot 5\cdot\sqrt{2}$
$= 10\sqrt{2}$

13. $\sqrt{12x^2} = \sqrt{2\cdot 2\cdot 3\cdot x\cdot x}$
$= 2\cdot x\cdot\sqrt{3}$
$= 2x\sqrt{3}$

15. $\sqrt{50x^2} = \sqrt{2\cdot 5\cdot 5\cdot x\cdot x}$
$= 5\cdot x\cdot\sqrt{2}$
$= 5x\sqrt{2}$

17. $\sqrt{75x^3} = \sqrt{3\cdot 5\cdot 5\cdot x\cdot x\cdot x}$
$= 5\cdot x\cdot\sqrt{3\cdot x}$
$= 5x\sqrt{3x}$

19. $\sqrt{50x^3} = \sqrt{2\cdot 5\cdot 5\cdot x\cdot x\cdot x}$
$= 5\cdot x\cdot\sqrt{2\cdot x}$
$= 5x\sqrt{2x}$

21. $\sqrt{32x^2y^3} = \sqrt{2\cdot 4\cdot 4\cdot x\cdot x\cdot y\cdot y\cdot y}$
$= 4\cdot x\cdot y\cdot\sqrt{2\cdot y}$
$= 4xy\sqrt{2y}$

23. $\sqrt{243x^4} = \sqrt{3\cdot 9\cdot 9\cdot x\cdot x\cdot x\cdot x}$
$= 9\cdot x\cdot x\cdot\sqrt{3}$
$= 9x^2\sqrt{3}$

25. $\sqrt{72x^2y^4} = \sqrt{2\cdot 2\cdot 2\cdot 3\cdot 3\cdot x\cdot x\cdot y\cdot y\cdot y\cdot y}$
$= 2\cdot 3\cdot x\cdot y\cdot y\cdot\sqrt{2}$
$= 6xy^2\sqrt{2}$

27. $\sqrt{12x^3y^3} = \sqrt{2\cdot 2\cdot 3\cdot x\cdot x\cdot x\cdot y\cdot y\cdot y}$
$= 2\cdot x\cdot y\cdot\sqrt{3\cdot x\cdot y}$
$= 2xy\sqrt{3xy}$

29. $c = \sqrt{a^2 + b^2}$
$= \sqrt{3^2 + 3^2}$
$= \sqrt{9+9}$
$= \sqrt{18}$
$= \sqrt{2\cdot 3\cdot 3}$
$= 3\sqrt{2}$ in.

31. $c = \sqrt{a^2 + b^2}$
$= \sqrt{4^2 + 4^2}$
$= \sqrt{16 + 16}$
$= \sqrt{32}$
$= \sqrt{2 \cdot 4 \cdot 4}$
$= 4\sqrt{2}$ in.

33. $c = \sqrt{a^2 + b^2}$
$= \sqrt{x^2 + x^2}$
$= \sqrt{2x^2}$
$= \sqrt{2 \cdot x \cdot x}$
$= x\sqrt{2}$
$= x\sqrt{2}$ in.

Applying the Concepts

35. $t = \sqrt{\frac{h}{16}}$ Let $h = 25$
$= \sqrt{\frac{25}{16}}$
$= \frac{5}{4}$
$= 1.25$
It will hit the ground in 1.25 seconds.

37. See "The Spiral of Roots" in chapter introduction.

39. $\sqrt{72} \approx 8.485281$ $\quad 6\sqrt{2} \approx 6(1.414214)$
$= 8.485284$
Answers will vary depending on how many decimal places are carried. However, the decimal approximations should support that the two roots are equal.

41. $x\sqrt{x} = 5\sqrt{5}$ Let $x = 5$
$\approx 5(2.236068)$
$= 11.18034$

43. $x^2\sqrt{x} = 5^2\sqrt{5}$ Let $x = 5$
$\approx 25(2.236068)$
$= 55.9017$

45. The table is as follows:

x	$\sqrt{x}$	$2\sqrt{x}$	$\sqrt{4x}$
1	$\sqrt{1} = 1$	$2\sqrt{1} = 2 \cdot 1$ $= 2$	$\sqrt{4 \cdot 1} = \sqrt{4}$ $= 2$
2	$\sqrt{2} \approx 1.414$	$2\sqrt{2} \approx 2 \cdot 1.414$ $= 2.828$	$\sqrt{4 \cdot 2} = \sqrt{8}$ ≈ 2.828
3	$\sqrt{3} \approx 1.732$	$2\sqrt{3} \approx 2 \cdot 1.732$ $= 3.464$	$\sqrt{4 \cdot 3} = \sqrt{12}$ ≈ 3.464
4	$\sqrt{4} = 2$	$2\sqrt{4} = 2 \cdot 2$ $= 4$	$\sqrt{4 \cdot 4} = \sqrt{16}$ $= 4$

47. The table is as follows:

x	$\sqrt{x}$	$3\sqrt{x}$	$\sqrt{9x}$
1	$\sqrt{1}=1$	$3\sqrt{1} = 3 \cdot 1$ $= 3$	$\sqrt{9 \cdot 1} = \sqrt{9}$ $= 3$
2	$\sqrt{2} \approx 1.414$	$3\sqrt{2} \approx 3 \cdot 1.414$ ≈ 4.243	$\sqrt{9 \cdot 2} = \sqrt{18}$ ≈ 4.243
3	$\sqrt{3} \approx 1.732$	$3\sqrt{3} \approx 3 \cdot 1.732$ $= 5.196$	$\sqrt{9 \cdot 3} = \sqrt{27}$ ≈ 5.196
4	$\sqrt{4} \approx 2$	$3\sqrt{4} = 3 \cdot 2$ $= 6$	$\sqrt{9 \cdot 4} = \sqrt{36}$ $= 6$

Review Problems

49. When $x = 0$:
$y = \frac{1}{2}x - 3$
$y = \frac{1}{2}(0) - 3$
$y = 0 - 3$
$y = -3$

51. When $x=-4$:

$$y=\frac{1}{2}x-3$$
$$y=\frac{1}{2}(-4)-3$$
$$y=-2-3$$
$$y=-5$$

53. When $x=2$:

$$y=\frac{1}{2}x-3$$
$$y=\frac{1}{2}(2)-3$$
$$y=1-3$$
$$y=-2$$

55. When $x=0$:

$$2x+5y=10$$
$$2(0)+5y=10$$
$$0+5y=10$$
$$5y=10$$
$$y=2$$

57. When $x=5$:

$$2x+5y=10$$
$$2(5)+5y=10$$
$$10+5y=10$$
$$5y=0$$
$$y=0$$

59. When $y=0$:

$$2x+5y=10$$
$$2x+5(0)=10$$
$$2x+0=10$$
$$2x=10$$
$$x=5$$

61. When $y=-2$:

$$2x+5y=10$$
$$2x+5(-2)=10$$
$$2x+(-10)=10$$
$$2x=20$$
$$x=10$$

Section 5.9

1. $2\sqrt{3}+8\sqrt{3}=(2+8)\sqrt{3}$
$=10\sqrt{3}$

3. $7\sqrt{5}-3\sqrt{5}=(7-3)\sqrt{5}$
$=4\sqrt{5}$

5. $9\sqrt{x}+3\sqrt{x}-5\sqrt{x}=(9+3-5)\sqrt{x}$
$=7\sqrt{x}$

7. $8\sqrt{7}+\sqrt{7}=8\sqrt{7}+1\sqrt{7}$
$=(8+1)\sqrt{7}$
$=9\sqrt{7}$

9. $2\sqrt{y}+\sqrt{y}+3\sqrt{y}=2\sqrt{y}+1\sqrt{y}+3\sqrt{y}$
$=(2+1+3)\sqrt{y}$
$=6\sqrt{y}$

11. $\sqrt{18}+\sqrt{32}=\sqrt{3\cdot 3\cdot 2}+\sqrt{4\cdot 4\cdot 2}$
$=3\sqrt{2}+4\sqrt{2}$
$=(3+4)\sqrt{2}$
$=7\sqrt{2}$

13. $\sqrt{75}+\sqrt{27}=\sqrt{5\cdot 5\cdot 3}+\sqrt{3\cdot 3\cdot 3}$
$=5\sqrt{3}+3\sqrt{3}$
$=(5+3)\sqrt{3}$
$=8\sqrt{3}$

15. $2\sqrt{75}-4\sqrt{27}=2\sqrt{5\cdot 5\cdot 3}-4\sqrt{3\cdot 3\cdot 3}$
$=2\cdot 5\sqrt{3}-4\cdot 3\sqrt{3}$
$=10\sqrt{3}-12\sqrt{3}$
$=(10-12)\sqrt{3}$
$=-2\sqrt{3}$

17. $2\sqrt{90}+3\sqrt{40}-4\sqrt{10}$
$=2\sqrt{3\cdot 3\cdot 10}+3\sqrt{2\cdot 2\cdot 10}-4\sqrt{10}$
$=2\cdot 3\sqrt{10}+3\cdot 2\sqrt{10}-4\sqrt{10}$
$=6\sqrt{10}+6\sqrt{10}-4\sqrt{10}$
$=(6+6-4)\sqrt{10}$
$=8\sqrt{10}$

19. $\sqrt{72x^2}-\sqrt{50x^2}$
$=\sqrt{2\cdot 2\cdot 2\cdot 3\cdot 3\cdot x\cdot x}-\sqrt{5\cdot 5\cdot 2\cdot x\cdot x}$
$=2\cdot 3\cdot x\sqrt{2}-5\cdot x\sqrt{2}$
$=6x\sqrt{2}-5x\sqrt{2}$
$=(6x-5x)\sqrt{2}$
$=x\sqrt{2}$

21. $4\sqrt{20x^3}+3\sqrt{45x^3}$
$=4\sqrt{2\cdot 2\cdot 5\cdot x\cdot x\cdot x}+3\sqrt{3\cdot 3\cdot 5\cdot x\cdot x\cdot x}$
$=4\cdot 2\cdot x\sqrt{5x}+3\cdot 3\cdot x\sqrt{5x}$
$=8x\sqrt{5x}+9x\sqrt{5x}$
$=(8x+9x)\sqrt{5x}$
$=17x\sqrt{5x}$

23.

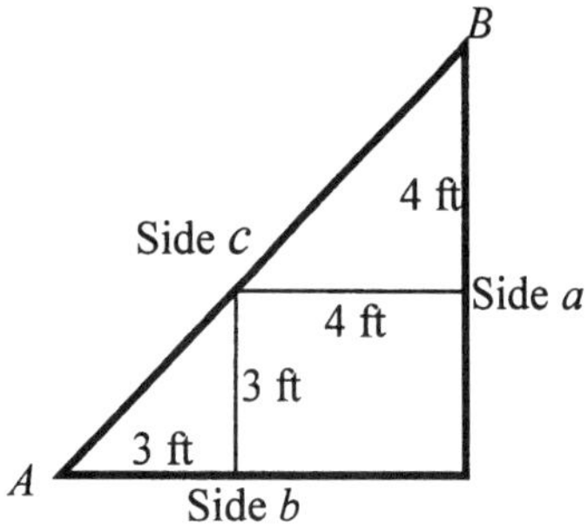

The distance from A to B is given by the hypotenuse of the larger triangle. First, find the lengths of the legs:

Side $a = 3 \text{ ft} + 4 \text{ ft} = 7 \text{ ft}$

Side $b = 3 \text{ ft} + 4 \text{ ft} = 7 \text{ ft}$

$$\begin{aligned} c &= \sqrt{a^2 + b^2} \\ &= \sqrt{7^2 + 7^2} \\ &= \sqrt{49 + 49} \\ &= \sqrt{98} \\ &= \sqrt{7 \cdot 7 \cdot 2} \\ &= 7\sqrt{2} \end{aligned}$$

The distance from A to B is $7\sqrt{2}$ ft.

Calculator Problems

25. $\sqrt{2} + \sqrt{3} \approx 3.1463$

$\sqrt{5} \approx 2.2361$

$\sqrt{2} + \sqrt{3} \neq \sqrt{5}$

27. The table is as follows:

x	$\sqrt{x^2 + 9}$	$x + 3$
1	$\sqrt{1^2 + 9} = \sqrt{10} \approx 3.162$	$1 + 3 = 4$
2	$\sqrt{2^2 + 9} = \sqrt{13} \approx 3.606$	$2 + 3 = 5$
3	$\sqrt{3^2 + 9} = \sqrt{18} \approx 4.243$	$3 + 3 = 6$
4	$\sqrt{4^2 + 9} = \sqrt{25} = 5$	$4 + 3 = 7$
5	$\sqrt{5^2 + 9} = \sqrt{34} \approx 5.831$	$5 + 3 = 8$
6	$\sqrt{6^2 + 9} = \sqrt{45} \approx 6.708$	$6 + 3 = 9$

The table entries prove that $\sqrt{x^2 + 9} \neq x + 3$

29. The table is as follows:

x	$\sqrt{x+3}$	$\sqrt{x} + \sqrt{3}$
1	$\sqrt{1+3} = \sqrt{4} = 2$	$\sqrt{1} + \sqrt{3} = 1 + \sqrt{3} \approx 2.732$
2	$\sqrt{2+3} = \sqrt{5} \approx 2.236$	$\sqrt{2} + \sqrt{3} \approx 3.146$
3	$\sqrt{3+3} = \sqrt{6} \approx 2.449$	$\sqrt{3} + \sqrt{3} \approx 3.464$
4	$\sqrt{4+3} = \sqrt{7} \approx 2.646$	$\sqrt{4} + \sqrt{3} = 2 + \sqrt{3} \approx 3.732$
5	$\sqrt{5+3} = \sqrt{8} = 2.828$	$\sqrt{5} + \sqrt{3} \approx 3.968$
6	$\sqrt{6+3} = \sqrt{9} = 3$	$\sqrt{6} + \sqrt{3} \approx 4.182$

The table entries prove $\sqrt{x+3} \neq \sqrt{x} + \sqrt{3}$

Review Problems

31. Let $x = 0$: Let $y = 0$:

Let $x = 0$:	Let $y = 0$:
$x + y = 3$	$x + y = 3$
$0 + y = 3$	$x + 0 = 3$
$y = 3$	$x = 3$

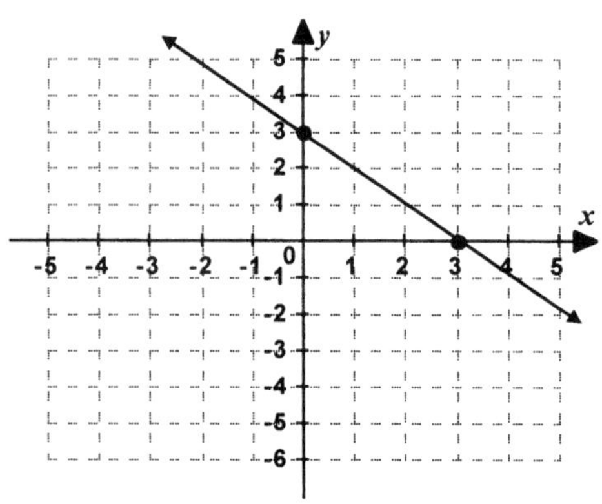

33. Let $x = 0$:

$y = \frac{1}{2}x - 3$

$y = \frac{1}{2}(0) - 3$

$y = 0 - 3$

$y = -3$

Let $y = 0$:

$y = \frac{1}{2}x - 3$

$0 = \frac{1}{2}x - 3$

$3 = \frac{1}{2}x$

$6 = x$

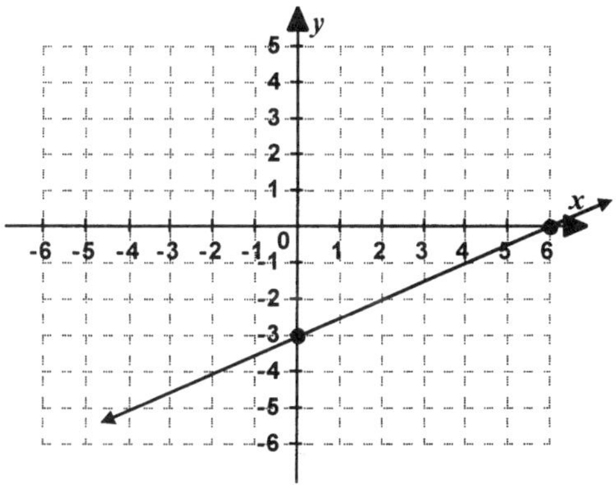

35. Let $x = 0$:

$2x + 5y = 10$

$2(0) + 5y = 10$

$0 + 5y = 10$

$5y = 10$

$y = 2$

Let $y = 0$:

$2x + 5y = 10$

$2x + 5(0) = 10$

$2x + 0 = 10$

$2x = 10$

$x = 5$

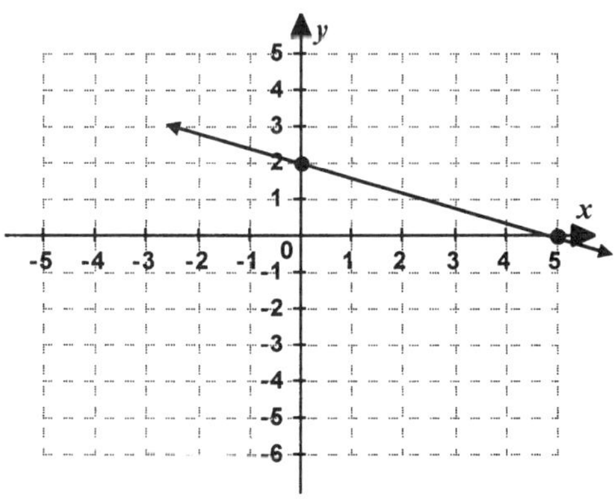

Chapter 5 Review

1. 36.007

$\uparrow$

thousandths

3. Thirty-seven and forty-two ten thousandths is 37.0042

5. Look at the digit to the right of the hundredths place: 98.76**54**
$5 \geq 5$ so add 1 to the hundredths place and delete all digits to the right 98.77

7.
$$\begin{array}{r} 11.076 \\ -\ 3.297 \\ \hline 7.779 \end{array}$$

9.
$$\begin{array}{rl} 24.24 & \textit{2 decimal places} \\ \times\ 0.89 & \textit{2 decimal places} \\ \hline 21816 & \\ 193920 & \\ \hline 21.5736 & \textit{4 decimal places} \end{array}$$

$$-0.89(24.24) = -21.5736$$

11. $0.58\overline{)0.7134} \rightarrow 58\overline{)71.34}$

$$\begin{array}{r} 1.23 \\ 58\overline{)71.34} \\ 58\downarrow \\ \hline 133 \\ 116\downarrow \\ \hline 174 \\ 174 \\ \hline 0 \end{array}$$

13. $0.705 = \dfrac{705}{1,000} = \dfrac{141}{200}$

15.
$$\begin{aligned} 3.3 - 4(0.22) &= 3.3 - 0.88 \\ &= 2.42 \end{aligned}$$

17.
$$\begin{aligned} 125\left(\frac{3}{5}\right) + 4 &= 75 + 4 \\ &= 79 \end{aligned}$$

19.
$$\begin{aligned} x + 9.8 &= 3.9 \\ x + 9.8 + (-9.8) &= 3.9 + (-9.8) \\ x &= -5.9 \end{aligned}$$

21.
$$\begin{aligned} 0.5y - 0.2 &= 3 \\ 0.5y - 0.2 + 0.2 &= 3 + 0.2 \\ 0.5y &= 3.2 \\ \frac{0.5y}{0.5} &= \frac{3.2}{0.5} \\ y &= 6.4 \end{aligned}$$

23. $3\sqrt{25} = 3 \cdot 5 = 15$

25.
$$\begin{aligned} 4\sqrt{25} + 3\sqrt{81} &= 4 \cdot 5 + 3 \cdot 9 \\ &= 20 + 27 \\ &= 47 \end{aligned}$$

27.
$$\begin{aligned} P &= 2l + 2w \\ &= 2(6.14 \text{ in}) + 2(2.56 \text{ in}) \\ &= 12.28 \text{ in} + 5.12 \text{ in} \\ &= 17.4 \text{ in} \end{aligned}$$

29. $r = \dfrac{d}{t} = \dfrac{3,106.3 \text{ mi}}{137.05 \text{ hr}} = 22.7\dfrac{\text{mi}}{\text{hr}}$

31a. Complete the table as follows:

Class	Units		Value	Grade Points
Algebra	4	C	2	$4\times2=8$
Speech	3	A	4	$3\times4=12$
Accting	3	B	3	$3\times3=9$
Mrktng	3	B	3	$3\times3=9$
Real Estate	2	C	2	$2\times2=4$
Total	15			42

$$\frac{\text{Total grade points}}{\text{Total units}}=\text{GPA}$$

$$\frac{42}{15}=2.80$$

Her GPA is 2.80.

31b. The table is as follows:

Class	Units		Value	Grade Points
Algebra	4	B	3	$4\times3=12$
Speech	3	A	4	$3\times4=12$
Accting	3	B	3	$3\times3=9$
Markting	3	B	3	$3\times3=9$
Real Estate	2	C	2	$2\times2=4$
Total	15			46

$$\frac{\text{Total grade points}}{\text{Total units}}=\text{GPA}$$

$$\frac{46}{15}=3.0\overline{6}$$

The difference between $3.0\overline{6}$ and 2.8 ≈ 0.27

33. $c=\sqrt{a^2+b^2}$
$=\sqrt{5^2+2^2}$
$=\sqrt{25+4}$
$=\sqrt{29}$
≈ 5.4
The hypotenuse is 5.4 ft long.

35. $P=$ sum of all sides
$\approx 5\text{ ft}+2\text{ ft}+5.4\text{ ft}$
$=12.4\text{ ft}$

$A=\frac{1}{2}bh$
$=\frac{1}{2}(5\text{ ft})(2\text{ ft})$
$=\frac{1}{2}(10\text{ ft}^2)$
$=5\text{ ft}^2$

37. Note: The radius is half the diameter:
$r=\frac{1}{2}d=\frac{1}{2}(32\text{ in})=16\text{ in.}$
$C=2\pi r$
$=2(3.14)(16\text{ in})$
$=100.48\text{ in}$
He will need 100.48 inches of pipe.

39. $\sqrt{12}=\sqrt{2\cdot2\cdot3}=2\sqrt{3}$

41. $\sqrt{20x^2}=\sqrt{2\cdot2\cdot5\cdot x\cdot x}$
$=2\cdot x\sqrt{5}$
$=2x\sqrt{5}$

43. $\sqrt{18x^2y^3}=\sqrt{3\cdot3\cdot2\cdot x\cdot x\cdot y\cdot y\cdot y}$
$=3\cdot x\cdot y\sqrt{2\cdot y}$
$=3xy\sqrt{2y}$

45. $8\sqrt{3}+2\sqrt{3}=(8+2)\sqrt{3}=10\sqrt{3}$

47. $4\sqrt{3}+\sqrt{3}=4\sqrt{3}+1\sqrt{3}$
$=(4+1)\sqrt{3}$
$=5\sqrt{3}$

49. $\sqrt{24}+\sqrt{54}=\sqrt{2\cdot 2\cdot 6}+\sqrt{3\cdot 3\cdot 6}$
$=2\sqrt{6}+3\sqrt{6}$
$=(2+3)\sqrt{6}$
$=5\sqrt{6}$

51. $3\sqrt{75}-8\sqrt{27}=3\sqrt{5\cdot 5\cdot 3}-8\sqrt{3\cdot 3\cdot 3}$
$=3\cdot 5\cdot\sqrt{3}-8\cdot 3\cdot\sqrt{3}$
$=15\sqrt{3}-24\sqrt{3}$
$=(15-25)\sqrt{3}$
$=-9\sqrt{3}$

Cumulative Review Chapters 1-5

1. $6+\frac{5}{x}=\frac{6\cdot x}{x}+\frac{5}{x}=\frac{6x+5}{x}$

3.
$$\begin{array}{r} 287 \\ \times\ 56 \\ \hline 1,722 \\ 14,350 \\ \hline 16,072 \end{array}$$

5.
$$\begin{aligned} -\frac{2}{5}(5x+10)-4 &= -\frac{2}{5}(5x)+\left(-\frac{2}{5}\right)(10)-4 \\ &= -2x-4-4 \\ &= -2x-8 \end{aligned}$$

7.
$$\begin{array}{r} 12.96 \\ \times\ 4.3 \\ \hline 3888 \\ 51840 \\ \hline 55.728 \end{array}$$
$(-12.96)(-4.3)=55.728$

9. $\frac{5}{14}\div\frac{15}{21}=\frac{5}{14}\cdot\frac{21}{15}=\frac{\cancel{5}\cdot\cancel{3}\cdot\cancel{7}}{\cancel{7}\cdot 2\cdot\cancel{3}\cdot\cancel{5}}=\frac{1}{2}$

11.
$$\begin{array}{r} 15 \\ 4\overline{)63} \\ 4\downarrow \\ \hline 23 \\ 20 \\ \hline 3 \end{array}$$
$\frac{63}{4}=15+\frac{3}{4}=15\frac{3}{4}$

13. Substitute $x=-2$:
$$\begin{aligned} 3x-5&=1 \\ 3(-2)-5&=1 \\ -6-5&=1 \\ -11&=1 \text{ false} \end{aligned}$$
$x=-2$ is not a solution.

15. $\frac{72}{-8}=-9$

17. $\underbrace{\text{Three times}}$ $\underbrace{\text{the sum of 13 and 4}}$ is 51

$\downarrow$ $\qquad\downarrow\qquad$ $\downarrow\downarrow$

$3\times$ $\qquad(13+4)\qquad$ $=51$
$$\begin{aligned} 3(13+4)&=51 \\ 3(17)&=51 \\ 51&=51 \end{aligned}$$

19. When $y=-3$:
$$\begin{aligned} &5x-10y=15 \\ &5x-10(-3)=15 \\ &5x+30=15 \\ &5x=-15 \\ &x=-3 \end{aligned}$$

21. $\frac{-6+2(-4)}{8-10}=\frac{-6+(-8)}{-2}=\frac{-14}{-2}=7$

23.
$$\begin{aligned} \frac{2}{3}(0.45)-\frac{4}{5}(0.8) &= \frac{2(0.45)}{3}-0.8(0.8) \\ &= \frac{0.90}{3}-0.64 \\ &= 0.30-0.64 \\ &= -0.34 \end{aligned}$$

25. $\left(3\frac{1}{3}-\frac{1}{2}\right)\left(4\frac{1}{2}+\frac{3}{4}\right)=\left(\frac{10}{3}-\frac{1}{2}\right)\left(\frac{9}{2}+\frac{3}{4}\right)$

$$=\left(\frac{20}{6}-\frac{3}{6}\right)\left(\frac{18}{4}+\frac{3}{4}\right)$$
$$=\left(\frac{17}{6}\right)\left(\frac{21}{4}\right)$$
$$=\frac{17\cdot\cancel{3}\cdot 7}{\cancel{3}\cdot 2\cdot 4}$$
$$=\frac{119}{8}$$
$$=14\frac{7}{8}$$

27. $\frac{3}{4}x=21$

$$4\left(\frac{3}{4}x\right)=4(21)$$
$$3x=84$$
$$x=28$$

29. $2(x-6)=-10$

$$2x-12=-10$$
$$2x=2$$
$$x=1$$

31. $a-\frac{3}{4}=\frac{3}{8}$ LCD = 8

$$8\left(a-\frac{3}{4}\right)=8\left(\frac{3}{8}\right)$$
$$8(a)-8\left(\frac{3}{4}\right)=8\left(\frac{3}{8}\right)$$
$$8a-6=3$$
$$8a=9$$
$$a=\frac{9}{8}=1\frac{1}{8}$$

33. Average (Mean) $=\frac{\text{Sum of all values}}{\text{Number of values}}$

$$=\frac{83+85+79+93+80}{5}$$
$$=\frac{420}{5}$$
$$=84$$

Her mean score is 84.

35. $P=2l+2w$

$$=2(12.7\text{cm})+2(6.7\text{cm})$$
$$=25.4\text{ cm}+13.4\text{ cm}$$
$$=38.8\text{ cm}$$

37. $V=lwh$

$$=(5\text{ in})(3\text{ in})(7\text{ in})$$
$$=105\text{ in}^3$$

39. $\frac{4}{5}$ of $40=\frac{4}{5}(40)=\frac{4\cdot\cancel{5}\cdot 8}{\cancel{5}}=32$

32 students passed.

Chapter 5 Test

1. 5.053 in words:
Five and fifty-three thousandths

3. Seventeen and four hundred six ten thousandths is 17.0406

5. Note: $7 = 7.00$

$0.6 = 0.60$

$$\begin{array}{r} 7.00 \\ 0.60 \\ +\ 0.58 \\ \hline 8.18 \end{array}$$

7.

$$\begin{array}{r} 6.24 \\ \times\ 5.7 \\ \hline 4368 \\ 31200 \\ \hline 35.568 \end{array}$$

9.

$$\begin{array}{r} 0.92 \\ 25\overline{)23.00} \\ 225\downarrow \\ \hline 50 \\ 50 \\ \hline 0 \end{array}$$

11. $5.2(2.8+0.2) = 5.2(2.82)$
$= 14.664$

13. $23.852 - 3(2.01+0.231)$
$= 23.852 - 3(2.241)$
$= 23.852 - 6.723$
$= 17.129$

15.

$$\begin{aligned} 6a - 0.18 &= a + 0.77 \\ 6a - 0.18 + 0.18 &= a + 0.77 + 0.18 \\ 6a &= a + 0.95 \\ 6a + (-a) &= a + 0.95 + (-a) \\ 5a &= 0.95 \\ \frac{5a}{5} &= \frac{0.95}{5} \\ a &= 0.19 \end{aligned}$$

17. $\sqrt{\frac{25}{81}} = \frac{5}{9}$, because $\left(\frac{5}{9}\right)^2 = \frac{25}{81}$

19.

Cost per pound:	\$5.44
× Total pounds:	× 3.5
Total cost:	2 720
	16 320
	\$19.040

The total is \$19.04

21. $c = \sqrt{4^2 + 3^2}$
$= \sqrt{16+9}$
$= \sqrt{25}$
$= 5$ in

23. $\sqrt{48x^2y} = \sqrt{4 \cdot 4 \cdot 3 \cdot x \cdot x \cdot y}$
$= 4\ x \cdot \sqrt{3 \cdot y}$
$= 4x\sqrt{3y}$

25. $6\sqrt{12} - 5\sqrt{48} = 6\sqrt{2 \cdot 2 \cdot 3} - 5\sqrt{4 \cdot 4 \cdot 3}$
$= 6 \cdot 2\sqrt{3} - 5 \cdot 4\sqrt{3}$
$= 12\sqrt{3} - 20\sqrt{3}$
$= (12-20)\sqrt{3}$
$= -8\sqrt{3}$

Chapter 6 Preview

1. $\frac{16}{48}=\frac{\cancel{16}}{\cancel{16}\cdot 3}=\frac{1}{3}$

3. $\begin{array}{r} 0.25 \\ 4\overline{)1.00} \\ \underline{8\downarrow} \\ 20 \\ \underline{20} \\ 0 \end{array}$ $\frac{1}{4}=0.25$

5. $3(0.4)=1.2$

7. $\frac{2}{3}\cdot 6=\frac{2\cdot 6}{3}=\frac{2\cdot 2\cdot \cancel{3}}{\cancel{3}}=4$

9. $\begin{array}{r} 0.695 \\ 2\overline{)1.390} \\ \underline{12\downarrow} \\ 19 \\ \underline{18\downarrow} \\ 10 \\ \underline{10} \\ 0 \end{array}$ $\frac{1.39}{2}=0.695$

11. $\frac{\frac{2}{3}}{\frac{4}{9}}=\frac{2}{3}\cdot\frac{9}{4}=\frac{\cancel{2}\cdot\cancel{3}\cdot 3}{\cancel{3}\cdot\cancel{2}\cdot 2}=\frac{3}{2}$

13. $4\cdot 6=3\cdot 8$
$24=24$
true

15. $65=y\cdot 10$
$\frac{65}{10}=\frac{y\cdot 10}{10}$
$6.5=y$

17. $4=5\cdot x$
$\frac{4}{5}=\frac{5\cdot x}{5}$
$\frac{4}{5}=x$

19. $18x=360$
$\frac{18x}{18}=\frac{360}{18}$
$x=20$

Chapter 6 Pretest

1. $\frac{15}{25}=\frac{3}{5}$

3. $\frac{\frac{5}{4}}{\frac{7}{4}}=\frac{5}{4}\cdot\frac{4}{7}=\frac{5}{7}$

5. a. $\frac{\text{fees}}{\text{personal expenses}}=\frac{\$6{,}000}{\$1{,}600}=\frac{15}{4}$

b. $\frac{\text{books/supplies}}{\text{fees}}=\frac{\$1{,}450}{\$6{,}000}=\frac{29}{120}$

7. $\frac{5\text{ gallons}}{40\text{ seconds}}=\frac{1}{8}\frac{\text{gal}}{\text{sec}}=0.125\text{ gal per sec}$

9. $\frac{399¢}{12\text{ cans}}=33.25¢\text{ per can}$

11. $\frac{2}{y}=\frac{4}{10}$

$2\cdot10=y\cdot4$

$20=4y$

$5=y$

13. $\frac{n}{8}=\frac{\frac{1}{6}}{\frac{1}{9}}$

$n\cdot\frac{1}{9}=8\cdot\frac{1}{6}$

$\frac{n}{9}=\frac{8}{6}$

$9\cdot\frac{n}{9}=9\cdot\frac{8}{6}$

$n=\frac{72}{6}=12$

15. Let x = ml of essential oil

$\frac{7\text{ ml oil}}{40\text{ ml alcohol}}=\frac{x\text{ ml oil}}{20\text{ ml alcohol}}$

$7(20)=40x$

$140=40x$

$3.5=x$

3.5 ml of essential oil.

17. Let x = amount reimbursed for 150 miles

$\frac{36.5¢}{1\text{ mi}}=\frac{x\ ¢}{150\text{ mi}}$

$36.5(150)=1(x)$

$5{,}475=x$

They will be reimbursed 5,475¢, or \$54.75

19. $\frac{6}{15}=\frac{20}{BC}$

$6\cdot BC=300$

$BC=50$

Section 6.1

1. $\dfrac{8}{6}=\dfrac{4}{3}$

3. $\dfrac{64}{12}=\dfrac{16}{3}$

5. $\dfrac{100}{250}=\dfrac{2}{5}$

7. $\dfrac{13}{26}=\dfrac{1}{2}$

9. $\dfrac{\frac{3}{4}}{\frac{1}{4}}=\dfrac{3}{4}\cdot\dfrac{4}{1}=\dfrac{3}{1}$

11. $\dfrac{\frac{7}{3}}{\frac{6}{3}}=\dfrac{7}{3}\cdot\dfrac{3}{6}=\dfrac{7}{6}$

13. $\dfrac{\frac{6}{5}}{\frac{6}{7}}=\dfrac{6}{5}\cdot\dfrac{7}{6}=\dfrac{7}{5}$

15. $\dfrac{2\frac{1}{2}}{3\frac{1}{2}}=\dfrac{\frac{5}{2}}{\frac{7}{2}}=\dfrac{5}{2}\cdot\dfrac{2}{7}=\dfrac{5}{7}$

17. $\dfrac{2\frac{2}{3}}{\frac{5}{3}}=\dfrac{\frac{8}{3}}{\frac{5}{3}}=\dfrac{8}{3}\cdot\dfrac{3}{5}=\dfrac{8}{5}$

19. $\dfrac{0.05}{0.15}=\dfrac{0.05\times100}{0.15\times100}=\dfrac{5}{15}=\dfrac{1}{3}$

21. $\dfrac{0.3}{3}=\dfrac{0.3\times10}{3\times10}=\dfrac{3}{30}=\dfrac{1}{10}$

23. $\dfrac{1.2}{10}=\dfrac{1.2\times10}{10\times10}=\dfrac{12}{100}=\dfrac{3}{25}$

25. a. $\dfrac{\text{shaded squares}}{\text{nonshaded squares}}=\dfrac{20}{40}=\dfrac{1}{2}$

b. $\dfrac{\text{shaded squares}}{\text{total squares}}=\dfrac{20}{60}=\dfrac{1}{3}$

c. $\dfrac{\text{nonshaded squares}}{\text{total squares}}=\dfrac{40}{60}=\dfrac{2}{3}$

Applying the Concepts

27. a. $\dfrac{\$650}{\$400}=\dfrac{650}{400}=\dfrac{13}{8}$

b. $\dfrac{\$100}{\$400}=\dfrac{100}{400}=\dfrac{1}{4}$

c. $\dfrac{\$150}{\$400}=\dfrac{150}{400}=\dfrac{3}{8}$

d. $\dfrac{\$650}{\$150}=\dfrac{650}{150}=\dfrac{13}{3}$

29. a. $\dfrac{\text{Q1 revenue}}{\text{Q1 profit}}=\dfrac{\$6{,}000}{\$1{,}000}=\dfrac{6}{1}$

b. $\dfrac{\text{Q2 revenue}}{\text{Q2 profit}}=\dfrac{\$7{,}500}{\$1{,}500}=\dfrac{5}{1}$

c. $\dfrac{\text{Q3 revenue}}{\text{Q3 profit}}=\dfrac{\$8{,}400}{\$2{,}100}=\dfrac{4}{1}$

d. $\dfrac{\text{Q4 revenue}}{\text{Q4 profit}}=\dfrac{\$10{,}500}{\$3{,}500}=\dfrac{3}{1}$

e. $\dfrac{\text{annual revenue}}{\text{annual profit}}=\dfrac{\$32{,}400}{\$8{,}100}=\dfrac{4}{1}$

Calculator Problems

31. $\frac{2{,}314}{2{,}408} \approx 0.96$

33. $\frac{2{,}314}{4{,}722} \approx 0.49$

Review Problems

35. $\frac{90}{5} = \frac{90 \div 5}{5 \div 5} = \frac{18}{1} = 18$

37. $2\overline{)125.0}$ = 62.5

$$\begin{array}{r} 62.5 \\ 2\overline{)125.0} \\ \underline{12\downarrow} \\ 5 \\ \underline{4\downarrow} \\ 10 \\ \underline{10} \\ 0 \end{array}$$

39.

$$\begin{array}{r} 0.615 \\ 2\overline{)1.230} \\ \underline{12\downarrow} \\ 3 \\ \underline{2\downarrow} \\ 10 \\ \underline{10} \\ 0 \end{array}$$

41. $0.25\overline{)46.0} \rightarrow 25\overline{)4600}$

$$\begin{array}{r} 184 \\ 25\overline{)4600} \\ \underline{25\downarrow} \\ 210 \\ \underline{200\downarrow} \\ 100 \\ \underline{100} \\ 0 \end{array}$$

Improving Your Quantitative Literacy

43. IBM $\frac{P}{E} = \frac{\$146.05}{\$6.35} = \frac{14{,}605}{635} = \frac{23}{1} = 23$

AO $\frac{P}{E} = \frac{\$139.69}{\$0.61} = \frac{13{,}969}{61} = \frac{229}{1} = 229$

DIS $\frac{P}{E} = \frac{\$30.03}{\$0.91} = \frac{3{,}003}{91} = \frac{33}{1} = 33$

KM $\frac{P}{E} = \frac{\$15.64}{\$0.68} = \frac{1{,}564}{68} = \frac{23}{1} = 23$

GE $\frac{P}{E} = \frac{\$90.75}{\$2.75} = \frac{9{,}075}{275} = \frac{33}{1} = 33$

TOY $\frac{P}{E} = \frac{\$19.92}{\$1.66} = \frac{1{,}992}{166} = \frac{12}{1} = 12$

Undervalued stocks: IBM, KM, TOY

Section 6.2

1. $\dfrac{220 \text{ miles}}{4 \text{ hours}} = 55\dfrac{\text{miles}}{\text{hour}} = 55$ miles per hour

3. $\dfrac{252 \text{ kilometers}}{3 \text{ hours}} = 84\dfrac{\text{kilometers}}{\text{hour}}$
$= 84$ kilometers per hour

5. $\dfrac{3 \text{ gallons}}{15 \text{ seconds}} = 0.2\dfrac{\text{gallons}}{\text{second}}$
$= 0.2$ gallons per second

7. $\dfrac{56 \text{ liters}}{4 \text{ minutes}} = 14\dfrac{\text{liters}}{\text{minute}}$
$= 14$ liters per minute

9. $\dfrac{95 \text{ miles}}{5 \text{ gallons}} = 19\dfrac{\text{miles}}{\text{gallon}} = 19$ miles per gallon

11. $\dfrac{325 \text{ miles}}{75 \text{ liters}} = 4\dfrac{1}{3}\dfrac{\text{miles}}{\text{liter}} = 4\dfrac{1}{3}$ miles per liter

13. $\dfrac{96¢}{6 \text{ ounces}} = 16¢$ per ounce

15. $\dfrac{99¢}{20 \text{ ounces}} = 4.95¢$ per ounce

17. Huggies:
$\dfrac{\$12.49}{36 \text{ diapers}} \approx \dfrac{\$0.347}{\text{diaper}} = 34.7¢$ per diaper
Cruisers:
$\dfrac{\$11.99}{38 \text{ diapers}} \approx \dfrac{\$0.316}{\text{diaper}} = 31.6¢$ per diaper
Cruisers are a better buy.

19. $\dfrac{675.4 \text{ miles}}{12\frac{1}{2} \text{ hours}} = \dfrac{675.4 \text{ miles}}{12.5 \text{ hours}}$
$\approx 54.03\dfrac{\text{miles}}{\text{hour}}$
$= 54.03$ miles per hour

21. $\dfrac{128.4 \text{ miles}}{13.8 \text{ gallons}} \approx 9.3\dfrac{\text{miles}}{\text{gallon}}$
$= 9.3$ miles per gallon

23. Daily rate of pay $= \dfrac{\$64}{\text{day}} = \$64$ per day

25. Annual rate of pay
$= \dfrac{\$320}{\text{week}} \times \dfrac{50 \text{ weeks}}{\text{year}} = \dfrac{\$16{,}000}{\text{year}}$

Review Problems

27. The solution to $2 \cdot n = 12$ is $n = 6$, because $2 \cdot 6 = 12$.

29. The solution to $6 \cdot n = 24$ is $n = 4$, because $6 \cdot 4 = 24$.

31. The solution to $20 = 5 \cdot n$ is $n = 4$, because $20 = 5 \cdot 4$.

33. The solution to $650 = 10 \cdot n$ is $n = 65$, because $650 = 10 \cdot 65$.

Improving Your Quantitative Literacy

35. Old: $\dfrac{\$6.99}{100 \text{ oz}} = \dfrac{\$6.99}{3.12 \text{ quart}} \approx \dfrac{\$2.24}{\text{quart}}$

New: $\dfrac{\$5.75}{80 \text{ oz}} = \dfrac{\$5.75}{2.5 \text{ quart}} = \dfrac{\$2.30}{\text{quart}}$

The 100 ounce jug is the better buy.

Section 6.3

1. Means: 3,5
Extremes: 1,15

$$\frac{1}{3}=\frac{5}{15}$$
$$1\cdot 15=3\cdot 5$$
$$15=15 \text{ True}$$

3. Means: 25,2
Extremes: 10,5

$$\frac{10}{25}=\frac{2}{5}$$
$$10\cdot 5=25\cdot 2$$
$$50=50 \text{ True}$$

5. Means: $\frac{1}{2}, 4$

Extremes: $\frac{1}{3}, 6$

$$\frac{\frac{1}{3}}{\frac{1}{2}}=\frac{4}{6}$$
$$\frac{1}{2}\cdot 4=\frac{1}{3}\cdot 6$$
$$2=2 \text{ True}$$

7. Means: 5,1
Extremes: 0.5,10

$$\frac{0.5}{5}=\frac{1}{10}$$
$$0.5(10)=5(1)$$
$$5=5 \text{ True}$$

9. $\frac{2}{5}=\frac{4}{x}$

$$2\cdot x=5\cdot 4$$
$$2x=20$$
$$x=10$$

11. $\frac{1}{y}=\frac{5}{12}$

$$1\cdot 12=y\cdot 5$$
$$12=5y$$
$$\frac{12}{5}=y$$

13. $\frac{x}{4}=\frac{3}{8}$

$$x\cdot 8=4\cdot 3$$
$$8x=12$$
$$x=\frac{12}{8}=\frac{3}{2}$$

15. $\frac{5}{9}=\frac{x}{2}$

$$5\cdot 2=9\cdot x$$
$$10=9x$$
$$\frac{10}{9}=x$$

17. $\frac{3}{7}=\frac{3}{x}$

$$3\cdot x=7\cdot 3$$
$$3x=21$$
$$x=7$$

19. $\frac{25}{100}=\frac{x}{4}$

$$25\cdot 4=100\cdot x$$
$$100=100x$$
$$1=x$$

21. $$\frac{\frac{1}{2}}{y}=\frac{\frac{1}{3}}{12}$$
$$\frac{1}{2}\cdot 12=y\cdot\frac{1}{3}$$
$$6=\frac{1}{3}y$$
$$18=y$$

23. $$\frac{n}{12}=\frac{\frac{1}{4}}{\frac{1}{2}}$$
$$n\cdot\frac{1}{2}=12\cdot\frac{1}{4}$$
$$\frac{1}{2}n=3$$
$$n=6$$

25. $$\frac{10}{20}=\frac{20}{n}$$
$$10\cdot n=20\cdot 20$$
$$10n=400$$
$$n=40$$

27. $$\frac{x}{10}=\frac{10}{2}$$
$$x\cdot 2=10\cdot 10$$
$$2x=100$$
$$x=50$$

29. $$\frac{1}{100}=\frac{y}{50}$$
$$1\cdot 50=100\cdot y$$
$$50=100y$$
$$\frac{50}{100}=y$$
$$\frac{1}{2}=y$$

31. $$\frac{0.4}{1.2}=\frac{1}{x}$$
$$0.4(x)=1.2(1)$$
$$0.4x=1.2$$
$$x=\frac{1.2}{0.4}$$
$$x=3$$

33. $$\frac{0.3}{0.18}=\frac{n}{0.6}$$
$$0.3(0.6)=0.18(n)$$
$$0.18=0.18n$$
$$1=n$$

35. $$\frac{0.5}{x}=\frac{1.4}{0.7}$$
$$0.5(0.7)=x(1.4)$$
$$0.35=1.4x$$
$$\frac{0.35}{1.4}=x$$
$$0.25=x$$
$$\frac{1}{4}=x$$

37. $$\frac{168}{324}=\frac{56}{x}$$
$$168\cdot x=324\cdot 56$$
$$168x=18{,}144$$
$$x=108$$

39. $$\frac{429}{y}=\frac{858}{130}$$
$$429\cdot 130=y\cdot 858$$
$$55{,}770=858y$$
$$65=y$$

41. $$\frac{n}{39} = \frac{533}{507}$$
$$n \cdot 507 = 39 \cdot 533$$
$$507n = 20,787$$
$$n = 41$$

43. $$\frac{756}{903} = \frac{x}{129}$$
$$756 \cdot 129 = 903x$$
$$97,524 = 903x$$
$$108 = x$$

Review Problems

45. 250.14
↑
tens

47.
$$\begin{array}{r} 2.300 \\ 0.180 \\ +\ 24.036 \\ \hline 26.516 \end{array}$$

49.
$$\begin{array}{r} 3.18 \\ -\ 2.79 \\ \hline 0.39 \end{array}$$

Improving Your Quantitative Literacy

51. Answers will vary.

Section 6.4

1. Let x = distance traveled in 7 hours

$$\frac{235 \text{ miles}}{5 \text{ hours}} = \frac{x \text{ miles}}{7 \text{ hours}}$$
$$235 \cdot 7 = 5 \cdot x$$
$$1{,}645 = 5x$$
$$329 = x$$

She will travel 329 miles.

3. Let x = points scored in 20 games

$$\frac{162 \text{ points}}{9 \text{ games}} = \frac{x \text{ points}}{20 \text{ games}}$$
$$162 \cdot 20 = 9 \cdot x$$
$$3{,}240 = 9x$$
$$360 = x$$

At that rate, he will score 360 points.

5. Let x = pints of water

$$\frac{8 \text{ pts antifreeze}}{5 \text{ pts water}} = \frac{24 \text{ pts antifreeze}}{x \text{ pints water}}$$
$$8 \cdot x = 5 \cdot 24$$
$$8x = 120$$
$$x = 15$$

You need 15 pints of water to get the same concentration.

7. Let x = actual distance between the cities

$$\frac{1 \text{ in}}{95 \text{ miles}} = \frac{4.5 \text{ in}}{x \text{ miles}}$$
$$1(x) = 95(4.5)$$
$$x = 427.5$$

The distance between the cities is 427.5 miles.

9. Let x = number of marketable eggs

$$\frac{50 \text{ laid}}{45 \text{ marketable}} = \frac{1{,}000 \text{ laid}}{x \text{ marketable}}$$
$$50 \cdot x = 45 \cdot 1{,}000$$
$$50x = 45{,}000$$
$$x = 900$$

900 eggs will be marketable.

11. Let x = length of actual box car

$$\frac{1 \text{ in model}}{87 \text{ in actual}} = \frac{5 \text{ in model}}{x \text{ in actual}}$$
$$1 \cdot x = 87 \cdot 5$$
$$x = 435$$

The actual box car is 435 inches, which is 36.25 feet

13. Let x = cost for a week

$$\frac{21¢}{1 \text{ mile}} = \frac{x\ ¢}{570 \text{ miles}}$$
$$21 \cdot 570 = 1 \cdot x$$
$$11{,}970 = x$$

It will cost him 11,970¢, or $119.70

15. Let x = grams of water in 10 ounces

$$\frac{6 \text{ oz serving}}{159 \text{ g of water}} = \frac{10 \text{ oz serving}}{x \text{ g of water}}$$
$$6 \cdot x = 159 \cdot 10$$
$$6x = 1{,}590$$
$$x = 265$$

There are 265 grams of water in a ten ounce serving.

17. Let x = liters of gas for 692 miles

$$\frac{378.9 \text{ miles}}{50 \text{ liters}} = \frac{692 \text{ miles}}{x \text{ liters}}$$
$$378.9(x) = 50(692)$$
$$378.9x = 34{,}600$$
$$x = 91.3 \text{ (to the nearest tenth)}$$

You would need 91.3 liters of gas.

19. Let x = number of people who voted

$$\frac{47 \text{ voters}}{100 \text{ registered}} = \frac{x \text{ voted}}{127{,}900 \text{ registered}}$$
$$47 \cdot 127{,}900 = 100 \cdot x$$
$$6{,}011{,}300 = 100x$$
$$60{,}113 = x$$

You would expect 60,113 people voted.

Review Problems

21.
$$\begin{array}{r} 2.7 \\ \times\ 0.5 \\ \hline 1.35 \end{array}$$

23.
$$\begin{array}{r} 3.18 \\ \times\ 1.2 \\ \hline 636 \\ 3{,}180 \\ \hline 3.816 \end{array}$$

25. $\dfrac{2.8}{0.7} = \dfrac{2.8 \times 10}{0.7 \times 10} = \dfrac{28}{7} = 4$

27. $2.33\overline{)24} \rightarrow 15\overline{)2400}$

$$\begin{array}{r} 160 \\ 15\overline{)2400} \\ \underline{15\downarrow} \\ 90 \\ \underline{90\downarrow} \\ 00 \end{array}$$

Improving Your Quantitative Literacy

29. Answers will vary.

Section 6.5

1. $\frac{6}{h}=\frac{4}{6}$

$36=4h$

$9=h$

3. $\frac{y}{8}=\frac{21}{12}$

$12y=168$

$y=14$

5. $\frac{12}{16}=\frac{9}{x}$

$12x=144$

$x=12$

7. $\frac{5}{a}=\frac{3}{15}$

$3a=75$

$a=25$

9. $\frac{50}{40}=\frac{40}{y}$

$50y=1{,}600$

$y=32$

11. $\frac{BA}{AC}=\frac{ED}{DF}$

$\frac{4}{3}=\frac{ED}{6}$

$24=3\cdot EL$

$8=ED$

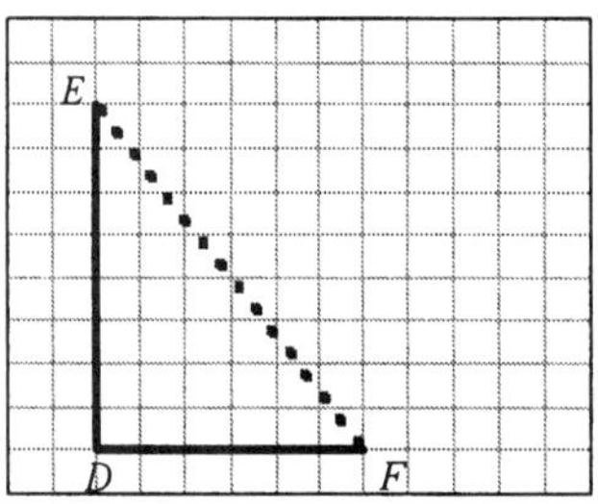

13. Let D be the midpoint of BC, and let H be the midpoint of EF.

$\frac{BD}{DA}=\frac{EH}{HG}$

$\frac{3}{6}=\frac{4}{HG}$

$3\cdot HG=24$

$HG=8$

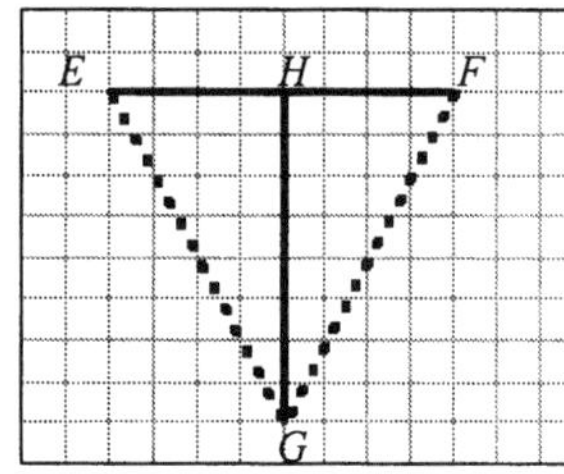

15. $\frac{AD}{DC}=\frac{EH}{HG}$

$\frac{3}{4}=\frac{EH}{8}$

$4\cdot EH=24$

$EH=6$

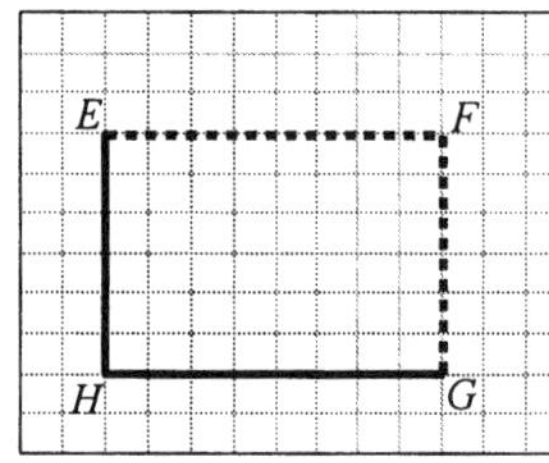

17. $\frac{AB}{AE}=\frac{FG}{FJ}$

$\frac{2}{4}=\frac{4}{FJ}$

$2\cdot FJ=16$

$FJ=8$

Let L be the midpoint of AE, and let K be the midpoint of FJ.

$\frac{AL}{LC}=\frac{FK}{KH}$

$\frac{2}{4}=\frac{4}{KH}$

$2\cdot KH=16$

$KH=8$

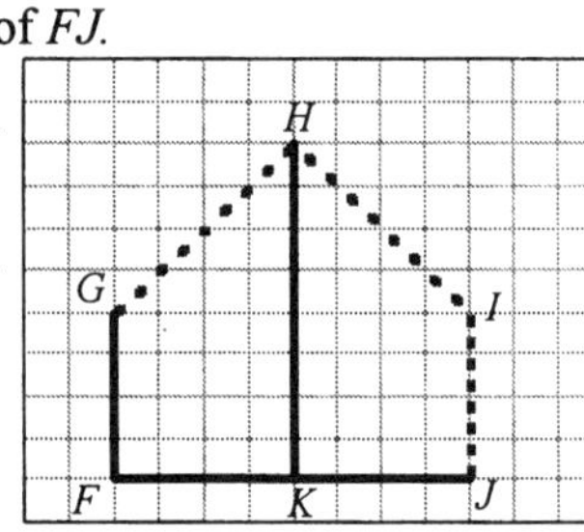

Applying the Concepts

19. Let x = body length of the Bass

$$\frac{\text{body length viola}}{\text{body length Bass}} = \frac{\text{total length viola}}{\text{total length Bass}}$$

$$\frac{15}{x} = \frac{24}{72}$$

$$24x = 1{,}080$$

$$x = 45$$

The body length of the Bass is 45 inches.

21. Let x = body length of the viola

$$\frac{\text{body length viola}}{\text{body length cello}} = \frac{\text{total length viola}}{\text{total length cello}}$$

$$\frac{x}{30} = \frac{26}{48}$$

$$48x = 780$$

$$x = 16.25$$

The body length of the viola is 16.25 inches.

23. Let x = new vertical resolution

$$\frac{\text{old horizontal}}{\text{new horizontal}} = \frac{\text{old vertical}}{\text{new vertical}}$$

$$\frac{800}{1{,}280} = \frac{600}{x}$$

$$800x = 768{,}000$$

$$x = 960$$

The new vertical resolution is 960 pixels.

25. Let x = new horizontal resolution

$$\frac{20\text{" horizontal}}{17\text{" horizontal}} = \frac{20\text{" vertical}}{17\text{" vertical}}$$

$$\frac{1{,}680}{1{,}050} = \frac{x}{900}$$

$$1{,}050x = 1{,}512{,}000$$

$$x = 1{,}440$$

The horizontal resolution of 1,440 pixels.

27. Let x = height of tree

$$\frac{\text{height of tree}}{\text{height of man}} = \frac{\text{tree shadow}}{\text{man shadow}}$$

$$\frac{x}{6} = \frac{38}{4}$$

$$4x = 228$$

$$x = 57$$

The tree is 57 feet tall.

29. Let x = height of the child

$$\frac{\text{height of tower}}{\text{height of child}} = \frac{\text{tower shadow}}{\text{child shadow}}$$

$$\frac{36}{x} = \frac{54}{6}$$

$$216 = 54x$$

$$4 = x$$

The child is 4 feet tall.

Review Problems

31. $4\overline{)3.00}$ with quotient 0.75; $28\downarrow$, 40, 40, 0 $\qquad \frac{3}{4} = 0.75$

33. $\frac{1}{2} \rightarrow 2\overline{)1.0}$ with quotient 0.5; 10, 0 $\qquad 5\frac{1}{2} = 5.5$

35. $\frac{3}{100} = 0.03$

37. $0.34 = \frac{34}{100} = \frac{17}{50}$

39. $2.4 = 2\frac{4}{10} = 2\frac{2}{5}$

41. $1.75 = 1\frac{75}{100} = 1\frac{3}{4}$

Extending the Concepts

43. a. $\frac{\text{shaded}}{\text{non-shaded}} = \frac{8}{16} = \frac{1}{2}$

b. Let x= number of shaded rectangles in the larger rectangle.

$$\frac{8 \text{ shaded}}{24 \text{ total}} = \frac{x \text{ shaded}}{54 \text{ total}}$$

$$24x = 432$$

$$x = 18$$

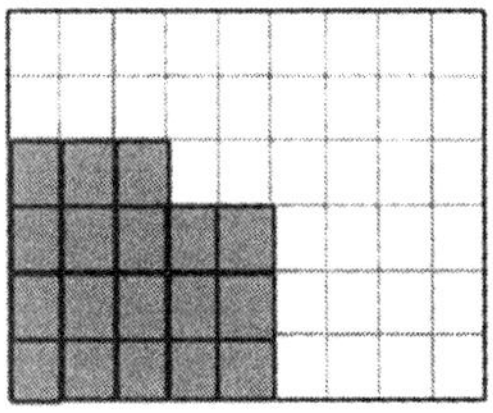

c. small rectangle: $\frac{\text{shaded}}{\text{total}} = \frac{8}{24} = \frac{1}{3}$

large rectangle: $\frac{\text{shaded}}{\text{total}} = \frac{18}{54} = \frac{1}{3}$

Chapter 6 Review

1. $\frac{9}{30} = \frac{3}{10}$

3. $\frac{\frac{3}{7}}{\frac{4}{7}} = \frac{3}{\cancel{7}} \cdot \frac{\cancel{7}}{4} = \frac{3}{4}$

5. $\frac{2\frac{1}{3}}{1\frac{2}{3}} = \frac{\frac{7}{3}}{\frac{5}{3}} = \frac{7}{\cancel{3}} \cdot \frac{\cancel{3}}{5} = \frac{7}{5}$

7. $\frac{0.6}{1.2} = \frac{0.6 \times 10}{1.2 \times 10} = \frac{6}{12} = \frac{1}{2}$

9. $\frac{\frac{1}{5}}{\frac{3}{5}} = \frac{1}{\cancel{5}} \cdot \frac{\cancel{5}}{3} = \frac{1}{3}$

11. $\frac{\text{taxes}}{\text{oil}} = \frac{18¢}{4¢} = \frac{9}{2}$

13. $\frac{\text{profits}}{\text{costs}} = \frac{4¢}{16¢} = \frac{1}{4}$

15. $\frac{285 \text{ miles}}{15 \text{ gallons}} = 19 \frac{\text{mi}}{\text{gal}}$

17. $\frac{138¢}{6 \text{ ounces}} = 23 \frac{¢}{\text{oz}} = 23¢$ per ounce

19. $\frac{5}{7} = \frac{35}{x}$

$5 \cdot x = 7 \cdot 35$

$5x = 245$

$x = 49$

21. $\frac{\frac{1}{2}}{10} = \frac{y}{2}$

$\frac{1}{2} \cdot 2 = 10 \cdot y$

$1 = 10y$

$\frac{1}{10} = y$

23. Let x = mL of solution

$$\frac{2{,}000 \text{ mL of solution}}{24 \text{ mL of drug}} = \frac{x \text{ mL of solution}}{18 \text{ mL of drug}}$$

$2{,}000 \cdot 18 = 24 \cdot x$

$36{,}000 = 24x$

$1{,}500 = x$

1,500 mL of solution are required.

25. Let x = time to lose 20 pounds (in weeks)

$$\frac{8 \text{ lbs}}{2 \text{ weeks}} = \frac{20 \text{ lbs}}{x \text{ weeks}}$$

$8 \cdot x = 2 \cdot 20$

$8x = 40$

$x = 5$

It will take him 5 weeks.

27. $\frac{8}{6} = \frac{12}{x}$

$8x = 6 \cdot 12$

$8x = 72$

$x = 9$

Chapters 1-6 Cumulative Review

1. $-4\left(\frac{x}{2}-3\right)=-4\left(\frac{x}{2}\right)-4(-3)$

$=-2x+12$

3. $\begin{array}{r} \overset{5}{\cancel{6}}\,\overset{10}{\cancel{1}}\,\overset{13}{\cancel{3}} \\ -2\,9\,7 \\ \hline 3\,1\,6 \end{array}$

5. $-\frac{2}{3}(3x+9)=-\frac{2}{3}\cdot 3x-\frac{2}{3}\cdot 9$

$=-2x-6$

7. $\frac{5}{8}\div(-10)=\frac{5}{8}\cdot\left(-\frac{1}{10}\right)=-\frac{\cancel{5}\cdot 1}{8\cdot 2\cdot \cancel{5}}=-\frac{1}{16}$

9. $4\frac{7}{8}=\frac{8\cdot 4+7}{8}=\frac{39}{8}$

11. $5(x+9)=5(x)+5(9)$

Distributive Property

13. $\sqrt{\frac{36}{49}}=\frac{6}{7}$, because $\left(\frac{6}{7}\right)^2=\frac{36}{49}$

15. $\frac{9(-4)+8(3)}{-2^2}$ *exponents first*

$=\frac{9(-4)+8(3)}{-4}$ *multiply left to right*

$=\frac{-36+24}{-4}$ *add*

$=\frac{-12}{-4}$ *divide*

$=3$

17. $-(-6)=6$

The opposite of –6 is 6.

19. $\frac{\frac{2}{3}}{\frac{3}{4}}=\frac{2}{3}\cdot\frac{4}{3}=\frac{2\cdot 4}{3\cdot 3}=\frac{8}{9}$

21. 4 times x | subtracted from | twice | the sum of x and 9

$2\times\ (x+9)\ -4x$

$2(x+9)-4x$

23.
$$\begin{aligned}
7x+0.21&=x-0.69\\
7x+0.21+(-0.21)&=x-0.69+(-0.21)\\
7x&=x+(-0.9)\\
7x+(-x)&=x+(-0.9)+(-x)\\
6x&=-0.9\\
x&=-\frac{0.9}{6}\\
x&=-0.15
\end{aligned}$$

25.
$$\begin{aligned}
5(2x-7)&=8x-5-4x\\
10x-35&=4x-5\\
10x-35+35&=4x-5+35\\
10x&=4x+30\\
10x+(-4x)&=4x+30+(-4x)\\
6x&=30\\
x&=5
\end{aligned}$$

27.

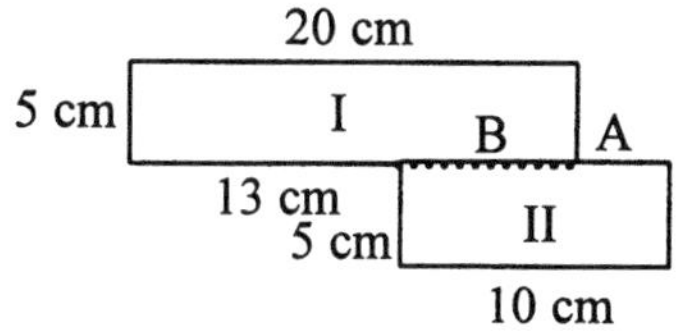

Find the lengths of the missing sides:

Side B $= 20 - 13 = 7$ cm

Side A $= 10 - 7 = 6$ cm

$P =$ sum of all sides

$= 20 + 5 + 3 + 5 + 10 + 5 + 13 + 5 = 66$ cm

Area: Separate into 2 areas.

$$\text{Area I} = lw = (20 \text{ cm})(5 \text{ cm}) = 100 \text{ cm}^2$$

$$\text{Area II} = lw = (5 \text{ cm})(10 \text{ cm}) = 50 \text{ cm}^2$$

$$\text{Area} = \text{Area I} + \text{Area II} = 100 \text{ cm}^2 + 50 \text{ cm}^2 = 150 \text{ cm}^2$$

29. When $x = -2$:

$$2x - 7y = 10$$
$$2(-2) - 7y = 10$$
$$-4 - 7y = 10$$
$$-7y = 14$$
$$y = -2$$

31. Difference = High − Low

$$= 9° - (-8°) = 9° + 8° = 17°$$

33. Let x = number of women

$$\frac{3 \text{ men}}{4 \text{ women}} = \frac{15 \text{ men}}{x \text{ women}}$$
$$3 \cdot x = 4 \cdot 15$$
$$3x = 60$$
$$x = 20$$

35. $\text{Average (mean)} = \dfrac{\text{sum of the values}}{\text{number of values}}$

$$= \frac{72 + 113 + 108 + 95}{4} = \frac{388}{4} = 97$$

The mean distance is 97 miles.

37. $\dfrac{579¢}{64 \text{ ounces}} \approx 9.05\dfrac{¢}{\text{oz}} = 9.05¢$ per ounce

$\dfrac{269¢}{32 \text{ ounces}} = 8.41\dfrac{¢}{\text{oz}} = 8.41¢$ per ounce

The 32-oz carton is the better buy.

39. $\dfrac{12}{x} = \dfrac{8}{10}$

$$120 = 8x$$
$$15 = x$$

The length is $x = 15$ in.

Chapter 6 Test

1. $\frac{24}{18}=\frac{4}{3}$

3. $\frac{5}{3\frac{1}{3}}=\frac{5}{\frac{10}{3}}=5\cdot\frac{3}{10}=\frac{15}{10}=\frac{3}{2}$

5. $\frac{\frac{3}{11}}{\frac{5}{11}}=\frac{3}{\cancel{11}}\cdot\frac{\cancel{11}}{5}=\frac{3}{5}$

7. $\frac{\text{phone payment}}{\text{food payment}}=\frac{\$60}{\$250}=\frac{6}{25}$

9. 16 oz can:

$\frac{\$2.59}{16\text{ oz}}=\frac{259¢}{16\text{ oz}}\approx 16.19¢\text{ per ounce}$

12 oz can :

$\frac{\$1.89}{12\text{ oz}}=\frac{189¢}{12\text{ oz}}=15.75¢\text{ per ounce}$

The 12 ounce can is the better buy.

11. $\frac{1.8}{6}=\frac{2.4}{x}$

$1.8(x)=6(2.4)$

$1.8x=14.4$

$x=8$

13. Let x = actual distance between the cities

$\frac{1\text{ inches}}{60\text{ miles}}=\frac{2\frac{1}{4}\text{ in}}{x\text{ miles}}$

$1(x)=60(2.25)$

$x=135$

The cities are 135 miles apart.

15. $\frac{20}{15}=\frac{h}{12}$

$240=15h$

$16=h$

Chapter 7 Preview

1.
$$\begin{array}{r} 136.00 \\ +\ 5.44 \\ \hline 141.44 \end{array}$$

3.
$$\begin{array}{r} 1{,}793{,}000 \\ -\ 315{,}568 \\ \hline 1{,}477{,}432 \end{array}$$

5. $0.2 \times 100 = 20$

7.
$$\begin{array}{r} 0.15 \\ \times\ 63 \\ \hline 45 \\ 900 \\ \hline 9.45 \end{array}$$

9. $3.62 \div 100 = \dfrac{3.62}{100} = 0.0362$

11.
$$\begin{aligned} 600 \times 0.04 \times \frac{60}{360} &= \frac{600}{1} \cdot \frac{4}{100} \cdot \frac{60}{360} \\ &= \frac{\cancel{100} \cdot \cancel{6} \cdot 4 \cdot \cancel{60}}{1 \cdot \cancel{100} \cdot \cancel{60} \cdot \cancel{6}} \\ &= 4 \end{aligned}$$

13. $\dfrac{45}{1{,}000} = \dfrac{9 \cdot \cancel{5}}{200 \cdot \cancel{5}} = \dfrac{9}{200}$

15.
$$\begin{array}{r} 0.375 \\ 8\overline{)3.000} \\ 24\downarrow \\ \hline 60 \\ 56\downarrow \\ \hline 40 \\ 40 \\ \hline 0 \end{array} \qquad \frac{3}{8} = 0.375$$

17.
$$2\frac{1}{2} = \frac{5}{2} \rightarrow \begin{array}{r} 2.5 \\ 2\overline{)5.0} \\ 4\downarrow \\ \hline 10 \\ 10 \\ \hline 0 \end{array} \qquad 2\frac{1}{2} = 2.5$$

19.
$$\begin{aligned} 0.12 \cdot n &= 1{,}836 \\ \frac{0.\cancel{12} \cdot n}{0.\cancel{12}} &= \frac{1{,}836}{0.12} \\ n &= \frac{1{,}836}{0.12} \\ n &= 15{,}300 \end{aligned}$$

Chapter 7 Pretest

1. $68\% = \dfrac{68}{100} = 0.68$

3. $21.5\% = \dfrac{21.5}{100} = 0.215$

5. $0.386 = 0.386 \times 100\% = 38.6\%$

7. $33\% = \dfrac{33}{100}$

9. $8.5\% = \dfrac{8.5}{100} = \dfrac{85}{1{,}000} = \dfrac{17}{200}$

11. $\dfrac{4}{5} = \dfrac{4 \cdot 20}{5 \cdot 20} = \dfrac{80}{100} = 80\%$

13. What number is 5% of 24?

$n = 0.05 \cdot 24$

$n = 1.2$

15. 12 is 24% of what number?

$$12 = 0.24 \cdot n$$

$$\frac{12}{0.24} = \frac{\cancel{0.24} \cdot n}{\cancel{0.24}}$$

$$n = \frac{12}{0.24}$$

$$n = 50$$

17. What is 6% of \$150,000?

$n = 0.06 \cdot \$150{,}000$

$n = \$9{,}000$

19. $I = P \times R \times T$

$= \$2{,}000 \times 0.015 \times 1$

$= \$30$

Original Principal:	\$2,000
+ Interest:	+30
Total amount after 1 year:	\$2,030

Section 7.1

1. $20\% = \frac{20}{100}$

3. $60\% = \frac{60}{100}$

5. $24\% = \frac{24}{100}$

7. $65\% = \frac{65}{100}$

9. $23\% = \frac{23}{100} = 0.23$

11. $92\% = \frac{92}{100} = 0.92$

13. $9\% = \frac{9}{100} = 0.09$

15. $3.4\% = \frac{3.4}{100} = 0.034$

17. $6.34\% = \frac{6.34}{100} = 0.0634$

19. $0.9\% = \frac{0.9}{100} = 0.009$

21. Move decimal 2 places right, and attach the % symbol:
$0.23 = 23.0\% = 23\%$

23. Move decimal 2 places right, and attach the % symbol:
$0.92 = 92.0\% = 92\%$

25. Move decimal 2 places right, and attach the % symbol:
$0.45 = 45.0\% = 45\%$

27. Move decimal 2 places right, and attach the % symbol:
$0.03 = 3.0\% = 3\%$

29. Move decimal 2 places right, and attach the % symbol:
$0.6 = 60.0\% = 60\%$

31. Move decimal 2 places right, and attach the % symbol:
$0.8 = 80.0\% = 80\%$

33. Move decimal 2 places right, and attach the % symbol:
$0.27 = 27.0\% = 27\%$

35. Move decimal 2 places right, and attach the % symbol:
$1.23 = 123.0\% = 123\%$

37. $60\% = \frac{60}{100} = \frac{3}{5}$

39. $75\% = \frac{75}{100} = \frac{3}{4}$

41. $4\% = \frac{4}{100} = \frac{1}{25}$

43. $26.5\% = \frac{26.5}{100} = \frac{26.5 \times 10}{100 \times 10} = \frac{265}{1,000} = \frac{53}{200}$

45. $71.87\% = \frac{71.87}{100} = \frac{71.87 \times 100}{100 \times 100} = \frac{7,187}{10,000}$

47. $0.75\% = \frac{0.75}{100} = \frac{0.75 \times 100}{100 \times 100} = \frac{75}{10,000} = \frac{3}{400}$

49. $6\frac{1}{4}\% = \frac{6\frac{1}{4}}{100} = \frac{\frac{25}{4}}{100} = \frac{25}{4} \times \frac{1}{100} = \frac{25}{400} = \frac{1}{16}$

51. $33\frac{1}{3}\% = \dfrac{33\frac{1}{3}}{100} = \dfrac{\frac{100}{3}}{100} = \dfrac{100}{3} \times \dfrac{1}{100} = \dfrac{1}{3}$

53. *Change denominator to 100, then switch to percent:*
$\dfrac{1}{2} = \dfrac{1 \times 50}{2 \times 50} = \dfrac{50}{100} = 50\%$

55. *Change denominator to 100, then switch to percent:*
$\dfrac{3}{4} = \dfrac{3 \times 25}{4 \times 25} = \dfrac{75}{100} = 75\%$

57.
$$\begin{array}{r} 0.33 \\ 3\overline{)1.00} \\ 9\downarrow \\ \hline 10 \\ 9 \\ \hline 1 \end{array}$$
$\dfrac{1}{3} = 0.\overline{3} = 33\dfrac{1}{3}\%$

59. *Change denominator to 100, then switch to percent:*
$\dfrac{4}{5} = \dfrac{4 \times 20}{5 \times 20} = \dfrac{80}{100} = 80\%$

61.
$$\begin{array}{r} 0.875 \\ 8\overline{)7.000} \\ 64\downarrow \\ \hline 60 \\ 56\downarrow \\ \hline 40 \\ 40 \\ \hline 0 \end{array}$$
$\dfrac{7}{8} = 0.875 = 87.5\%$

63. *Change denominator to 100, then switch to percent:*
$\dfrac{7}{50} = \dfrac{7 \times 2}{50 \times 2} = \dfrac{14}{100} = 14\%$

65. *Change denominator to 100, then switch to percent:*
$3\dfrac{1}{4} = \dfrac{13}{4} = \dfrac{13 \times 25}{4 \times 25} = \dfrac{325}{100} = 325\%$

67. *Change denominator to 100, then switch to percent:*
$1\dfrac{1}{2} = \dfrac{3}{2} = \dfrac{3 \times 50}{2 \times 50} = \dfrac{150}{100} = 150\%$

69.
$$\begin{array}{r} 0.4883 \\ 43\overline{)21.0000} \\ 172\downarrow \\ \hline 380 \\ 344\downarrow \\ \hline 360 \\ 344 \\ \hline 160 \\ 129 \\ \hline 31 \end{array}$$
$\dfrac{21}{43} \approx 0.4883 \approx 48.8\%$

Applying the Concepts

71. $50\% = \dfrac{50}{100} = 0.50$

$75\% = \dfrac{75}{100} = 0.75$

73. a. $10\% = \frac{10}{100} = \frac{1}{10}$

$4\% = \frac{4}{100} = \frac{1}{25}$

b. $10\% = \frac{10}{100} = 0.1$

$4\% = \frac{4}{100} = 0.04$

c. $\frac{10\%}{4\%} = \frac{0.10}{0.04} = \frac{0.10\times 10}{0.04\times 10} = \frac{10}{4} = 2.5$

Boys are $2\frac{1}{2}$ times more likely to have ADHD than girls.

75. *Change denominator to 100, then switch to percent:*

$\frac{1}{5} = \frac{1\times 20}{5\times 20} = \frac{20}{100} = 20\%$

77. Science and Mathematics:

$\frac{3}{25} = \frac{3\cdot 4}{25\times 4} = \frac{12}{100} = 12\%$

Agriculture:

$\frac{11}{50} = \frac{11\times 2}{50\times 2} = \frac{22}{100} = 22\%$

Architecture and Environmental Design:

$\frac{1}{10} = \frac{1\times 10}{10\times 10} = \frac{10}{100} = 10\%$

Business:

$\frac{3}{20} = \frac{3\times 5}{20\times 5} = \frac{15}{100} = 15\%$

Engineering:

$\frac{1}{4} = \frac{1\times 25}{4\times 25} = \frac{25}{100} = 25\%$

Liberal Arts:

$\frac{4}{25} = \frac{4\times 4}{25\times 4} = \frac{16}{100} = 16\%$

Calculator Problems

79. $\frac{29}{37} \approx 0.7838 = 78.4\%$

81. $\frac{6}{51} \approx 0.1176 = 11.8\%$

83. $\frac{236}{327} \approx 0.7217 = 72.2\%$

85. $\frac{1}{12} \approx 0.08333 = 8.3\%$

$\frac{1}{450} \approx 0.0022 = 0.2\%$

Review Problems

87.
```
    300
 × 0.25
 ------
   1500
   6000
 ------
  75.00
```
The trailing 0's can be dropped: 75

89.
```
  0.53
  × 16
  ----
   318
   530
  ----
  8.48
```

91. $0.05\overline{)25} \rightarrow 5\overline{)2500}$ (quotient 500; 25↓, 00) $\quad \frac{25}{0.05} = 500$

93. $0.25\overline{)45} \rightarrow 25\overline{)4500}$ $\quad \frac{45}{0.25} = 180$

$$\begin{array}{r} 180 \\ 25\overline{)4500} \\ \underline{25\downarrow} \\ 200 \\ \underline{200\downarrow} \\ 00 \end{array}$$

95. $3 \cdot n = 12$

$$\frac{3n}{3} = \frac{12}{3}$$

$$n = 4$$

Improving your Quantitative Literacy

97. a. $100\% - 55\% = 45\%$

b. $100\% - 37\% = 63\%$

c. Answers will vary.

Section 7.2

1. What number is 25% of 32?

$n = 0.25 \cdot 32$

$n = 8$

3. What number is 20% of 120?

$n = 0.20 \cdot 120$

$n = 24$

5. What number is 54% of 38?

$n = 0.54 \cdot 38$

$n = 20.52$

7. What number is 11% of 67?

$n = 0.11 \cdot 67$

$n = 7.37$

9. What percent of 24 is 12?

$n \cdot 24 = 12$

$$\frac{n \cdot \cancel{24}}{\cancel{24}} = \frac{12}{24}$$

$$n = \frac{12}{24}$$

$n = 0.5 = 50\%$

11. What percent of 50 is 5?

$n \cdot 50 = 5$

$$\frac{n \cdot \cancel{50}}{\cancel{50}} = \frac{5}{50}$$

$$n = \frac{5}{50}$$

$n = 0.1 = 10\%$

13. What percent of 36 is 9?

$n \cdot 36 = 9$

$$\frac{n \cdot \cancel{36}}{\cancel{36}} = \frac{9}{36}$$

$$n = \frac{9}{36}$$

$n = 0.25 = 25\%$

15. What percent of 8 is 6?

$n \cdot 8 = 6$

$$\frac{n \cdot \cancel{8}}{\cancel{8}} = \frac{6}{8}$$

$$n = \frac{6}{8}$$

$n = 0.75 = 2 = 75\%$

17. 32 is 50% of what number?

$32 = 0.50 \cdot n$

$$\frac{32}{0.50} = \frac{\cancel{0.50} \cdot n}{\cancel{0.50}}$$

$$\frac{32}{0.50} = n$$

$64 = n$

19. 10 is 20% of what number?

$10 = 0.20 \cdot n$

$$\frac{10}{0.20} = \frac{\cancel{0.20} \cdot n}{\cancel{0.20}}$$

$$\frac{10}{0.20} = n$$

$50 = n$

21. 37 is 4% of what number?

$37 = 0.04 \cdot n$

$$\frac{37}{0.04} = \frac{\cancel{0.04} \cdot n}{\cancel{0.04}}$$

$$\frac{37}{0.04} = n$$

$925 = n$

23. 8 is 2% of what number?

$8 = 0.02 \cdot n$

$$\frac{8}{0.02} = \frac{\cancel{0.02} \cdot n}{\cancel{0.02}}$$

$$\frac{8}{0.02} = n$$

$400 = n$

25. What is 6.4% of 87?

$n = 0.064 \cdot 87$

$n = 5.568$

27. 25% of what number is 30?

$0.25 \cdot n = 30$

$\frac{\cancel{0.25} \cdot n}{\cancel{0.25}} = \frac{30}{0.25}$

$n = \frac{30}{0.25}$

$n = 120$

29. 28% of 49 is what number?

$0.28 \cdot 49 = n$

$13.72 = n$

31. 27 is 120% of what number?

$27 = 1.20 \cdot n$

$\frac{27}{1.20} = \frac{\cancel{1.20} \cdot n}{\cancel{1.20}}$

$\frac{27}{1.20} = n$

$22.5 = n$

33. 65 is what percent of 130?

$65 = n \cdot 130$

$\frac{65}{130} = \frac{n \cdot \cancel{130}}{\cancel{130}}$

$\frac{65}{130} = n$

$n = 0.5 = 50\%$

35. What is 0.4% of 235,671?

$n = 0.004 \cdot 235{,}671$

$n = 942.684$

37. 4.89% of 2,000 is what number?

$0.0489 \cdot 2{,}000 = n$

$97.8 = n$

Applying the Concepts

39. Total calories: 210
Fat calories: 10
10 is what percent of 210?

$10 = n \cdot 210$

$\frac{10}{210} = \frac{n \cdot \cancel{210}}{\cancel{210}}$

$\frac{10}{210} = n$

$n \approx 0.0476 = 4.8\%$ (to the nearest tenth)

Because the fat calories are less than 30% of total calories, the food is considered healthy.

41. Total calories: 20
Fat calories: 10
10 is what percent of 20?

$10 = n \cdot 20$

$\frac{10}{20} = \frac{n \cdot \cancel{20}}{\cancel{20}}$

$\frac{10}{20} = n$

$n \approx 0.5 = 50\%$

Because the fat calories are more than 30% of total calories, the food not considered healthy.

Calculator Problems

43. What number is 62.5% of 398?

$n = 0.625 \cdot 398$

$n = 248.75$

45. What percent of 789 is 204?

$n \cdot 789 = 204$

$\frac{n \cdot \cancel{789}}{\cancel{789}} = \frac{204}{789}$

$n = \frac{204}{789}$

$n \approx 0.2585 = 25.9\%$ (to the nearest tenth)

47. 19 is 38.4% of what number?

$$19 = 0.384 \cdot n$$

$$\frac{19}{0.384} = \frac{\cancel{0.384} \cdot n}{\cancel{0.384}}$$

$$\frac{19}{0.384} = n$$

$$n = 49.5 \text{ (to the nearest tenth)}$$

Review Problems

49. $45 = 9 \cdot 5$
$= 3 \cdot 3 \cdot 5$

51. $105 = 5 \cdot 21$
$= 5 \cdot 3 \cdot 7$
$= 3 \cdot 5 \cdot 7$

53. $36 = 4 \cdot 9$
$= 2 \cdot 2 \cdot 3 \cdot 3$

55. $210 = 10 \cdot 21$
$= 2 \cdot 5 \cdot 3 \cdot 7$
$= 2 \cdot 3 \cdot 5 \cdot 7$

Extending the Concepts

57.

$$\begin{array}{ccccc} n & = & 0.25 & \cdot & 350 \\ \downarrow & \downarrow & \downarrow & \downarrow & \downarrow \\ \text{What number} & \text{is} & 25\% & \text{of} & 350? \end{array}$$

59.

$$\begin{array}{ccccc} n & \cdot & 24 & = & 16 \\ \downarrow & \downarrow & \downarrow & \downarrow & \downarrow \\ \text{What percent} & \text{of} & 24 & \text{is} & 16? \end{array}$$

61.

$$\begin{array}{ccccc} 46 & = & 0.75 & \cdot & n \\ \downarrow & \downarrow & \downarrow & \downarrow & \downarrow \\ 46 & \text{is} & 75\% & \text{of} & \text{what number?} \end{array}$$

Improving Your Quantitative Literacy

63. Low estimate:
From eggs to hatchling:
What is 70% of 1,000?

$n = 0.70(1000)$

$n = 700$

From hatchling to fledgling:
What is 50% of 700?

$n = 0.50(700)$

$n = 350$

From hatchling to breeder:
What is 50% of 350?

$n = 0.50(350)$

$n = 175$

High estimate:
From eggs to hatchling:
What is 80% of 1,000?

$n = 0.80(1,000)$

$n = 800$

From hatchling to fledgling:
What is 70% of 800?

$n = 0.70(800)$

$n = 560$

From hatchling to breeder:
What is 50% of 350?

$n = 0.50(560)$

$n = 280$

Fewer than 175 to 280 gulls.

Section 7.3

1. 84 is what percent of 120?

$84 = n \cdot 120$

$\frac{84}{120} = \frac{n \cdot \cancel{120}}{\cancel{120}}$

$\frac{84}{120} = n$

$n = 0.7 = 70\%$

3. 63 is what percent of 75?

$63 = n \cdot 75$

$\frac{63}{75} = \frac{n \cdot \cancel{75}}{\cancel{75}}$

$\frac{63}{75} = n$

$n = 0.84 = 84\%$

5. What is 75% of 60 mL?

$n = 0.75 \cdot 60$

$n = 45$ mL

7. What is 65% of 28 acres?

$n = 0.65 \cdot 28$

$n = 18.2$ acres

18.2 acres are available for farming.

$28 - 18.2 = 9.8$

9.8 acres are not available for farming.

9. 1,440 is 48% of what number?

$1{,}440 = 0.48 \cdot n$

$\frac{1{,}440}{0.48} = \frac{\cancel{0.48} \cdot n}{\cancel{0.48}}$

$\frac{1{,}440}{0.48} = n$

$n = 3{,}000$ total students

11. 240 is 60% of what number?

$240 = 0.6 \cdot n$

$\frac{240}{0.6} = \frac{\cancel{0.6} \cdot n}{\cancel{0.6}}$

$\frac{240}{0.6} = n$

$n = 400$ total students

13. What is 52% of 3,200 students?

$n = 0.52 \cdot 3{,}200$

$n = 1{,}664$ female students

15. 10,000 is what percent of 32,000?

$10{,}000 = n \cdot 32{,}000$

$\frac{10{,}000}{32{,}000} = \frac{n \cdot \cancel{32{,}000}}{\cancel{32{,}000}}$

$\frac{10{,}000}{32{,}000} = n$

$n = 0.3125 = 31.25\%$

Calculator Problems

17. 3,972 is what percent of 7,892?

$3{,}972 = n \cdot 7{,}892$

$\frac{3{,}972}{7{,}892} = \frac{n \cdot \cancel{7{,}892}}{\cancel{7{,}892}}$

$\frac{3{,}972}{7{,}892} = n$

$n = 0.503 \approx 50\%$

19. a. $119{,}648 + 6{,}742 + 506 + 77 = 126{,}973$

b. 126,973 is 3% of what number?
$126{,}973 = 0.03 \cdot n$

$$\frac{126{,}973}{0.03} = \frac{\cancel{0.03} \cdot n}{\cancel{0.03}}$$

$$\frac{126{,}973}{0.03} = n$$

$n \approx 4{,}232{,}433$ children born

c. 119,648 is what percent of 126,973?
$119{,}648 = n \cdot 126{,}973$

$$\frac{119{,}648}{126{,}973} = \frac{n \cdot \cancel{126{,}973}}{\cancel{126{,}973}}$$

$$\frac{119{,}648}{126{,}973} = n$$

$n \approx 0.9423 \approx 94.23\%$

Review Problems

21. $\frac{1}{2} \cdot \frac{2}{5} = \frac{1 \cdot \cancel{2}}{\cancel{2} \cdot 5} = \frac{1}{5}$

23. $\frac{3}{4} \cdot \frac{5}{9} = \frac{\cancel{3} \cdot 5}{4 \cdot 3 \cdot \cancel{3}} = \frac{5}{12}$

25. $2 \cdot \frac{3}{8} = \frac{\cancel{2} \cdot 3}{\cancel{2} \cdot 4} = \frac{3}{4}$

27. $1\frac{1}{4} \cdot \frac{8}{15} = \frac{5}{4} \cdot \frac{8}{15} = \frac{\cancel{5} \cdot 2 \cdot \cancel{4}}{\cancel{4} \cdot 3 \cdot \cancel{5}} = \frac{2}{3}$

Extending the Concepts

29. 55 is what percent of 163?
$55 = n \cdot 163$

$$\frac{55}{163} = \frac{n \cdot \cancel{163}}{\cancel{163}}$$

$$\frac{55}{163} = n$$

$n \approx 0.3374 \approx 33.7\%$

31. Note: Batting 0.333 = 33.3%
What is 33.3% of 132 at bats?
$n = 0.333 \cdot 132$

$n = 44$ hits (to the nearest whole)

33. It is easier to work this problem using cumulative batting average versus the average for the next 50 at bats.

From the opening discussion on batting averages in the text, Suzuki had 72 hits in 200 at bats for an average of 0.360.

After the next 50 at bats, he would have a cumulative total of 200 + 50 = 250 at bats.

Find the number of hits needed to have 0.360 (36.0%) out of 250 at bats:
What is 36% of 250 at bats?
$n = 0.36 \cdot 250$

$n = 90$ hits
He needs 90 total hits. Because he already has 72 hits, he must have at least $90 - 72 = 18$ hits to maintain his average.

Section 7.4

1. What is 7% of \$750?

$n = 0.07 \cdot 750$

$n = 52.5$

The sales tax is \$52.50

3. What is 6% of \$45?

$n = 0.06 \cdot 45$

$n = 2.7$

The sales tax is \$2.70

$$\begin{aligned}\text{Total price} &= \text{purchase price} + \text{sales tax}\\ &= \$45 + \$2.70\\ &= \$47.70\end{aligned}$$

5. 4% of what number is \$6?

$0.04 \cdot n = 6$

$$\frac{\cancel{0.04}n}{\cancel{0.04}} = \frac{6}{0.04}$$

$n = \$150$

The purchase price is \$150.

$$\begin{aligned}\text{Total price} &= \text{purchase price} + \text{sales tax}\\ &= \$150 + \$6\\ &= \$156\end{aligned}$$

7. \$22.50 is what percent of \$450?

$22.50 = n \cdot 450$

$$\frac{22.50}{450} = \frac{n \cdot \cancel{450}}{\cancel{450}}$$

$$n = \frac{22.5}{450}$$

$n = 0.05 = 5\%$

9. What is 3% of \$94,000?

$n = 0.03 \cdot 94{,}000$

$n = 2{,}820$

The commission is \$2,820

11. 12% of what number is \$24?

$0.12 \cdot n = 24$

$$\frac{\cancel{0.12}n}{\cancel{0.12}} = \frac{24}{0.12}$$

$n = 200$

The purchase price is \$200.

13. \$112 is what percent of \$800?

$112 = n \cdot 800$

$$\frac{112}{800} = \frac{n \cdot \cancel{800}}{\cancel{800}}$$

$$n = \frac{112}{800}$$

$n = 0.14 = 14\%$

Calculator Problems

15. What is 5.5% of \$216.95?

$n = 0.055 \cdot 216.95$

$n = 11.9323$

The sales tax is \$11.93

17. \$10.33 is what percent of \$229.50?

$10.33 = n \cdot 229.50$

$$\frac{10.33}{229.50} = \frac{n \cdot \cancel{229.50}}{\cancel{229.50}}$$

$$n = \frac{10.33}{229.50}$$

$n = 0.04501 = 4.5\%$

19. 13% of what number is \$519.35?

$0.13 \cdot n = 519.35$

$$\frac{\cancel{0.13}n}{\cancel{0.13}} = \frac{519.35}{0.13}$$

$n = \$3{,}995$

The car sold for \$3,995

21. Let x = price of the car
Note: The sales tax on the car is
7.25% of purchase price $= 0.0725 \cdot x$

Purchase price + Sales tax = Total price

$$x \quad + 0.0725x = \$17,481.75$$

$$1x + 0.0725x = 17{,}481.75$$
$$1.0725x = 17{,}481.75$$
$$\frac{1.0725x}{1.0725} = \frac{17{,}481.75}{1.0725}$$
$$x = \frac{17{,}481.75}{1.0725}$$
$$x = 16{,}300$$

The purchase price is \$16,300

23. First find the total cash for the day:

$$\begin{array}{r} \$1058.60 \\ -\ 125.50 \\ \hline \$933.10 \end{array} \text{ total cash}$$

Second, find the amount of the \$933.10 that was from purchases (not including sales tax).
Let x = money from purchases (not tax)
Note: The sales tax on the purchases is
8.5% of purchases $= 0.085 \cdot x$

Purchase price + Sales tax = Total cash

$$x \quad + 0.085x = \$933.10$$

$$1x + 0.085x = 933.10$$
$$1.085x = 933.10$$
$$\frac{\cancel{1.085}x}{\cancel{1.085}} = \frac{933.10}{1.085}$$
$$x = \frac{933.10}{1.085}$$
$$x = 860$$

The purchases totaled \$860. Since each item in the store sells for \$1, 860 items were sold.

Review Problems

25. $\frac{1}{3} \div \frac{2}{3} = \frac{1}{3} \cdot \frac{3}{2} = \frac{1 \cdot \cancel{3}}{\cancel{3} \cdot 2} = \frac{1}{2}$

27. $2 \div \frac{3}{4} = 2 \cdot \frac{4}{3} = \frac{2 \cdot 4}{3} = \frac{8}{3}$ or $2\frac{2}{3}$

29. $\frac{3}{8} \div \frac{1}{4} = \frac{3}{8} \cdot \frac{4}{1} = \frac{3 \cdot \cancel{4}}{2 \cdot \cancel{4} \cdot 1} = \frac{3}{2}$ or $1\frac{1}{2}$

31. $2\frac{1}{4} \div \frac{1}{2} = \frac{9}{4} \cdot \frac{2}{1} = \frac{9 \cdot \cancel{2}}{\cancel{2} \cdot 2 \cdot 1} = \frac{9}{2} = 4\frac{1}{2}$

Extending the Concepts: Luxury Taxes

33. Sales tax:
What is 6% of \$53,000?
$n = 0.06 \cdot 53{,}000$
$n = \$3{,}180$
To find luxury tax:

Purchase Price:	\$53,000
−\$30,000:	−30,000
Amount over \$30,000:	\$23,000

Luxury tax:
What is 10% of \$23,000?
$n = 0.1 \cdot 23{,}000$
$n = \$2{,}300$

35. You would have saved the luxury tax of \$2,300

37.

$$\begin{array}{rr} \text{Sticker Price:} & \$31,500 \\ -\ \text{Re duced Price:} & -\ 29,900 \\ \hline \text{Sticker savings:} & \$1,600 \end{array}$$

Luxury Tax savings:

$$\begin{array}{rr} \text{Purchase Price:} & \$31,500 \\ -\ \$30,000: & -\ 30,0000 \\ \hline \text{Amount over \$30,000:} & \$1,500 \end{array}$$

Luxury tax:

What is 10% of $1,500

$n = 0.1 \cdot 1,500$

$n = \$150$

You save $1,600 on the sticker price and $150 in luxury tax. If you live in a state with a 6% sales tax, you saved an additional $96, as shown below:

Sales Tax savings:

What is 6% of $1,600

$n = 0.06 \cdot 1,600$

$n = \$96$

Section 7.5

1. Increase:
What is 7% of $23,000?
$n = 0.07 \cdot 23,000$
$n = \$1,610$

Old Salary:	$23,000
+ Increase:	+1,610
New Salary:	$24,610

3. Increase:
What is 17% of $3,000?
$n = 0.17 \cdot 3,000$
$n = \$510$

Present tuition:	$3,000
+ Increase:	+510
New tuition:	$3,510

5. Decrease:
What is 20% of $16,500?
$n = 0.2 \cdot 16,500$
$n = \$3,300$

New price:	$16,500
– Decrease:	–3,300
Value after 1 year:	$13,200

7. 350 is what percent of 3,500?
$350 = n \cdot 3,500$

$$\frac{350}{3,500} = \frac{n \cdot \cancel{3,500}}{\cancel{3,500}}$$

$$n = \frac{350}{3,500}$$

$n = 0.1 = 10\%$

9. Decrease = $25 – $20 = $5
Percent Decrease:
$5 is what percent of $25?
$5 = n \cdot 25$

$$\frac{5}{25} = \frac{n \cdot \cancel{25}}{\cancel{25}}$$

$$n = \frac{5}{25}$$

$n = 0.2 = 20\%$ decrease

11. a. Increase = $39.51 – $25.97 = $13.54
Percent Increase:
$13.54 is what percent of $25.97?
$13,54 = n \cdot 25.97$

$$\frac{13.54}{25.97} = \frac{n \cdot \cancel{25.97}}{\cancel{25.97}}$$

$$n = \frac{13.54}{25.97}$$

$n \approx 0.5214 \approx 52.1\%$ increase

b. Increase = $39.51 – $36.39 = $3.12
Percent Increase:
$3.12 is what percent of $36.39?
$3.12 = n \cdot 36.39$

$$\frac{3.12}{36.39} = \frac{n \cdot \cancel{36.39}}{\cancel{36.39}}$$

$$n = \frac{3.12}{36.39}$$

$n \approx 0.0857 \approx 8.6\%$ increase

13. Discount:
What is 15% of $300?
$n = 0.15 \cdot 300$
$n = \$45$
The discount is $45.

Original price:	$300
– Discount:	– 45
Discount Price:	$255

15. Discount:

What is 20% of \$450?

$n = 0.20 \cdot 450$

$n = \$90$

Original price:	\$450
− Discount:	− 90
Sale Price:	\$360

Sales tax:

What is 6% of \$360?

$n = 0.06 \cdot 360$

$n = \$21.60$

Sale price:	\$360.00
+ Sales tax:	+ 21.60
Total bill:	\$381.60

Calculator Problems

17. Increase:

What is 6.5% of \$43,752?

$n = 0.065 \cdot 43,752$

$n = \$2,843.88$

Old Salary:	\$43,752.00
+ Increase:	+ 2,843.88
New Salary:	\$46,595.88

19. Area of smallest field:

$$A = lw = (100 \text{ yd})(55 \text{ yd}) = 5,500 \text{ yd}^2$$

Area of Rose Bowl:

$$A = lw = (116 \text{ yd})(72 \text{ yd}) = 8,352 \text{ yd}^2$$

Area of largest field:

$$A = lw = (120 \text{ yd})(75 \text{ yd}) = 9,000 \text{ yd}^2$$

a. Increase $= 8,352 \text{ yd}^2 - 5,500 \text{ yd}^2$

$= 2,852 \text{ yd}^2$

Percent Increase:

$2,852 \text{ yd}^2$ is what percent of $5,500 \text{ yd}^2$?

$$2,852 = n \cdot 5,500$$

$$\frac{2,852}{5,500} = \frac{n \cdot \cancel{5,500}}{\cancel{5,500}}$$

$$n = \frac{2,852}{5,500}$$

$$n \approx 0.5185 \approx 51.9\% \text{ increase}$$

b. Increase $= 9,000 \text{ yd}^2 - 8,352 \text{ yd}^2$

$= 648 \text{ yd}^2$

Percent Increase:

648 yd^2 is what percent of $8,352 \text{ yd}^2$?

$$648 = n \cdot 8,352$$

$$\frac{648}{8,352} = \frac{n \cdot \cancel{8,352}}{\cancel{8,352}}$$

$$n = \frac{648}{8,352}$$

$$n \approx 0.0776 \approx 7.8\% \text{ increase}$$

Review Problems

21. $\frac{1}{3} + \frac{2}{3} = \frac{1+2}{3} = \frac{3}{3} = 1$

23. $\frac{1}{2} + \frac{1}{4} = \frac{1 \cdot 2}{2 \cdot 2} + \frac{1}{4} = \frac{2}{4} + \frac{1}{4} = \frac{2+1}{4} = \frac{3}{4}$

25. $\frac{3}{4} + \frac{2}{3} = \frac{3 \cdot 3}{4 \cdot 3} + \frac{2 \cdot 4}{3 \cdot 4} = \frac{9}{12} + \frac{8}{12} = \frac{17}{12} = 1\frac{5}{12}$

27.

$$\begin{array}{r} 2\frac{1}{2} \\ +\ 3\frac{1}{2} \\ \hline 5\frac{2}{2} = 5+1=6 \end{array}$$

Improving Your Quantitative Literacy

29. a. Increase = $5{,}458 - 517 = 4{,}941$

Percent Increase:

4,941 is what percent of 517?

$$4{,}941 = n \cdot 517$$

$$\frac{4{,}941}{517} = \frac{n \cdot \cancel{517}}{\cancel{517}}$$

$$n = \frac{4{,}941}{517}$$

$$n \approx 9.55705 \approx 956\% \text{ increase}$$

b. Decrease $= \$1{,}533.85 - \$359.00$

$= \$1{,}174.85$

Percent decrease:

\$1,174.85 is what percent of \$1,533.85?

$$1{,}174.85 = n \cdot 1{,}533.85$$

$$\frac{1{,}174.85}{1{,}533.85} = \frac{n \cdot \cancel{1{,}533.85}}{\cancel{1{,}533.85}}$$

$$n = \frac{1{,}174.85}{1{,}533.85}$$

$$n \approx 0.76594 \approx 76.59\% \text{ decrease}$$

c. No. A 100% decrease in price would mean the camcorders are free.

Section 7.6

1. $I = P \times R \times T$

$= \$2{,}000 \times 0.08 \times 1$

$= \$160$

Original Principal:	\$2,000
+ Interest:	+160
Total amount after 1 year:	\$2,160

3. $I = P \times R \times T$

$= \$9{,}500 \times 0.07 \times 1$

$= \$665$

5. $I = P \times R \times T$

$= \$8{,}000 \times 0.07 \times 1$

$= \$560$

Amount borrowed:	\$8,000
+ Interest:	+560
Total amount to pay back:	\$8,560

7. $I = P \times R \times T$

$= \$2{,}000 \times 0.08 \times 1$

$= \$160$

Amount borrowed:	\$2,000
+ Interest:	+160
Total amount to pay back:	\$2,160

9. $I = P \times R \times T$

$= \$600 \times 0.05 \times \frac{60}{360}$

$= \$600 \times 0.05 \times \frac{1}{6}$

$= \$5$

11. $I = P \times R \times T$

$= \$800 \times 0.05 \times \frac{120}{360}$

$= \$800 \times 0.05 \times \frac{1}{3}$

$= \$13.33$ (to nearest cent)

Original Principal:	\$800.00
+ Interest:	+ 13.33
Total amount to withdraw:	\$813.33

13. Interest after 1st year:

$I = P \times R \times T$

$= \$5{,}000 \times 0.06 \times 1$

$= \$300$

Original Principal:	\$5,000
+ Interest:	+300
Total amount after 1 year:	\$5,300

Interest after 2nd year:

$I = P \times R \times T$

$= \$5{,}300 \times 0.06 \times 1$

$= \$318$

Amount at start of 2nd year:	\$5,300
+ Interest:	+318
Total amount after 2 years:	\$5,618

15. 1st Quarter:

$$\begin{aligned} I &= P \times R \times T \\ &= \$8{,}000 \times 0.05 \times \frac{1}{4} \\ &= \$100 \end{aligned}$$

Original principal:	\$8,000
+ Interest:	+100
New principal:	\$8,100

2nd quarter:

$$\begin{aligned} I &= P \times R \times T \\ &= \$8{,}100 \times 0.05 \times \frac{1}{4} \\ &= \$101.25 \end{aligned}$$

Original principal:	\$8,100.00
+ Interest:	+101.25
New principal:	\$8,201.25

3rd Quarter:

$$\begin{aligned} I &= P \times R \times T \\ &= \$8{,}201.25 \times 0.05 \times \frac{1}{4} \\ &= \$102.52 \end{aligned}$$

Original principal:	\$8,201.25
+ Interest:	+102.52
New principal:	\$8,303.77

4th Quarter:

$$\begin{aligned} I &= P \times R \times T \\ &= \$8{,}303.77 \times 0.05 \times \frac{1}{4} \\ &= \$103.80 \end{aligned}$$

Original principal:	\$8,303.77
+ Interest:	+103.80
New principal:	\$8,407.57

Calculator Problems

17.

$$\begin{aligned} I &= P \times R \times T \\ &= \$917.26 \times 0.0625 \times 1 \\ &= \$57.33 \end{aligned}$$

Original Principal:	\$917.26
+ Interest:	+ 57.33
Total amount after 1 year:	\$974.59

19. a.

$$\begin{aligned} A &= P\left(1+\frac{r}{n}\right)^{n \cdot t} \\ &= 10{,}000\left(1+\frac{0.06}{4}\right)^{4 \cdot 5} \\ &= 10{,}000(1+0.015)^{20} \\ &= 10{,}000(1.015)^{20} \\ &= 10{,}000(1.346855) \\ &= \$13{,}468.55 \end{aligned}$$

b.

$$\begin{aligned} A &= P\left(1+\frac{r}{n}\right)^{n \cdot t} \\ &= 10{,}000\left(1+\frac{0.06}{12}\right)^{12 \cdot 5} \\ &= 10{,}000(1+0.005)^{60} \\ &= 10{,}000(1.005)^{60} \\ &= 10{,}000(1.3488501) \\ &= \$13{,}488.50 \end{aligned}$$

c.

$$\begin{aligned} A &= P\left(1+\frac{r}{n}\right)^{n \cdot t} \\ &= 10{,}000\left(1+\frac{0.05}{4}\right)^{4 \cdot 5} \\ &= 10{,}000(1+0.0125)^{20} \\ &= 10{,}000(1.0125)^{20} \\ &= 10{,}000(1.282037232) \\ &= \$12{,}820.37 \end{aligned}$$

19. d. $A = P\left(1+\dfrac{r}{n}\right)^{n\cdot t}$

$$= 10{,}000\left(1+\frac{0.05}{12}\right)^{12\cdot 5}$$
$$= 10{,}000(1+0.004166667)^{60}$$
$$= 10{,}000(1.00416666667)^{60}$$
$$= 10{,}000(1.283358679)$$
$$= \$12{,}833.59$$

Review Problems

21. $\dfrac{3}{4}-\dfrac{1}{4}=\dfrac{3-1}{4}=\dfrac{2}{4}=\dfrac{1}{2}$

23. $\dfrac{5}{8}-\dfrac{1}{4}=\dfrac{5}{8}-\dfrac{1\cdot 2}{4\cdot 2}=\dfrac{5}{8}-\dfrac{2}{8}=\dfrac{5-2}{8}=\dfrac{3}{8}$

25. $\dfrac{1}{3}-\dfrac{1}{4}=\dfrac{1\cdot 4}{3\cdot 4}-\dfrac{1\cdot 3}{4\cdot 3}=\dfrac{4}{12}-\dfrac{3}{12}=\dfrac{4-3}{12}=\dfrac{1}{12}$

27.
$$\begin{array}{r} 3\frac{1}{4} \\ -\ 2 \\ \hline 1\frac{1}{4} \end{array}$$

Extending the Concepts

29. Star Wars to The Empire Strikes Back:
Increase $= 18-11=7$
Percent Increase:
7 million is what percent of 11 million?
$$7 = n\cdot 11$$
$$\frac{7}{11}=\frac{n\cdot \cancel{11}}{\cancel{11}}$$
$$n=\frac{7}{11}$$
$n = 0.636 = 63.6\%$ (to nearest tenth)

The Empire Strikes Back to Return of the Jedi:
Increase $= 32.5-18=14.5$
Percent Increase:
14.5 million is what percent of 18 million?
$$14.5 = n\cdot 18$$
$$\frac{14.5}{18}=\frac{n\cdot \cancel{18}}{\cancel{18}}$$
$$n=\frac{14.5}{18}$$
$n = 0.8055 = 80.6\%$ (to nearest tenth)

Return of the Jedi to The Phantom Menace:
Increase $= 115-32.5=82.5$
Percent Increase:
82.5 million is what percent of 32.5 million?
$$82.5 = n\cdot 32.5$$
$$\frac{82.5}{32.5}=\frac{n\cdot \cancel{32.5}}{\cancel{32.5}}$$
$$n=\frac{82.5}{32.5}$$
$n = 2.5384 = 253.8\%$ (to nearest tenth)

Section 7.7

1. Note: each section of pie is worth 5%

Number of A's:

$\frac{5}{20} = 0.25 = 25\% = 5$ sections of the circle

Number of B's:

$\frac{8}{20} = 0.40 = 40\% = 8$ sections of the circle

Number of C's:

$\frac{7}{20} = 0.35 = 35\% = 7$ sections

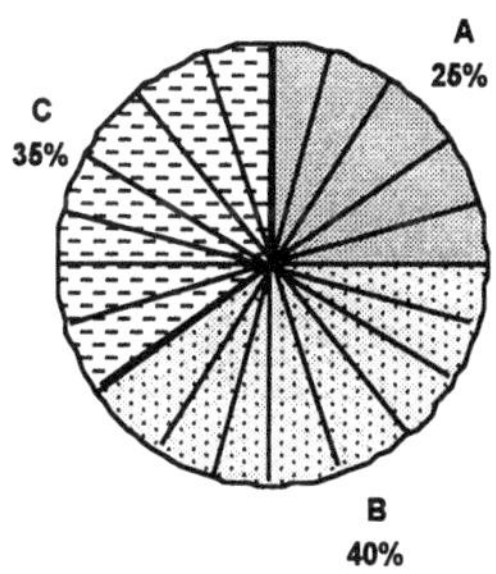

3. Note: each section of pie is worth 5%

Kitchen:

$\frac{400}{2{,}400} \approx 0.17 = 17\% =$ just over 3 sections

Dining Room:

$\frac{310}{2{,}400} \approx 0.13 = 13\% =$ just over 2 sections

Bedrooms:

$\frac{890}{2{,}400} \approx 0.37 = 37\% =$ just over 5 sections

Living Room:

$\frac{600}{2{,}400} = 0.25 = 25\% =$ 5 sections

Bathrooms:

$\frac{200}{2{,}400} = 0.08 = 8\% =$ just over 1 section

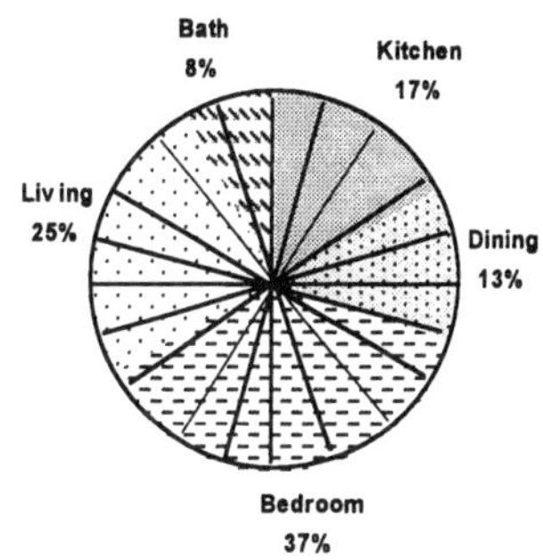

Review Problems

5. $8 \cdot \frac{1}{3} = \frac{8}{1} \cdot \frac{1}{3} = \frac{8 \cdot 1}{1 \cdot 3} = \frac{8}{3} = 2\frac{2}{3}$

7. $25 \cdot \frac{1}{1{,}000} = \frac{25}{1} \cdot \frac{1}{1{,}000} = \frac{\cancel{25} \cdot 1}{1 \cdot \cancel{25} \cdot 40} = \frac{1}{40}$

9. $36.5\cdot\frac{1}{100}\cdot 10 = 36\frac{1}{2}\cdot\frac{1}{100}\cdot 10$

$$= \frac{73}{2}\cdot\frac{1}{100}\cdot\frac{10}{1}$$

$$= \frac{73\cdot 1\cdot \cancel{10}}{2\cdot 10\cdot \cancel{10}}$$

$$= \frac{73}{20} \text{ or } 3\frac{13}{20}$$

11. $248\cdot\frac{1}{10}\cdot\frac{1}{10} = \frac{248}{10\cdot 10} = \frac{248}{100} = \frac{62}{25} = 2\frac{12}{25}$

13. $48\cdot\frac{1}{12}\cdot\frac{1}{3} = \frac{48\cdot 1\cdot 1}{12\cdot 3} = \frac{\cancel{12}\cdot 4}{\cancel{12}\cdot 3} = \frac{4}{3} = 1\frac{1}{3}$

Chapter 7 Review

1. $35\% = \frac{35}{100} = 0.35$

3. $5\% = \frac{5}{100} = 0.05$

5. Move the decimal point 2 places right and attach the % sign:
$0.95 = 95.0\% = 95\%$

7. Move the decimal point 2 places right and attach the % sign:
$0.495 = 49.5\%$

9. $75\% = \frac{75}{100} = \frac{3}{4}$

11. $145\% = \frac{145}{100} = \frac{29}{20} = 1\frac{9}{20}$

13. $\frac{3}{10} = \frac{3 \cdot 10}{10 \cdot 10} = \frac{30}{100} = 30\%$

15.
$$\begin{array}{r} 0.666 \\ 3\overline{)2.000} \\ \underline{18\downarrow} \\ 20 \\ \underline{18\downarrow} \\ 20 \\ \underline{18} \\ 2 \end{array}$$

$\frac{2}{3} = 0.666... = 66\frac{2}{3}\%$

17. What number is 60% of 28?
$n = 0.60 \cdot 28$
$n = 16.8$

19. 24 is 30% of what number?
$24 = 0.30 \cdot n$
$\frac{24}{0.30} = \frac{\cancel{0.30} \cdot n}{\cancel{0.30}}$
$\frac{24}{0.30} = n$
$80 = n$

21. a. What is 4.9% of 21,000 people?
$n = 0.049 \cdot 21{,}000$
$n = 1029$ people

b. What is 10.1% of 21,000 people?
$n = 0.101 \cdot 21{,}000$
$n = 2{,}121$ people

c. What is 74.1% of 280 million people?
$n = 0.741 \cdot 280{,}000{,}000$
$n = 207{,}480{,}000$ people

23. Discount:
What is 25% of $600?
$n = 0.25 \cdot 600$
$n = \$150$

Original price:	$600
− Discount:	−150
Sale Price:	$450

Sales tax:
What is 6% of $450?
$n = 0.06 \cdot 450$
$n = \$27$

Sale price:	$450
+ Sales tax:	+ 27
Total price:	$477

25. Increase:
What is 16% of $1.25?

$n = 0.16 \cdot 1.25$

$n = \$0.20$

Old price:	$1.25
+ increase:	+ 0.20
New price	$1.45

27. Increase = $906 − $583 = $323
Percent Increase:
$323 is what percent of $583?

$$\frac{323}{583} = \frac{n \cdot 583}{583}$$

$$n = \frac{323}{583} = 0.554 = 55.4\%$$

(to the nearest tenth of a percent)

29.
$$I = P \times R \times T$$
$$= \$1,800 \times 0.07 \times \frac{120}{360}$$
$$= \$1,800 \times 0.07 \times \frac{1}{3}$$
$$= \$42$$

Chapters 1-7 Cumulative Review

1.
$$\begin{array}{r} \overset{1\,1}{3,261} \\ 1,018 \\ +\ 297 \\ \hline 4,576 \end{array}$$

3.
$$\begin{array}{r} 240 \\ \times\ 21 \\ \hline 240 \\ 4800 \\ \hline 5,040 \end{array}$$

5. $6.31\overline{)265.02} \rightarrow 631\overline{)26502}$

$$\begin{array}{r} 42 \\ 631\overline{)26502} \\ 2524\downarrow \\ \hline 1262 \\ 1262 \\ \hline 0 \end{array}$$

7. $-\frac{2}{7}\cdot\frac{3}{8} = -\frac{\cancel{2}\cdot 3}{7\cdot\cancel{2}\cdot 4} = -\frac{3}{28}$

9. $5\cdot 3\frac{1}{2} = 5\cdot\frac{7}{2} = \frac{5}{1}\cdot\frac{7}{2} = \frac{35}{2} = 17\frac{1}{2}$

11.
$$\begin{array}{r} 2.01 \\ \times\ 4.8 \\ \hline 1608 \\ 8040 \\ \hline 9.648 \end{array}$$
$-4.8(2.01) = -9.648$

13. $\frac{9}{28} \div \frac{3}{16} = \frac{9}{28}\cdot\frac{16}{3} = \frac{\cancel{3}\cdot 3\cdot\cancel{4}\cdot 4}{\cancel{4}\cdot 7\cdot\cancel{3}} = \frac{12}{7}$ or $1\frac{5}{7}$

15. $6.237-(-1.38) = 6.237+1.38 = 7.617$

17. $3\sqrt{100}-\sqrt{81} = 3\cdot 10-9 = 30-9 = 21$

19.
$$\begin{aligned} 3(7x+4)-12 &= 3\cdot 7x+3\cdot 4-12 \\ &= 21x+12-12 \\ &= 21x \end{aligned}$$

21.
$$\frac{x}{20} = \frac{5}{4}$$
$$4x = 100$$
$$\frac{4x}{4} = \frac{100}{4}$$
$$x = 25$$

23.

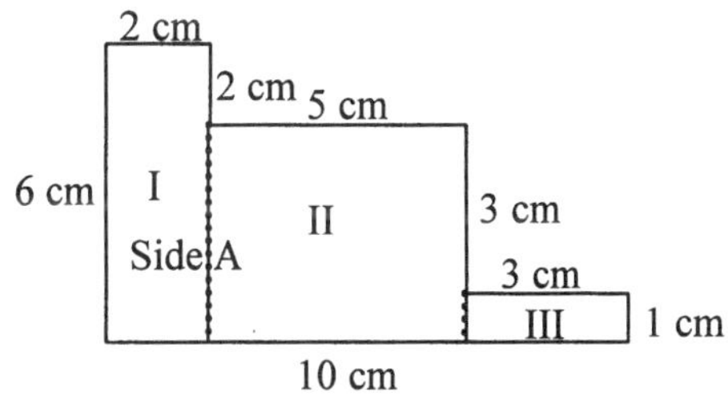

$P =$ sum of all sides

$= 2+2+5+3+3+1+10+6 = 32$ cm

To find area, split into 3 regions.

Area I $= (6\text{ cm})(2\text{ cm}) = 12\text{ cm}^2$

Note: Side A $= 3\text{ cm}+1\text{ cm} = 4\text{ cm}$

Area II $= (5\text{ cm})(4\text{ cm}) = 20\text{ cm}^2$

Area III $= (3\text{ cm})(1\text{ cm}) = 3\text{ cm}^2$

Total area $= 12+20+3 = 35\text{ cm}^2$

25.
$$\begin{aligned} 264 &= 2\cdot 132 \\ &= 2\cdot 2\cdot 66 = 2\cdot 2\cdot 2\cdot 33 \\ &= 2\cdot 2\cdot 2\cdot 3\cdot 11 \end{aligned}$$

27. $\frac{25}{20} = \frac{\cancel{5}\cdot 5}{\cancel{5}\cdot 4} = \frac{5}{4}$

29. $3.5 = 3.5\times 100\% = 350\%$

31. What number is 4% of 15?

$x = 0.04(15)$

$x = 0.6$

33. Discount: What is 15% of \$20?

$$x = 0.15(20)$$
$$x = \$3$$

$$\text{Sale price} = \text{Sticker price} - \text{discount}$$
$$= \$20 - \$3$$
$$= \$17$$

35. $\text{Total sales} = 32\ \cancel{\text{cups}} \cdot \dfrac{75¢}{\cancel{\text{cups}}}$

$$= 2400¢ \text{ or } \$24.00$$

37. 26.5 cm is $\frac{1}{8}$ of what length?

$$26.5 = \frac{1}{18} \cdot x$$
$$18(26.5) = 18\left(\frac{1}{18} \cdot x\right)$$
$$477 = x$$

The Buick is 477 cm long.

39. Let x = actual height of the tower

$$\frac{1 \text{ in model}}{29 \text{ in actual}} = \frac{15\frac{3}{4} \text{ in}}{x \text{ in actual}}$$
$$x = 29(15.75)$$
$$x = 456.75 \text{ in}$$

Chapter 7 Test

1. $18\% = \frac{18}{100} = 0.18$

3. $0.5\% = \frac{0.5}{100} = 0.005$

5. Move decimal point 2 places right and attach % sign:
$0.7 = 70.0\% = 70\%$

7. $65\% = \frac{65}{100} = \frac{13}{20}$

9. $3.5\% = \frac{3.5}{100} = \frac{3.5\times 10}{100\times 10} = \frac{35}{1,000} = \frac{7}{200}$

11. $8\overline{)3.000}$ with quotient 0.375

$$\begin{array}{r} 24\downarrow \\ \hline 60 \\ 56\downarrow \\ \hline 40 \\ 40 \\ \hline 0 \end{array}$$

$\frac{3}{8} = 0.375 = 37.5\%$

13. What number is 75% of 60?
$n = 0.75 \cdot 60$
$n = 45$

15. 16 is 20% of what number?
$16 = 0.20 \cdot n$
$\frac{16}{0.20} = \frac{\cancel{0.20} \cdot n}{\cancel{0.20}}$
$\frac{16}{0.20} = n$
$80 = n$

17. What is 8% of \$12,000?
$n = 0.08 \cdot 12,000$
$n = \$960$

19. Discount:
What is 25% of \$280?
$n = 0.25 \cdot 280$
$n = \$70$

Original price:	\$280
− Discount:	−70
Sale Price:	\$210

Sales tax:
What is 5% of \$210?
$n = 0.05 \cdot 210$
$n = \$10.50$

Sale price:	\$210.00
+ Sales tax:	+ 10.50
Total price:	\$220.50

21. At the end of 1st year:
$I = P \times R \times T$
$= \$12,000 \times 0.1 \times 1$
$= \$1,200$

Original Principal:	\$12,000
+ Interest:	+1,200
Total amount after 1 year:	\$13,200

At the end of 2nd year:
$I = P \times R \times T$
$= \$13,200 \times 0.1 \times 1$
$= \$1,320$

Interest from 1st year:	\$1,200
+ Interest from 2nd year:	+1,320
Total interest:	\$2,520

Chapter 8 Preview

1. $\frac{12}{30} = \frac{\not{6} \cdot 2}{\not{6} \cdot 5} = \frac{2}{5}$

3.
$$\begin{array}{r} 12 \\ \times\ 16 \\ \hline 72 \\ \underline{120} \\ 192 \end{array}$$

5.
$$\begin{array}{r} 43{,}560 \\ \times\ 75 \\ \hline 217{,}800 \\ 3{,}049{,}200 \\ \hline 3{,}267{,}000 \end{array}$$

7.
$$\begin{array}{r} 2.49 \\ \times\ 3.75 \\ \hline 1{,}245 \\ 17{,}430 \\ 74{,}700 \\ \hline 9.3375 \end{array}$$

9. $8 \times \frac{1}{3} = \frac{8}{1} \cdot \frac{1}{3} = \frac{8 \cdot 1}{1 \cdot 3} = \frac{8}{3} = 2\frac{2}{3}$

11. $\frac{1800}{4} = \frac{450}{1} = 450$

13. $\frac{80.5}{1.61} = \frac{80.5 \times 100}{1.61 \times 100} = \frac{8{,}050}{161} = \frac{50}{1} = 50$

15. $\frac{1{,}100 \times 60 \times 60}{5{,}280} = \frac{3{,}960000}{5{,}280} = 750$

17. $\frac{2 \cdot 1{,}000}{16.39} = \frac{2{,}000}{16.39} \approx 122$

19.
$$\begin{array}{r} 0.75 \\ 16\overline{)12.00} \\ \underline{112\downarrow} \\ 80 \\ \underline{80} \\ 0 \end{array}$$

Chapter 8 Pretest

1. $$\begin{aligned} 8 \text{ ft} &= 8 \cancel{\text{ft}} \times \frac{12 \text{ in}}{1 \cancel{\text{ft}}} \\ &= 8 \times 12 \text{ in} \\ &= 96 \text{ in} \end{aligned}$$

3. $$\begin{aligned} 32 \text{ m} &= 32 \cancel{\text{m}} \times \frac{100 \text{ cm}}{1 \cancel{\text{m}}} \\ &= 32 \times 100 \text{ cm} \\ &= 3,200 \text{ cm} \end{aligned}$$

5. $$\begin{aligned} 30 \text{ yd}^2 &= 30 \cancel{\text{yd}^2} \times \frac{9 \text{ ft}^2}{1 \cancel{\text{yd}^2}} \\ &= 30 \times 9 \text{ ft}^2 \\ &= 270 \text{ ft}^2 \end{aligned}$$

7. $$\begin{aligned} 3,840 \text{ acres} &= 3,840 \cancel{\text{acres}} \times \frac{1 \text{ mi}^2}{640 \cancel{\text{acres}}} \\ &= \frac{3,840}{640} \text{mi}^2 \\ &= 6 \text{ mi}^2 \end{aligned}$$

9. $$\begin{aligned} 3 \text{ gal} &= 3 \cancel{\text{gal}} \times \frac{4 \text{ qt}}{1 \cancel{\text{gal}}} \\ &= 3 \times 4 \text{ qt} \\ &= 12 \text{ qt} \end{aligned}$$

11. $$\begin{aligned} 251 \text{ mL} &= 251 \cancel{\text{mL}} \times \frac{1 \text{ L}}{1,000 \cancel{\text{mL}}} \\ &= \frac{251}{1,000} \text{L} \\ &= 0.251 \text{ L} \end{aligned}$$

13. $$\begin{aligned} 2,142 \text{ mg} &= 2,142 \cancel{\text{mg}} \times \frac{1 \text{ g}}{1,000 \cancel{\text{mg}}} \\ &= \frac{2,142}{1,000} \text{ g} \\ &= 2.142 \text{ g} \end{aligned}$$

15. $$\begin{aligned} 3 \text{ gal} &= 3 \cancel{\text{gal}} \times \frac{3.79 \text{ L}}{1 \cancel{\text{gal}}} \\ &= 3 \times 3.79 \text{ L} \\ &= 11.37 \text{ L} \end{aligned}$$

17. 45 miles per hour
$$\begin{aligned} &= 45 \frac{\cancel{\text{mi}}}{\cancel{\text{hr}}} \times \frac{5,280 \text{ ft}}{1 \cancel{\text{mi}}} \times \frac{1 \cancel{\text{hr}}}{60 \cancel{\text{min}}} \times \frac{1 \cancel{\text{min}}}{60 \text{ sec}} \\ &= \frac{45 \times 5,280}{60 \times 60} \frac{\text{ft}}{\text{sec}} \\ &= 66 \frac{\text{ft}}{\text{sec}} \end{aligned}$$

19. $$\begin{aligned} 5 \text{ gal} &= 5 \cancel{\text{gal}} \times \frac{128 \cancel{\text{fl oz}}}{1 \cancel{\text{gal}}} \times \frac{1 \text{ glass}}{8 \cancel{\text{fl oz}}} \\ &= \frac{5 \times 128}{8} \text{ glasses} \\ &= 80 \text{ glasses} \end{aligned}$$

Section 8.1

1. $5 \text{ ft} = 5 \cancel{\text{ft}} \times \frac{12 \text{ in}}{1 \cancel{\text{ft}}}$

$= 5 \times 12 \text{ in}$

$= 60 \text{ in}$

3. $10 \text{ ft} = 10 \cancel{\text{ft}} \times \frac{12 \text{ in}}{1 \cancel{\text{ft}}}$

$= 10 \times 12 \text{ in}$

$= 120 \text{ in}$

5. $2 \text{ yd} = 2 \cancel{\text{yd}} \times \frac{3 \text{ ft}}{1 \cancel{\text{yd}}}$

$= 2 \times 3 \text{ ft}$

$= 6 \text{ ft}$

7. $4.5 \text{ yd} = 4.5 \cancel{\text{yd}} \times \frac{3 \cancel{\text{ft}}}{1 \cancel{\text{yd}}} \times \frac{12 \text{ in}}{1 \cancel{\text{ft}}}$

$= 4.5 \times 3 \times 12 \text{ in}$

$= 162 \text{ in}$

9. $27 \text{ in} = 27 \cancel{\text{in}} \times \frac{1 \text{ ft}}{12 \cancel{\text{in}}}$

$= \frac{27}{12} \text{ ft}$

$= 2\frac{1}{4} \text{ ft}$

11. $19 \text{ ft} = 19 \cancel{\text{ft}} \times \frac{1 \text{ yd}}{3 \cancel{\text{ft}}}$

$= \frac{19}{3} \text{ yd}$

$= 6\frac{1}{3} \text{ yd}$

13. $48 \text{ in} = 48 \cancel{\text{in}} \times \frac{1 \cancel{\text{ft}}}{12 \cancel{\text{in}}} \times \frac{1 \text{ yd}}{3 \cancel{\text{ft}}}$

$= 48 \times \frac{1}{12} \times \frac{1}{3} \text{ yd}$

$= 4 \times \frac{1}{3} \text{ yd}$

$= \frac{4}{3} \text{ yd}$

$= 1\frac{1}{3} \text{ yd}$

15. $18 \text{ m} = 18 \cancel{\text{m}} \times \frac{100 \text{ cm}}{1 \cancel{\text{m}}}$

$= 18 \times 100 \text{ cm}$

$= 1{,}800 \text{ cm}$

17. $4.8 \text{ km} = 4.8 \cancel{\text{km}} \times \frac{1{,}000 \text{ m}}{1 \cancel{\text{km}}}$

$= 4.8 \times 1{,}000 \text{ m}$

$= 4{,}800 \text{ m}$

19. $5 \text{ dm} = 5 \cancel{\text{dm}} \times \frac{1 \cancel{\text{m}}}{10 \cancel{\text{dm}}} \times \frac{100 \text{ cm}}{1 \cancel{\text{m}}}$

$= \frac{5 \times 100}{10} \text{ cm}$

$= 50 \text{ cm}$

21. $248 \text{ m} = 248 \cancel{\text{m}} \times \frac{1 \text{ km}}{1{,}000 \cancel{\text{m}}}$

$= \frac{248}{1{,}000} \text{ km}$

$= 0.248 \text{ km}$

23. $67 \text{ cm} = 67 \cancel{\text{cm}} \times \frac{1 \cancel{\text{m}}}{100 \cancel{\text{cm}}} \times \frac{1{,}000 \text{ mm}}{1 \cancel{\text{m}}}$

$= \frac{67 \times 1{,}000}{100} \text{ mm}$

$= 670 \text{ mm}$

25. $3{,}498 \text{ cm} = 3{,}498 \text{ cm} \times \dfrac{1 \text{ m}}{100 \text{ cm}}$

$= \dfrac{3{,}498}{100} \text{m}$

$= 34.98 \text{ m}$

27. $63.4 \text{ cm} = 63.4 \text{ cm} \times \dfrac{1 \text{ m}}{100 \text{ cm}} \times \dfrac{10 \text{ dm}}{1 \text{ m}}$

$= \dfrac{63.4 \times 10}{100} \text{dm}$

$= 6.34 \text{ dm}$

Applying the Concepts

29. $60 \text{ ft} = 60 \text{ ft} \times \dfrac{1 \text{ yd}}{3 \text{ ft}}$

$= \dfrac{60}{3} \text{ yd}$

$= 20 \text{ yd}$

31. First convert 6 feet to inches:

$6 \text{ ft} = 6 \text{ ft} \times \dfrac{12 \text{ in}}{1 \text{ ft}}$

$= 6 \times 12 \text{ in}$

$= 72 \text{ in}$

6 feet 8 inches = 72 in + 8 in = 80 in

33. $2.44 \text{ m} = 2.44 \text{ m} \times \dfrac{100 \text{ cm}}{1 \text{ m}}$

$= 2.44 \times 100 \text{ cm}$

$= 244 \text{ cm}$

35. $6.5 \text{ cm} = 6.5 \text{ cm} \times \dfrac{1 \text{ m}}{100 \text{ cm}} \times \dfrac{1{,}000 \text{ mm}}{1 \text{ m}}$

$= \dfrac{6.5 \times 1{,}000}{100} \text{ mm}$

$= 65 \text{ mm}$

37. $37 \text{ mi} = 37 \text{ mi} \times \dfrac{80 \text{ chains}}{1 \text{ mi}}$

$= 37 \times 80 \text{ chains}$

$= 2{,}960 \text{ chains}$

39. 12 cm

$= 12 \text{ cm} \times \dfrac{1 \text{ m}}{100 \text{ cm}} \times \dfrac{1{,}000 \text{ mm}}{1 \text{ m}} \times \dfrac{1{,}000 \mu m}{1 \text{ mm}}$

$= \dfrac{12 \times 1{,}000 \times 1{,}000}{100} \mu\text{m}$

$= 120{,}000 \ \mu\text{m}$

41. 12 furlongs

$= 12 \text{ furlongs} \times \dfrac{220 \text{ yds}}{1 \text{ furlong}} \times \dfrac{3 \text{ ft}}{1 \text{ yd}}$

$= 12 \times 220 \times 3 \text{ ft}$

$= 7{,}920 \text{ ft}$

43. 55 miles/hour

$= 55 \dfrac{\text{mi}}{\text{hr}} \times \dfrac{5{,}280 \text{ft}}{1 \text{ mi}} \times \dfrac{1 \text{ hr}}{60 \text{ min}} \times \dfrac{1 \text{ min}}{60 \text{ sec}}$

$= \dfrac{55 \times 5{,}280 \text{ ft}}{60 \times 60 \text{ sec}}$

≈ 80.7 feet/second

45. 9.52 yards/second

$= 9.52 \dfrac{\text{yds}}{\text{sec}} \times \dfrac{3 \text{ ft}}{1 \text{ yd}} \times \dfrac{1 \text{ mi}}{5{,}280 \text{ ft}} \times \dfrac{60 \text{ sec}}{1 \text{ min}}$

$\times \dfrac{60 \text{ min}}{1 \text{ hr}}$

$= \dfrac{9.52 \times 3 \times 60 \times 60 \text{ mi}}{5{,}280 \text{ hr}}$

≈ 19.5 miles per hour

47. 1,500 feet/second

$$= 1{,}500\frac{\cancel{ft}}{\cancel{sec}} \times \frac{1\text{mi}}{5{,}280\ \cancel{ft}} \times \frac{60\ \cancel{sec}}{1\ \cancel{min}} \times \frac{60\ \cancel{min}}{1\text{ hr}}$$

$$= \frac{1{,}500 \times 60 \times 60\text{ mi}}{5{,}280\text{ hr}}$$

$$= 1{,}022.\overline{72}\text{ mi/hr}$$

$$\approx 1{,}023\text{ miles/hour}$$

49.

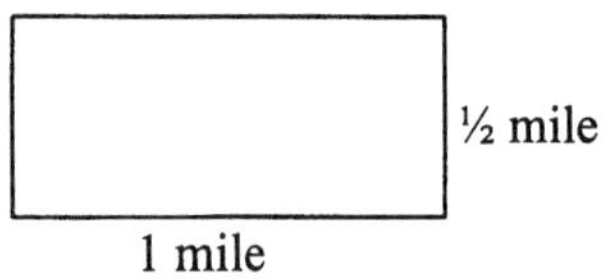

1st, find the perimeter:

Perimeter $= 2l + 2w$

$$= 2(1\text{ mi}) + 2\left(\frac{1}{2}\text{mi}\right)$$

$$= 2\text{ mi} + 1\text{ mi}$$

$$= 3\text{ mi}$$

Convert miles to dollars:

$$3\text{ miles} = 3\ \cancel{mi} \times \frac{5{,}280\ \cancel{ft}}{1\ \cancel{mi}} \times \frac{\$1.15}{1\ \cancel{ft}}$$

$$= 3 \times 5{,}280 \times 1.15\text{ dollars}$$

$$= \$18{,}216$$

It will cost $18,216.

51. $120\text{ days} = 120\ \cancel{days} \times \frac{1\ \cancel{bale}}{8\ \cancel{day}} \times \frac{\$10.50}{1\ \cancel{bale}}$

$$= \frac{120 \times 10.50\text{ dollars}}{8}$$

$$= \$157.50$$

Calculator Problems

53. $751\text{ miles} = 751\ \cancel{mi} \times \frac{5{,}280\text{ ft}}{1\ \cancel{mi}}$

$$= 751 \times 5{,}280\text{ ft}$$

$$= 3{,}965{,}280\text{ ft}$$

55. $4{,}982\text{ yd} = 4{,}982\ \cancel{yd} \times \frac{3\ \cancel{ft}}{1\ \cancel{yd}} \times \frac{12\text{ in}}{1\ \cancel{ft}}$

$$= 4{,}982 \times 3 \times 12\text{ in}$$

$$= 179{,}352\text{ in}$$

57. $14{,}494\text{ feet} = 14{,}494\ \cancel{ft} \times \frac{1\text{ mi}}{5{,}280\ \cancel{ft}}$

$$= \frac{14{,}494}{5{,}280}\text{ ft}$$

$$\approx 2.7\text{ mi}$$

59. $3{,}427\text{ miles} = 3{,}427\ \cancel{mi} \times \frac{5{,}280\text{ ft}}{1\ \cancel{mi}}$

$$= 3{,}427 \times 5{,}280\text{ ft}$$

$$= 18{,}094{,}560\text{ ft}$$

Review Problems

61. $\frac{2}{3} \cdot \frac{1}{2} = \frac{\cancel{2} \cdot 1}{3 \cdot \cancel{2}} = \frac{1}{3}$

63. $-8\left(-\frac{3}{4}\right) = \frac{2 \cdot \cancel{4} \cdot 3}{\cancel{4}} = 6$

65. $1\frac{1}{2} \cdot 2\frac{1}{3} = \frac{\cancel{3}}{2} \cdot \frac{7}{\cancel{3}} = \frac{7}{2} = 3\frac{1}{2}$

67. $\frac{3}{4} \div \frac{1}{8} = \frac{3}{4} \cdot \frac{8}{1} = \frac{3 \cdot 2 \cdot \cancel{4}}{\cancel{4} \cdot 1} = 6$

69. $-4 \div \left(-\frac{2}{3}\right) = 4 \cdot \frac{3}{2} = \frac{\cancel{2} \cdot 2 \cdot 3}{\cancel{2}} = 6$

71. $1\frac{3}{4} \div 2\frac{1}{2} = \frac{7}{4} \div \frac{5}{2} = \frac{7}{4} \cdot \frac{2}{5} = \frac{7 \cdot \cancel{2}}{\cancel{2} \cdot 2 \cdot 5} = \frac{7}{10}$

Improving Your Quantitative Literacy

73. 2,000 steps =

$$= 2{,}000 \ \cancel{\text{steps}} \times \frac{2.5 \ \cancel{\text{ft}}}{1 \ \cancel{\text{step}}} \times \frac{1 \text{ mi}}{5{,}280 \ \cancel{\text{ft}}}$$

$$= \frac{2{,}000 \times 2.5}{5{,}280} \text{ mi}$$

$$\approx 0.95 \text{ mi} \approx 1 \text{ mi}$$

10,000 steps =

$$= 10{,}000 \ \cancel{\text{steps}} \times \frac{2.5 \ \cancel{\text{ft}}}{1 \ \cancel{\text{step}}} \times \frac{1 \text{ mi}}{5{,}280 \ \cancel{\text{ft}}}$$

$$= \frac{10{,}000 \times 2.5}{5{,}280} \text{ mi}$$

$$\approx 4.73 \text{ mi} \approx 5 \text{ mi}$$

Both statements are correct.

Section 8.2

1. $3\text{ ft}^2 = 3\ \cancel{\text{ft}^2} \times \dfrac{144\text{ in}^2}{1\ \cancel{\text{ft}^2}}$

$= 3 \times 144\text{ in}^2$

$= 432\text{ in}^2$

3. $288\text{ in}^2 = 288\ \cancel{\text{in}^2} \times \dfrac{1\text{ ft}^2}{144\ \cancel{\text{in}^2}}$

$= \dfrac{288}{144}\text{ ft}^2$

$= 2\text{ ft}^2$

5. $30\text{ acres} = 30\ \cancel{\text{acres}} \times \dfrac{43{,}560\text{ ft}^2}{1\ \cancel{\text{acre}}}$

$= 30 \times 43{,}560\text{ ft}^2$

$= 1{,}306{,}800\text{ ft}^2$

7. $2\text{ mi}^2 = 2\ \cancel{\text{mi}^2} \times \dfrac{640\text{ acres}}{1\ \cancel{\text{mi}^2}}$

$= 2 \times 640\text{ acres}$

$= 1{,}280\text{ acres}$

9. $1{,}920\text{ acres} = 1{,}920\ \cancel{\text{acres}} \times \dfrac{1\text{ mi}^2}{640\ \cancel{\text{acres}}}$

$= \dfrac{1{,}920}{640}\text{mi}^2$

$= 3\text{ mi}^2$

11. $12\text{ yd}^2 = 12\ \cancel{\text{yd}^2} \times \dfrac{9\text{ ft}^2}{1\ \cancel{\text{yd}^2}}$

$= 12 \times 9\text{ ft}^2$

$= 108\text{ ft}^2$

13. $17\text{ cm}^2 = 17\ \cancel{\text{cm}^2} \times \dfrac{100\text{ mm}^2}{1\ \cancel{\text{cm}^2}}$

$= 17 \times 100\text{ mm}^2$

$= 1{,}700\text{ mm}^2$

15. $2.8\text{ m}^2 = 2.8\ \cancel{\text{m}^2} \times \dfrac{100\ \cancel{\text{dm}^2}}{1\ \cancel{\text{m}^2}} \times \dfrac{100\text{ cm}^2}{1\ \cancel{\text{dm}^2}}$

$= 2.8 \times 100 \times 100\text{ cm}^2$

$= 28{,}000\text{ cm}^2$

17. $1{,}200\text{ mm}^2$

$= 1{,}200\ \cancel{\text{mm}^2} \times \dfrac{1\ \cancel{\text{cm}^2}}{100\ \cancel{\text{mm}^2}} \times \dfrac{1\ \cancel{\text{dm}^2}}{100\ \cancel{\text{cm}^2}}$

$\times \dfrac{1\text{ m}^2}{100\ \cancel{\text{dm}^2}}$

$= \dfrac{1{,}200}{100 \times 100 \times 100}\text{m}^2$

$= 0.0012\text{ m}^2$

19. $5\text{ a} = 5\ \cancel{\text{a}} \times \dfrac{100\text{ m}^2}{1\ \cancel{\text{a}}}$

$= 5 \times 100\text{ m}^2$

$= 500\text{ m}^2$

21. $7\text{ ha} = 7\ \cancel{\text{ha}} \times \dfrac{100\text{ a}}{1\ \cancel{\text{ha}}}$

$= 7 \times 100\text{ a}$

$= 700\text{ a}$

23. $342\text{ a} = 342\ \cancel{\text{a}} \times \dfrac{1\text{ ha}}{100\ \cancel{\text{a}}}$

$= \dfrac{342}{100}\text{ ha}$

$= 3.42\text{ ha}$

Applying the Concepts

25. NFL field:

$$A = lw$$
$$= (100)\left(53\frac{1}{3}\right)$$
$$= 100\left(\frac{160}{3}\right)$$
$$= \frac{16,000}{3}$$
$$= 5,333\frac{1}{3}\text{ yd}^2;$$

$$5,333\frac{1}{3}\text{yd}^2$$
$$= 5,333\frac{1}{3}\cancel{\text{yd}^2} \times \frac{9\ \cancel{\text{ft}^2}}{1\ \cancel{\text{yd}^2}} \times \frac{1\text{ acre}}{43,560\ \cancel{\text{ft}^2}}$$
$$= \frac{5,333\frac{1}{3} \times 9}{43,560}\text{ acres}$$
$$\approx 1.10\text{ acres}$$

The NFL field is approximately 1.10 acres.

Canadian field:

$$A = lw = (110)(65) = 7,150\text{ yd}^2$$
$$7,150\text{ yd}^2 = 7,150\ \cancel{\text{yd}^2} \times \frac{9\ \cancel{\text{ft}^2}}{1\ \cancel{\text{yd}^2}} \times \frac{1\text{ acre}}{43,560\ \cancel{\text{ft}^2}}$$
$$= \frac{7,150 \times 9}{43,560}\text{ acres}$$
$$\approx 1.48\text{ acres}$$

The Canadian field is approximately 1.48 acres.

25. (continued)

Arena Football:

$$A = lw$$
$$= (50)\left(28\frac{1}{3}\right)$$
$$= 50\left(\frac{85}{3}\right)$$
$$= 1,416\frac{2}{3}\text{ yd}^2$$

$$1,416\frac{2}{3}\text{yd}^2$$
$$= 1,416\frac{2}{3}\cancel{\text{yd}^2} \times \frac{9\ \cancel{\text{ft}^2}}{1\ \cancel{\text{yd}^2}} \times \frac{1\text{ acre}}{43,560\ \cancel{\text{ft}^2}}$$
$$= \frac{1,416\frac{2}{3} \times 9}{43,560}\text{ acres}$$
$$\approx 0.29\text{ acres}$$

The Arena is approximately 0.29 acres.

27. $A = lw = (100\text{ m})(30\text{ m}) = 3,000\text{ m}^2$

$$3,000\text{ m}^2 = 3,000\ \cancel{\text{m}^2} \times \frac{1\text{ a}}{100\ \cancel{\text{m}^2}}$$
$$= \frac{3,000}{100}\text{a}$$
$$= 30\text{ a}$$

The area of the pool is 30 a.

29. Find the area of 1 brick:

$$A = (10\text{ cm})(20\text{ cm}) = 200\text{ cm}^2$$

Convert the area of the patio to bricks:

$$110\text{ m}^2$$
$$= 110\cancel{\text{m}^2} \times \frac{100\cancel{\text{dm}^2}}{1\cancel{\text{m}^2}} \times \frac{100\cancel{\text{cm}^2}}{1\cancel{\text{dm}^2}} \times \frac{1\text{ brick}}{200\cancel{\text{cm}^2}}$$
$$= \frac{110 \times 100 \times 100}{200}\text{ bricks}$$
$$= 5,500\text{ bricks}$$

It will take 5,500 bricks to make the patio.

31. $5\text{ yd}^3 = 5\ \cancel{\text{yd}^3} \times \dfrac{27\text{ ft}^3}{1\ \cancel{\text{yd}^3}}$

$= 5 \times 27\text{ ft}^3$

$= 135\text{ ft}^3$

33. $3\text{ pt} = 3\ \cancel{\text{pt}} \times \dfrac{16\text{ fl oz}}{1\ \cancel{\text{pt}}}$

$= 3 \times 16\text{ fl oz}$

$= 48\text{ fl oz}$

35. $2\text{ gal} = 2\ \cancel{\text{gal}} \times \dfrac{4\text{ qt}}{1\ \cancel{\text{gal}}}$

$= 2 \times 4\text{ qt}$

$= 8\text{ qt}$

37. $2.5\text{ gal} = 2.5\ \cancel{\text{gal}} \times \dfrac{4\ \cancel{\text{qt}}}{1\ \cancel{\text{gal}}} \times \dfrac{2\text{ pt}}{1\ \cancel{\text{qt}}}$

$= 2.5 \times 4 \times 2\text{ pt}$

$= 20\text{ pt}$

39. $15\text{ qt} = 15\ \cancel{\text{qt}} \times \dfrac{2\ \cancel{\text{pt}}}{1\ \cancel{\text{qt}}} \times \dfrac{16\text{ fl oz}}{1\ \cancel{\text{pt}}}$

$= 15 \times 2 \times 16\text{ fl oz}$

$= 480\text{ fl oz}$

41. $64\text{ pt} = 64\ \cancel{\text{pt}} \times \dfrac{1\ \cancel{\text{qt}}}{2\ \cancel{\text{pt}}} \times \dfrac{1\text{ gal}}{4\ \cancel{\text{qt}}}$

$= \dfrac{64}{2 \times 4}\text{ gal}$

$= 8\text{ gal}$

43. $12\text{ pt} = 12\ \cancel{\text{pt}} \times \dfrac{1\text{ qt}}{2\ \cancel{\text{pt}}}$

$= \dfrac{12}{2}\text{ qt}$

$= 6\text{ qt}$

45. $243\text{ ft}^3 = 243\ \cancel{\text{ft}^3} \times \dfrac{1\text{ yd}^3}{27\ \cancel{\text{ft}^3}}$

$= \dfrac{243}{27}\text{ yd}^3$

$= 9\text{ yd}^3$

47. $5\text{ L} = 5\ \cancel{\text{L}} \times \dfrac{1{,}000\text{ mL}}{1\ \cancel{\text{L}}}$

$= 5 \times 1{,}000\text{ mL}$

$= 5{,}000\text{ mL}$

49. $127\text{ mL} = 127\ \cancel{\text{mL}} \times \dfrac{1\text{ L}}{1{,}000\ \cancel{\text{mL}}}$

$= \dfrac{127}{1{,}000}\text{L}$

$= 0.127\text{ L}$

51. $4\text{ kL} = 4\ \cancel{\text{kL}} \times \dfrac{1{,}000\ \cancel{\text{L}}}{1\ \cancel{\text{kL}}} \times \dfrac{1{,}000\text{ mL}}{1\ \cancel{\text{L}}}$

$= 4 \times 1{,}000 \times 1{,}000\text{ mL}$

$= 4{,}000{,}000\text{ mL}$

53. $14.92\text{ kL} = 14.92\ \cancel{\text{kL}} \times \dfrac{1{,}000\text{ L}}{1\ \cancel{\text{kL}}}$

$= 14.92 \times 1{,}000\text{ L}$

$= 14{,}920\text{ L}$

Applying the Concepts

55. Convert 1 gallon to pints, then cups:

$1\text{ gal} = 1\text{ gal} \times \dfrac{4\text{ qt}}{1\text{ gal}} \times \dfrac{2\text{ pt}}{1\text{ qt}} \times \dfrac{1\text{ cup}}{\frac{1}{2}\text{ pt}}$

$= \dfrac{4 \times 2}{\frac{1}{2}}\text{ cups}$

$= 16\text{ cups}$

The coffee make can fill 16 cups.

57. $20 \text{ ft}^3 = 20 \cancel{\text{ft}^3} \times \frac{1{,}728 \text{ in}^3}{1 \cancel{\text{ft}^3}}$

$= 20 \times 1{,}728 \text{ in}^3$

$= 34{,}560 \text{ in}^3$

59. Convert gallons to ounces, then glasses:
3 gal

$= 3 \cancel{\text{gal}} \times \frac{4 \cancel{\text{qt}}}{1 \cancel{\text{gal}}} \times \frac{2 \cancel{\text{pt}}}{1 \cancel{\text{qt}}} \times \frac{16 \cancel{\text{fl oz}}}{1 \cancel{\text{pt}}} \times \frac{1 \text{ glass}}{8 \cancel{\text{fl oz}}}$

$= \frac{3 \times 4 \times 2 \times 16}{8}$ glasses

$= 48$ glasses

Calculator Problems

61. $31{,}700 \text{ mi}^2 = 31{,}700 \cancel{\text{mi}^2} \times \frac{640 \text{ acres}}{1 \cancel{\text{mi}^2}}$

$= 31{,}700 \times 640$ acres

$= 20{,}288{,}000$ acres

63. 2,067,795 acres

$= 2{,}067{,}795 \cancel{\text{acres}} \times \frac{1 \text{ mi}^2}{640 \cancel{\text{acres}}}$

$= \frac{2{,}067{,}795}{640} \text{ mi}^2$

$= 3{,}230.93 \text{ mi}^2$

65. $93.4 \text{ qt} = 93.4 \cancel{\text{qt}} \times \frac{1 \text{ gal}}{4 \cancel{\text{qt}}}$

$= \frac{93.4}{4}$ gal

$= 23.35$ gal

67. $796 \text{ yd}^3 = 796 \cancel{\text{yd}^3} \times \frac{27 \text{ ft}^3}{1 \cancel{\text{yd}^3}}$

$= 796 \times 27 \text{ ft}^3$

$= 21{,}492 \text{ ft}^3$

69. $10{,}585{,}000 \text{ yd}^3 = 10{,}585{,}000 \cancel{\text{yd}^3} \times \frac{27 \text{ ft}^3}{1 \cancel{\text{yd}^3}}$

$= 10.585{,}000 \times 27 \text{ ft}^3$

$= 285{,}795{,}000 \text{ ft}^3$

Review Problems

71. $\frac{3}{8} + \frac{1}{4} = \frac{3}{8} + \frac{1 \cdot 2}{4 \cdot 2} = \frac{3}{8} + \frac{2}{8} = \frac{3+2}{8} = \frac{5}{8}$

73.

$$\begin{array}{r} 3\frac{1}{2} \\ +\ 5\frac{1}{2} \\ \hline 8\frac{2}{2} = 8 + 1 = 9 \end{array}$$

75. $\frac{7}{15} - \frac{2}{15} = \frac{7-2}{15} = \frac{5}{15} = \frac{1}{3}$

77. $\left.\begin{array}{l} 36 = 2 \cdot 2 \cdot 3 \cdot 3 \\ 48 = 2 \cdot 2 \cdot 2 \cdot 2 \cdot 3 \end{array}\right\} \text{LCD} = 2 \cdot 2 \cdot 2 \cdot 2 \cdot 3 \cdot 3 = 144$

$\frac{5}{36} - \frac{1}{48} = \frac{5 \cdot 4}{36 \cdot 4} - \frac{1 \cdot 3}{48 \cdot 4} = \frac{20}{144} - \frac{3}{144} = \frac{17}{144}$

79.

$$\begin{array}{rcrcr} 5\frac{1}{6} & = & 5\frac{1 \cdot 3}{6 \cdot 3} & = & 5\frac{3}{18} \\ -\ 2\frac{1}{9} & = & -\ 2\frac{1 \cdot 2}{9 \cdot 2} & = & -\ 2\frac{2}{18} \\ \hline & & & & 3\frac{1}{18} \end{array}$$

Section 8.3

1. $8\text{ lb} = 8\ \cancel{\text{lb}} \times \frac{16\text{ oz}}{1\ \cancel{\text{lb}}}$

$= 8 \times 16\text{ oz}$

$= 128\text{ oz}$

3. $2\text{ T} = 2\ \cancel{\text{T}} \times \frac{2{,}000\text{ lb}}{1\ \cancel{\text{T}}}$

$= 2 \times 2{,}000\text{ lb}$

$= 4{,}000\text{ lb}$

5. $192\text{ oz} = 192\ \cancel{\text{oz}} \times \frac{1\text{ lb}}{16\ \cancel{\text{oz}}}$

$= \frac{192}{16}\text{lb}$

$= 12\text{ lb}$

7. $1{,}800\text{ lb} = 1{,}800\cancel{\text{lb}} \times \frac{1\text{ T}}{2{,}000\ \cancel{\text{lb}}}$

$= \frac{1{,}800}{2{,}000}\text{T}$

$= 0.9\text{ T}$

9. $1\text{ T} = 1\ \cancel{\text{T}} \times \frac{2{,}000\cancel{\text{lb}}}{1\ \cancel{\text{T}}} \times \frac{16\text{ oz}}{1\cancel{\text{lb}}}$

$= 1 \times 2{,}000 \times 16\text{ oz}$

$= 32{,}000\text{ oz}$

11. $3\frac{1}{2}\text{ lb} = 3\frac{1}{2}\cancel{\text{lb}} \times \frac{16\text{ oz}}{1\ \cancel{\text{lb}}}$

$= 3\frac{1}{2} \times 16\text{ oz}$

$= \frac{7}{2} \times 16\text{ oz}$

$= 56\text{ oz}$

13. $6\frac{1}{2}\text{ T} = 6\frac{1}{2}\ \cancel{\text{T}} \times \frac{2{,}000\text{ lb}}{1\ \cancel{\text{T}}}$

$= \frac{13}{2} \times 2{,}000\text{ lb}$

$= 13{,}000\text{ lb}$

15. $2\text{ kg} = 2\ \cancel{\text{kg}} \times \frac{1{,}000\text{ g}}{1\ \cancel{\text{kg}}}$

$= 2 \times 1{,}000\text{ g}$

$= 2{,}000\text{ g}$

17. $4\text{ cg} = 4\ \cancel{\text{cg}} \times \frac{1\ \cancel{\text{g}}}{100\ \cancel{\text{cg}}} \times \frac{1{,}000\text{ mg}}{1\ \cancel{\text{g}}}$

$= \frac{4 \times 1{,}000}{100}\text{ mg}$

$= 40\text{ mg}$

19. $2\text{ kg} = 2\ \cancel{\text{kg}} \times \frac{1{,}000\ \cancel{\text{g}}}{1\ \cancel{\text{kg}}} \times \frac{100\text{ cg}}{1\ \cancel{\text{g}}}$

$= 2 \times 1{,}000 \times 100\text{ cg}$

$= 200{,}000\text{ cg}$

21. $5.08\text{ g} = 5.08\ \cancel{\text{g}} \times \frac{100\text{ cg}}{1\ \cancel{\text{g}}}$

$= 5.08 \times 100\text{ cg}$

$= 508\text{ cg}$

23. $450\text{ cg} = 450\ \cancel{\text{cg}} \times \frac{1\text{ g}}{100\ \cancel{\text{cg}}}$

$= \frac{450}{100}\text{ g}$

$= 4.5\text{ g}$

25. 478.95 mg

$$= 478.95\ \cancel{mg} \times \frac{1\ \cancel{g}}{1{,}000\ \cancel{mg}} \times \frac{100\ cg}{1\ \cancel{g}}$$

$$= \frac{478.95 \times 100}{1{,}000}\ cg$$

$$= 47.895\ cg$$

27. $1{,}578\ mg = 1{,}578\ \cancel{mg} \times \frac{1\ g}{1{,}000\ \cancel{mg}}$

$$= \frac{1{,}578}{1{,}000}\ g$$

$$= 1.578\ g$$

29. $42{,}000\ cg = 42{,}000\ \cancel{cg} \times \frac{1\ \cancel{g}}{100\ \cancel{cg}} \times \frac{1\ kg}{1{,}000\ \cancel{g}}$

$$= \frac{42{,}000}{100 \times 1{,}000}\ kg$$

$$= 0.42\ kg$$

31. First, find the total mg in the bottle:

$$\text{Total mg} = 60\ \cancel{\text{softgels}} \cdot \frac{800\ mg}{\cancel{\text{softgel}}} = 48{,}000\ mg$$

Then convert mg to g:

$$48{,}000\ mg = 48{,}000\ \cancel{mg} \times \frac{1\ g}{1{,}000\ \cancel{mg}}$$

$$= \frac{48{,}000}{1{,}000}\ g$$

$$= 48\ g$$

The bottle contains 48 grams of fatty acid.

33. First, find the total mg in the bottle:

$$\text{Total mg} = 80\ \cancel{\text{vitamins}} \cdot \frac{50\ mg}{\cancel{\text{vitamin}}} = 4{,}000\ mg$$

Then convert mg to g:

$$4{,}000\ mg = 4{,}000\ \cancel{mg} \times \frac{1\ g}{1{,}000\ \cancel{mg}}$$

$$= \frac{4{,}000}{1{,}000}\ g$$

$$= 4\ g$$

The bottle contains 4 grams of Riboflavin.

35. First, find the total mg in the bottle:

$$\text{Total mg} = 120\ \cancel{\text{tablets}} \cdot \frac{81\ mg}{\cancel{\text{tablet}}} = 9{,}720\ mg$$

Then convert mg to g:

$$9{,}720\ mg = 9{,}720\ \cancel{mg} \times \frac{1\ g}{1{,}000\ \cancel{mg}}$$

$$= \frac{9{,}720}{1{,}000}\ g$$

$$= 9.72\ g$$

The bottle contains 9.72 grams of aspirin.

37. First, find the total mg in the bottle:

$$\text{Total mg} = 240\ \cancel{\text{vitamins}} \cdot \frac{500\ mg}{\cancel{\text{vitamin}}} = 120{,}000\ mg$$

Then convert mg to g:

$$120{,}000\ mg = 120{,}000\ \cancel{mg} \times \frac{1\ g}{1{,}000\ \cancel{mg}}$$

$$= \frac{120{,}000}{1{,}000}\ g$$

$$= 120\ g$$

The bottle contains 120 grams of Vitamin C.

39. First, find the total mL in a six pack:

$$\text{Total ml} = 6\ \cancel{\text{bottles}} \cdot \frac{500\text{ mL}}{\cancel{\text{bottle}}} = 3{,}000\text{ mL}$$

Then convert mL to L

$$3{,}000\text{ mL} = 3{,}000\ \cancel{\text{mL}} \times \frac{1\text{ L}}{1{,}000\ \cancel{\text{mL}}}$$

$$= \frac{3{,}000}{1{,}000}\text{ L}$$

$$= 3\text{ L}$$

There are 3 L in a six pack.

41. First, find the total mL in a six pack:

$$\text{Total ml} = 6\ \cancel{\text{bottles}} \cdot \frac{250\text{ mL}}{\cancel{\text{bottle}}} = 1{,}500\text{ mL}$$

Then convert mL to L

$$1{,}500\text{ mL} = 1{,}500\ \cancel{\text{mL}} \times \frac{1\text{ L}}{1{,}000\ \cancel{\text{mL}}}$$

$$= \frac{1{,}500}{1{,}000}\text{ L}$$

$$= 1.5\text{ L}$$

There are 1.5 L in a six pack.

Review Problems

43. $0.18 = \frac{18}{100} = \frac{9}{50}$

45. $0.09 = \frac{9}{100}$

47. $0.8 = \frac{8}{10} = \frac{4}{5}$

49.

$$\begin{array}{r} 0.9 \\ 10\overline{)9.0} \\ \underline{90} \\ 0 \end{array}$$

$\frac{9}{10} = 0.9$

51.

$$\begin{array}{r} 0.125 \\ 8\overline{)1.000} \\ \underline{8\downarrow} \\ 20 \\ \underline{16\downarrow} \\ 40 \\ \underline{40} \\ 0 \end{array}$$

$\frac{1}{8} = 0.125$

Section 8.4

1. $6\text{ in} = 6\cancel{\text{ in}} \times \frac{2.54\text{ cm}}{1\cancel{\text{ in}}}$
$= 6 \times 2.54\text{ cm}$
$= 15.24\text{ cm}$

3. $4\text{ m} = 4\cancel{\text{ m}} \times \frac{3.28\text{ ft}}{1\cancel{\text{ m}}}$
$= 4 \times 3.28\text{ ft}$
$= 13.12\text{ ft}$

5. $6\text{ m} = 6\cancel{\text{ m}} \times \frac{3.28\cancel{\text{ ft}}}{1\cancel{\text{ m}}} \times \frac{1\text{ yd}}{3\cancel{\text{ ft}}}$
$= \frac{6 \times 3.28}{3}\text{ yd}$
$= 6.56\text{ yd}$

7. $20\text{ mi} = 20\cancel{\text{ mi}} \times \frac{1.61\cancel{\text{ km}}}{1\cancel{\text{ mi}}} \times \frac{1{,}000\text{ m}}{1\cancel{\text{ km}}}$
$= 20 \times 1.61 \times 1{,}000\text{ m}$
$= 32{,}200\text{ m}$

9. $5\text{ m}^2 = 5\cancel{\text{ m}^2} \times \frac{1.196\text{ yd}^2}{1\cancel{\text{ m}^2}}$
$= 5 \times 1.196\text{ yd}^2$
$= 5.98\text{ yd}^2$

11. $10\text{ ha} = 10\cancel{\text{ ha}} \times \frac{2.47\text{ acres}}{1\cancel{\text{ ha}}}$
$= 10 \times 2.47\text{ acres}$
$= 24.7\text{ acres}$

13. $500\text{ in}^3 = 500\cancel{\text{ in}^3} \times \frac{16.39\text{ mL}}{1\cancel{\text{ in}^3}}$
$= 500 \times 16.39\text{ mL}$
$= 8{,}195\text{ mL}$

15. $2\text{ L} = 2\cancel{\text{ L}} \times \frac{1.06\text{ qt}}{1\cancel{\text{ L}}}$
$= 2 \times 1.06\text{ qt}$
$= 2.12\text{ qt}$

17. $20\text{ gal} = 20\cancel{\text{ gal}} \times \frac{3.79\text{ L}}{1\cancel{\text{ gal}}}$
$= 20 \times 3.79\text{ L}$
$= 75.8\text{ L}$

19. $12\text{ oz} = 12\cancel{\text{ oz}} \times \frac{28.3\text{ g}}{1\cancel{\text{ oz}}}$
$= 12 \times 28.3\text{ g}$
$= 339.6\text{ g}$

21. $15\text{ kg} = 15\cancel{\text{ kg}} \times \frac{2.20\text{ lb}}{1\cancel{\text{ kg}}}$
$= 15 \times 2.20\text{ lb}$
$= 33\text{ lb}$

23. When $C = 185$:
$F = \frac{9}{5}C + 32$
$= \frac{9}{5}(185) + 32$
$= 333 + 32$
$= 365$
185°C is equivalent to 365°F.

25. When $F = 86$:
$C = \frac{5(F - 32)}{9}$
$= \frac{5(86 - 32)}{9}$
$= \frac{5(54)}{9}$
$= 30$
86°F is equivalent to 30°C.

Calculator Problems

27. $10\text{ cm} = 10\ \cancel{\text{cm}} \times \dfrac{1\text{ in}}{2.54\ \cancel{\text{cm}}}$

$= \dfrac{10}{2.54}\text{ in}$

$\approx 3.94\text{ in}$

29. $25\text{ ft} = 25\ \cancel{\text{ft}} \times \dfrac{1\text{ m}}{3.28\ \cancel{\text{ft}}}$

$= \dfrac{25}{3.28}\text{ m}$

$\approx 7.62\text{ m}$

31. $49\text{ qt} = 49\ \cancel{\text{qt}} \times \dfrac{1\text{ L}}{1.06\ \cancel{\text{qt}}}$

$= \dfrac{49}{1.06}\text{ L}$

$\approx 46.23\text{ L}$

33. $500\text{ g} = 500\ \cancel{\text{g}} \times \dfrac{1\text{ oz}}{28.3\ \cancel{\text{g}}}$

$= \dfrac{500}{28.3}\text{ oz}$

$\approx 17.67\text{ oz}$

35. Answers will vary.

37. $100\text{ yd} = 100\ \cancel{\text{yd}} \times \dfrac{3\ \cancel{\text{ft}}}{1\ \cancel{\text{yd}}} \times \dfrac{1\text{ m}}{3.28\ \cancel{\text{ft}}}$

$= \dfrac{100 \times 3}{3.28}\text{ m}$

$\approx 91.46\text{ m}$

39. $25\text{ yd}^2 = 25\ \cancel{\text{yd}^2} \times \dfrac{1\text{ m}^2}{1.196\ \cancel{\text{yd}^2}}$

$= \dfrac{25}{1.196}\text{ m}^2$

$\approx 20.90\text{ m}^2$

41. $55\text{ miles per hour} = 55\dfrac{\cancel{\text{mi}}}{\text{hr}} \times \dfrac{1.61\text{ km}}{1\ \cancel{\text{mi}}}$

$= 55 \times 1.61\dfrac{\text{km}}{\text{hr}}$

$= 88.55\dfrac{\text{km}}{\text{hr}}$

43. First convert 8 in to ft:

$8\text{ in} = 8\ \cancel{\text{in}} \times \dfrac{1\text{ ft}}{12\ \cancel{\text{in}}}$

$= \dfrac{8}{12}\text{ ft}$

$= \dfrac{2}{3}\text{ ft}$

$6\text{ ft }8\text{ in} = 6\dfrac{2}{3}\text{ ft}$

Second, convert ft to m:

$6\dfrac{2}{3}\text{ ft} = 6\dfrac{2}{3}\ \cancel{\text{ft}} \times \dfrac{1\text{ m}}{3.28\ \cancel{\text{ft}}}$

$= \dfrac{6\frac{2}{3}}{3.28}\text{ m}$

$\approx 2.03\text{ m}$

6 ft 8 in is equivalent to 2.03 m.

45. When $F = 101°$:

$C = \dfrac{5(F-32)}{9}$

$= \dfrac{5(101-32)}{9}$

$= \dfrac{5(69)}{9}$

≈ 38.3

101°F is equivalent to 38.3°C.

Review Problems

47. $4\overline{)3.00}$

$$\begin{array}{r} 0.75 \\ 4\overline{)3.00} \\ \underline{28\downarrow} \\ 20 \\ \underline{20} \\ 0 \end{array}$$

$\frac{3}{4} = 0.75$

49. $5\frac{1}{2} = \frac{11}{2}$:

$$\begin{array}{r} 5.5 \\ 2\overline{)11.0} \\ \underline{10\downarrow} \\ 10 \\ \underline{10} \\ 0 \end{array}$$

$5\frac{1}{2} = 5.5$

51. $\frac{3}{100} = 0.03$

53. $0.34 = \frac{34}{100} = \frac{17}{50}$

55. $2.4 = 2\frac{4}{10} = 2\frac{2}{5}$

57. $1.75 = 1\frac{75}{100} = 1\frac{3}{4}$

Section 8.5

1. a. $4\text{ hr }30\text{ min} = 4\ \cancel{\text{hr}} \times \dfrac{60\text{ min}}{1\ \cancel{\text{hr}}} + 30\text{ min}$

$= 240\text{ min} + 30\text{ min}$

$= 270\text{ min}$

b. $4\text{ hr }30\text{ min} = 4\text{ hr} + 30\ \cancel{\text{min}} \times \dfrac{1\text{ hr}}{60\ \cancel{\text{min}}}$

$= 4\text{ hr} + 0.5\text{ hr}$

$= 4.5\text{ hr}$

3. a. $5\text{ hr }20\text{ min} = 5\ \cancel{\text{hr}} \times \dfrac{60\text{ min}}{1\ \cancel{\text{hr}}} + 20\text{ min}$

$= 300\text{ min} + 20\text{ min}$

$= 320\text{ min}$

b. $5\text{ hr }20\text{ min} = 5\text{ hr} + 20\ \cancel{\text{min}} \times \dfrac{1\text{ hr}}{60\ \cancel{\text{min}}}$

$= 5\text{ hr} + 0.333\text{ hr}$

$= 5.333\text{ hr}$

5. a. $6\text{ min }30\text{ sec} = 6\ \cancel{\text{min}} \times \dfrac{60\text{ sec}}{1\ \cancel{\text{min}}} + 30\text{ sec}$

$= 360\text{ sec} + 30\text{ sec}$

$= 390\text{ sec}$

b. $6\text{ min }30\text{ sec} = 6\text{ min} + 30\ \cancel{\text{sec}} \times \dfrac{1\text{ min}}{60\ \cancel{\text{sec}}}$

$= 6\text{ min} + 0.5\text{ min}$

$= 6.5\text{ min}$

7. a. $5\text{ min }20\text{ sec} = 5\ \cancel{\text{min}} \times \dfrac{60\text{ sec}}{1\ \cancel{\text{min}}} + 20\text{ sec}$

$= 300\text{ sec} + 20\text{ sec}$

$= 320\text{ sec}$

b. $5\text{ min }20\text{ sec} = 5\text{ min} + 20\ \cancel{\text{sec}} \times \dfrac{1\text{ min}}{60\ \cancel{\text{sec}}}$

$= 5\text{ min} + 0.33\text{ min}$

$= 5.33\text{ min}$

9. a. $2\text{ lbs }8\text{ oz} = 2\ \cancel{\text{lb}} \times \dfrac{16\text{ oz}}{1\ \cancel{\text{lb}}} + 8\text{ oz}$

$= 32\text{ oz} + 8\text{ oz}$

$= 40\text{ oz}$

b. $2\text{ lb }8\text{ oz} = 2\text{ lb} + 8\ \cancel{\text{oz}} \times \dfrac{1\text{ lb}}{16\ \cancel{\text{oz}}}$

$= 2\text{ lb} + 0.5\text{ lb}$

$= 2.5\text{ lb}$

11. a. $4\text{ lbs }12\text{ oz} = 4\ \cancel{\text{lb}} \times \dfrac{16\text{ oz}}{1\ \cancel{\text{lb}}} + 12\text{ oz}$

$= 64\text{ oz} + 12\text{ oz}$

$= 76\text{ oz}$

b. $4\text{ lb }12\text{ oz} = 4\text{ lb} + 12\ \cancel{\text{oz}} \times \dfrac{1\text{ lb}}{16\ \cancel{\text{oz}}}$

$= 4\text{ lb} + 0.75\text{ lb}$

$= 4.75\text{ lb}$

13. a. $4\text{ ft }6\text{ in} = 4\ \cancel{\text{ft}} \times \dfrac{12\text{ in}}{1\ \cancel{\text{ft}}} + 6\text{ in}$

$= 48\text{ in} + 6\text{ in}$

$= 54\text{ in}$

b. $4\text{ ft }6\text{ in} = 4\text{ ft} + 6\ \cancel{\text{in}} \times \dfrac{1\text{ ft}}{12\ \cancel{\text{in}}}$

$= 4\text{ ft} + 0.5\text{ ft}$

$= 4.5\text{ ft}$

15. a. $5\text{ ft }9\text{ in} = 5\ \cancel{\text{ft}} \times \dfrac{12\text{ in}}{1\ \cancel{\text{ft}}} + 9\text{ in}$

$= 60\text{ in} + 9\text{ in}$

$= 69\text{ in}$

b. $5\text{ ft }9\text{ in} = 5\text{ ft} + 9\ \cancel{\text{in}} \times \dfrac{1\text{ ft}}{12\ \cancel{\text{in}}}$

$= 5\text{ ft} + 0.75\text{ ft}$

$= 5.75\text{ ft}$

17. a. $2 \text{ gal } 1 \text{ qt} = 2 \cancel{\text{gal}} \times \frac{4 \text{ qt}}{1 \cancel{\text{gal}}} + 1 \text{ qt}$

$= 8 \text{ qt} + 1 \text{ qt}$

$= 9 \text{ qt}$

b. $2 \text{ gal } 1 \text{ qt} = 2 \text{ gal} + 1 \cancel{\text{qt}} \times \frac{1 \text{ gal}}{4 \cancel{\text{qt}}}$

$= 2 \text{ gal} + 0.25 \text{ gal}$

$= 2.25 \text{ gal}$

19.
$$\begin{array}{r} 4 \text{ hr } 47 \text{ min} \\ + \; 6 \text{ hr } 13 \text{ min} \\ \hline 10 \text{ hr } 60 \text{ min} \end{array} = 10 \text{ hr} + 1 \text{ hr} = 11 \text{ hr}$$

21.
$$\begin{array}{r} 8 \text{ ft } 10 \text{ in} \\ + \; 13 \text{ ft } 6 \text{ in} \\ \hline 21 \text{ ft } 16 \text{ in} \end{array} = 21 \text{ ft} + 1 \text{ ft } 4 \text{ in} = 22 \text{ ft } 4 \text{ in}$$

23.
$$\begin{array}{r} 4 \text{ lb } 12 \text{ oz} \\ + \; 6 \text{ lb } 4 \text{ oz} \\ \hline 10 \text{ lb } 16 \text{ oz} \end{array} = 10 \text{ lb} + 1 \text{ lb} = 11 \text{ lb}$$

25.
$$\begin{array}{r} 8 \text{ hr } 15 \text{ min} \\ - \; 2 \text{ hr } 35 \text{ min} \\ \hline \end{array} = \begin{array}{r} 7 \text{ hr } 75 \text{ min} \\ - \; 2 \text{ hr } 35 \text{ min} \\ \hline 5 \text{ hr } 40 \text{ min} \end{array}$$

27.
$$\begin{array}{r} 7 \text{ hr } 30 \text{ min} \\ - \; 3 \text{ hr } 43 \text{ min} \\ \hline \end{array} = \begin{array}{r} 6 \text{ hr } 90 \text{ min} \\ - \; 3 \text{ hr } 43 \text{ min} \\ \hline 3 \text{ hr } 47 \text{ min} \end{array}$$

29.
$$\begin{array}{r} 5 \text{ hr } 9 \text{ min} \\ - \; 4 \text{ hr } 17 \text{ min} \\ \hline \end{array} = \begin{array}{r} 4 \text{ hr } 69 \text{ min} \\ - \; 4 \text{ hr } 17 \text{ min} \\ \hline 52 \text{ min} \end{array}$$

Applying the Concepts

31.
$$\begin{array}{r} 50 \text{ min } 36 \text{ sec} \\ 4 \text{ hr } 40 \text{ min } 4 \text{ sec} \\ + \; 2 \text{ hr } 47 \text{ min } 38 \text{ sec} \\ \hline 6 \text{ hr } 137 \text{ min } 78 \text{ sec} \end{array}$$

$= 6 \text{ hr} + 2 \text{ hr } 17 \text{ min} + 1 \text{ min } 18 \text{ sec}$

$= 8 \text{ hr } 18 \text{ min } 18 \text{ sec}$

Reid's total time is 8:18:18

$$\begin{array}{r} 56 \text{ min } 51 \text{ sec} \\ 5 \text{ hr } 9 \text{ min} \\ + \; 3 \text{ hr } 2 \text{ min } 10 \text{ sec} \\ \hline 8 \text{ hr } 67 \text{ min } 61 \text{ sec} \end{array}$$

$= 8 \text{ hr} + 1 \text{ hr } 7 \text{ min} + 1 \text{ min } 1 \text{ sec}$

$= 9 \text{ hr } 8 \text{ min } 1 \text{sec}$

Bowden's total time is 9:08:01

33.
$$\begin{array}{r} 56 \text{ min } 51 \text{ sec} \\ - \; 50 \text{ min } 36 \text{ sec} \\ \hline 6 \text{ min } 15 \text{ sec} \end{array}$$

Peter was 00:06:15 faster.

35. Total weight:
$$\begin{array}{r} 6 \text{ lb } 8 \text{ oz} \\ \times \quad 4 \\ \hline 24 \text{ lb } 32 \text{ oz} \end{array}$$

$24 \text{ lb } 32 \text{ oz} = 24 \text{ lb} + 32 \text{ oz} \times \frac{1 \text{ lb}}{16 \text{ oz}}$

$= 24 \text{ lb} + 2 \text{ lb}$

$= 26 \text{ lb}$

Total cost: $26 \cancel{\text{lb}} \times \frac{\$4.00}{\cancel{\text{lb}}} = \$104$

37. Time per week:
$$\begin{array}{r} 1 \text{ hr } 15 \text{ min} \\ \times \quad 4 \\ \hline 4 \text{ hr } 60 \text{ min} \end{array} = 5 \text{ hr}$$

Total time: $5 \frac{\text{hr}}{\cancel{\text{wk}}} \times 2 \cancel{\text{wk}} = 10 \text{ hr}$

39. Total fabric:

$$\begin{array}{r} 3\text{ yd }1\text{ ft} \\ \times \quad 6 \\ \hline 18\text{ yd }6\text{ ft} \end{array}$$

$$18\text{ yd }6\text{ ft} = 18\text{ yd} + 6\cancel{\text{ ft}} \times \frac{1\text{ yd}}{3\cancel{\text{ ft}}}$$
$$= 18\text{ yd} + 2\text{ yd}$$
$$= 20\text{ yd}$$

Total cost: $20\cancel{\text{ yd}} \times \frac{\$7.50}{\cancel{\text{yd}}} = \$150$

41. Total weight of 6 avocados (in lbs)

$$6\cancel{\text{ avoc}} \times \frac{8\cancel{\text{ oz}}}{\cancel{\text{avoc}}} \times \frac{1\text{ lb}}{16\cancel{\text{ oz}}} = \frac{6\times 8}{16}\text{ lb} = 3\text{ lb}$$

Total cost: $3\cancel{\text{ lb}} \times \frac{\$2.00}{\cancel{\text{lb}}} = \$6$

51.

$$\begin{array}{rr} 2\text{ min }1.19\text{ sec} = & 1\text{ min }61.19\text{ sec} \\ -\ 1\text{ min }59.4\text{ sec} = & -\ 1\text{ min }59.40\text{ sec} \\ \hline & 1.79\text{ sec} \end{array}$$

Secretariat was 1.79 seconds faster.

Review Problems

43. $(9x)^2 = (9x)(9x) = (9\cdot 9)(x\cdot x) = 81x^2$

45. $(3xy)^3 = (3xy)(3xy)(3xy)$
$= (3\cdot 3\cdot 3)(x\cdot x\cdot x)(y\cdot y\cdot y)$
$= 27x^3y^3$

47. $(2xy)^2(3x)^3$
$= (2xy)(2xy)(3x)(3x)(3x)$
$= (2\cdot 2\cdot 3\cdot 3\cdot 3)(x\cdot x\cdot x\cdot x\cdot x)(y\cdot y)$
$= 108x^5y^2$

49. $(10a)^2(3ab)^3$
$= (10a)(10a)(3ab)(3ab)(3ab)$
$= (10\cdot 10\cdot 3\cdot 3\cdot 3)(a\cdot a\cdot a\cdot a\cdot a)(b\cdot b\cdot b)$
$= 2{,}700a^5b^3$

Extending the Concepts

Chapter 8 Review

1. $12 \text{ ft} = 12 \cancel{\text{ft}} \times \dfrac{12 \text{ in}}{1 \cancel{\text{ft}}}$

$= 12 \times 12 \text{ in}$

$= 144 \text{ in}$

3. $49 \text{ cm} = 49 \cancel{\text{cm}} \times \dfrac{1 \text{ m}}{100 \cancel{\text{cm}}}$

$= \dfrac{49}{100} \text{m}$

$= 0.49 \text{ m}$

5. $10 \text{ acres} = 10 \cancel{\text{acres}} \times \dfrac{43{,}560 \text{ ft}^2}{1 \cancel{\text{acre}}}$

$= 10 \times 43{,}560 \text{ ft}^2$

$= 435{,}600 \text{ ft}^2$

7. $4 \text{ ft}^2 = 4 \cancel{\text{ft}^2} \times \dfrac{144 \text{ in}^2}{1 \cancel{\text{ft}^2}}$

$= 4 \times 144 \text{ in}^2$

$= 576 \text{ in}^2$

9. $24 \text{ qt} = 24 \cancel{\text{qt}} \times \dfrac{1 \text{ gal}}{4 \cancel{\text{qt}}}$

$= \dfrac{24}{4} \text{gal}$

$= 6 \text{ gal}$

11. $8 \text{ lb} = 8 \cancel{\text{lb}} \times \dfrac{16 \text{ oz}}{1 \cancel{\text{lb}}}$

$= 8 \times 16 \text{ oz}$

$= 128 \text{ oz}$

13. $5 \text{ kg} = 5 \cancel{\text{kg}} \times \dfrac{1{,}000 \text{ g}}{1 \cancel{\text{kg}}}$

$= 5 \times 1{,}000 \text{ g}$

$= 5{,}000 \text{ g}$

15. $4 \text{ in} = 4 \cancel{\text{in}} \times \dfrac{2.54 \text{ cm}}{1 \cancel{\text{in}}}$

$= 4 \times 2.54 \text{ cm}$

$= 10.16 \text{ cm}$

17. $7 \text{ L} = 7 \cancel{\text{L}} \times \dfrac{1.06 \text{ qts}}{1 \cancel{\text{L}}}$

$= 7 \times 1.06 \text{ qt}$

$= 7.42 \text{ qt}$

19. $5 \text{ oz} = 5 \cancel{\text{oz}} \times \dfrac{28.3 \text{ g}}{1 \cancel{\text{oz}}}$

$= 5 \times 28.3 \text{ g}$

$= 141.5 \text{ g}$

21. When $C = 120$:

$F = \dfrac{9}{5}C + 32$

$= \dfrac{9}{5}(120) + 32$

$= 216 + 32$

$= 248$

120°C is equivalent to 248°F.

23. $24 \text{ cans} = 24 \text{ cans} \times \dfrac{355 \cancel{\text{mL}}}{1 \cancel{\text{can}}} \times \dfrac{1 \text{ L}}{1{,}000 \cancel{\text{mL}}}$

$= \dfrac{24 \times 355}{1{,}000} \text{ L}$

$= 8.52 \text{ L}$

25. 2,805,269 acres

$= 2{,}805{,}269 \cancel{\text{acres}} \times \dfrac{1 \text{ mi}^2}{640 \cancel{\text{acre}}}$

$= \dfrac{2{,}805{,}269}{640} \text{ mi}^2$

$\approx 4{,}383.23 \text{ mi}^2$

27. $250\text{ mi} = 250\ \cancel{\text{mi}} \times \dfrac{1.61\text{ km}}{1\ \cancel{\text{mi}}}$

$= 250 \times 1.61\text{ km}$

$= 402.5\text{ km}$

29. Find the area of 1 brick in m^2:

$A = (20\text{ cm})(10\text{ cm})$

$= 200\ \cancel{\text{cm}}^2 \times \dfrac{1\text{ m}}{100\ \cancel{\text{cm}}} \times \dfrac{1\text{ m}}{100\ \cancel{\text{cm}}}$

$= \dfrac{200}{100 \times 100}\text{m}^2$

$= 0.02\text{ m}^2$

Convert the area of the entryway to bricks:

$12\text{ m}^2 = 12\ \cancel{\text{m}^2} \times \dfrac{1\text{ brick}}{0.02\ \cancel{\text{m}^2}} = \dfrac{12}{0.02}\text{ bricks}$

$= 600\text{ bricks}$

It will take 600 bricks to cover the patio.

31. Convert 5 gal to oz, then glasses:

$5\text{ gal} = 5\ \cancel{\text{gal}} \times \dfrac{128\ \cancel{\text{fl oz}}}{1\ \cancel{\text{gal}}} \times \dfrac{1\text{ glass}}{8\ \cancel{\text{fl oz}}}$

$= \dfrac{5 \times 128}{8}\text{ glasses}$

$= 80\text{ glasses}$

You can fill 80 glasses.

33. $188\text{ km per hour} = 188 \dfrac{\cancel{\text{km}}}{\text{hr}} \times \dfrac{1\text{ mi}}{1.61\ \cancel{\text{km}}}$

$= \dfrac{188}{1.61}\dfrac{\text{mi}}{\text{hr}}$

$\approx 117\dfrac{\text{mi}}{\text{hr}}$

35. a. $4\text{ hr }45\text{ min} = 4\ \cancel{\text{hr}} \times \dfrac{60\text{ min}}{1\ \cancel{\text{hr}}} + 45\text{ min}$

$= 240\text{ min} + 45\text{ min}$

$= 285\text{ min}$

b. $4\text{ hr }45\text{ min} = 4\text{ hr} + 45\ \cancel{\text{min}} \times \dfrac{1\text{ hr}}{60\ \cancel{\text{min}}}$

$= 4\text{ hr} + 0.75\text{ hr}$

$= 4.75\text{ hr}$

37. Total weight:

$$\begin{array}{r} 12\text{ lb }8\text{ oz} \\ \times\quad 2 \\ \hline 24\text{ lb }16\text{ oz} = 25\text{ lbs} \end{array}$$

Total cost: $25\ \cancel{\text{lb}} \times \dfrac{\$5.00}{\cancel{\text{lb}}} = \$125$

Chapters 1-8 Cumulative Review

1.
$$\begin{array}{r} \overset{1\,1}{230} \\ 4{,}976 \\ +\ 349 \\ \hline 5{,}555 \end{array}$$

3.
$$\begin{array}{r} 73.8 \\ 25\overline{)1845.0} \\ \underline{175\downarrow} \\ 95 \\ \underline{75\downarrow} \\ 200 \\ \underline{200} \\ 0 \end{array}$$

5. $\left(\frac{4}{5}\right)^3 = \left(\frac{4}{5}\right)\left(\frac{4}{5}\right)\left(\frac{4}{5}\right) = \frac{64}{125}$

7.
$$\begin{array}{r} 3.4 \\ \times\ 0.34 \\ \hline 136 \\ \underline{1020} \\ 1.156 \end{array}$$

9. $\frac{7}{12} \div \left(-\frac{21}{4}\right) = -\frac{7}{12} \cdot \frac{4}{21} = -\frac{\cancel{7} \cdot \cancel{4}}{3 \cdot \cancel{4} \cdot \cancel{7} \cdot 3} = -\frac{1}{9}$

11.
$$\begin{aligned} 3\frac{1}{5} - 1\frac{1}{10} &= \frac{16}{5} - \frac{11}{10} \\ &= \frac{16 \cdot 2}{5 \cdot 2} - \frac{11}{10} \\ &= \frac{32}{10} - \frac{11}{10} \\ &= \frac{21}{10} \text{ or } 2\frac{1}{10} \end{aligned}$$

13.
$$\begin{aligned} 4y - 16 &= -2y + 26 \\ 4y - 16 + 2y &= -2y + 26 + 2y \\ 6y - 16 &= 26 \\ 6y - 16 + 16 &= 26 + 16 \\ 6y &= 42 \\ \frac{6y}{6} &= \frac{42}{6} \\ y &= 7 \end{aligned}$$

15.
$$\begin{aligned} \frac{2}{x} + \frac{1}{3} &= \frac{2}{3} \\ \frac{2}{x} + \frac{1}{3} - \frac{1}{3} &= \frac{2}{3} - \frac{1}{3} \\ \frac{2}{x} &= \frac{1}{3} \\ x \cdot 1 &= 2 \cdot 3 \\ x &= 6 \end{aligned}$$

17. $\frac{3}{4} \cdot 1\frac{1}{2} = \frac{3}{4} \cdot \frac{3}{2} = \frac{3 \cdot 3}{4 \cdot 2} = \frac{9}{8} = 1\frac{1}{8}$

19. $\frac{12}{36} = \frac{\cancel{12}}{\cancel{12} \cdot 3} = \frac{1}{3}$

21.
$$\begin{aligned} 2.5 \text{ mi}^2 &= 2.5\ \cancel{\text{mi}^2} \times \frac{640 \text{ acres}}{1\ \cancel{\text{mi}^2}} \\ &= 2.5 \times 640 \text{ acres} \\ &= 1{,}600 \text{ acres} \end{aligned}$$

23. $55\% = \frac{55}{100} = \frac{11}{20}$

25. $30{,}405 = 30{,}000 + 400 + 5$

27. 26 is what percent of 39?

$$\begin{aligned} 26 &= x \cdot 39 \\ \frac{26}{39} &= \frac{x \cdot 39}{39} \\ x &= 0.\overline{6} = 66\frac{2}{3}\% \end{aligned}$$

29. $10 \text{ kg} = 10 \cancel{\text{kg}} \times \frac{2.2 \text{ lbs}}{1 \cancel{\text{kg}}}$

$= 10 \times 2.2 \text{ lbs}$

$= 22 \text{ lbs}$

31. When $C = 50$:

$F = \frac{9}{5}C + 32$

$= \frac{9}{5}(50) + 32$

$= 90 + 32$

$= 122$

50°C is equivalent to 122°F.

33. Total weight of 400 avocados (in lbs)

$400 \cancel{\text{avoc}} \times \frac{12 \cancel{\text{oz}}}{\cancel{\text{avoc}}} \times \frac{1 \text{ lb}}{16 \cancel{\text{oz}}} = \frac{400 \times 12}{16} \text{ lb} = 300 \text{ lb}$

Total sales: $300 \cancel{\text{lb}} \times \frac{\$1.50}{\cancel{\text{lb}}} = \$450$

35. a. Total people $= 36 \cancel{\text{carriages}} \times \frac{60 \text{ people}}{\cancel{\text{carriage}}}$

$= 2{,}160 \text{ people}$

b. $C = 2\pi r = \pi d = 3.14(250 \text{ ft}) = 785 \text{ ft}$

c. Speed is $\frac{\text{distance}}{\text{time}}$. For 1 revolution, the distance is C, and the time is 20 min:

$\frac{\text{distance}}{\text{time}} = \frac{758 \text{ ft}}{20 \text{ min}} = 39.25 \text{ ft/min}$

d. 39.25 ft per min

$= 39.25 \frac{\cancel{\text{ft}}}{\cancel{\text{min}}} \times \frac{1 \text{ mi}}{5{,}280 \cancel{\text{ft}}} \times \frac{60 \cancel{\text{min}}}{1 \text{ hr}}$

$= \frac{39.25 \times 60}{5{,}280} \frac{\text{mi}}{\text{hr}}$

$\approx 0.45 \frac{\text{mi}}{\text{hr}}$

Chapter 8 Test

1. $7 \text{ yd} = 7 \cancel{\text{yd}} \times \frac{3 \text{ ft}}{1 \cancel{\text{yd}}}$

$= 7 \times 3 \text{ ft}$

$= 21 \text{ ft}$

3. $3 \text{ acres} = 3 \cancel{\text{acres}} \times \frac{43{,}560 \text{ ft}^2}{1 \cancel{\text{acre}}}$

$= 3 \times 43{,}560 \text{ ft}^2$

$= 130{,}680 \text{ ft}^2$

5. $10 \text{ L} = 10 \cancel{\text{L}} \times \frac{1{,}000 \text{ mL}}{1 \cancel{\text{L}}}$

$= 10 \times 1{,}000 \text{ mL}$

$= 10{,}000 \text{ mL}$

7. $10 \text{ L} = 10 \cancel{\text{L}} \times \frac{1.06 \text{ qts}}{1 \cancel{\text{L}}}$

$= 10 \times 1.06 \text{ qt}$

$= 10.6 \text{ qt}$

9. $1 \text{ L} = 1 \cancel{\text{L}} \times \frac{1.06 \cancel{\text{qt}}}{1 \cancel{\text{L}}} \times \frac{1 \text{ gal}}{4 \cancel{\text{qt}}}$

$= \frac{1 \times 1.06}{4} \text{ gal}$

$= 0.27 \text{ gal}$

11. $409 \text{ in}^3 = 409 \cancel{\text{in}^3} \times \frac{1 \text{ ft}^3}{1{,}728 \cancel{\text{in}^3}}$

$= \frac{409}{1{,}728} \text{ ft}^3$

$= 0.24 \text{ ft}^3$

13. $245 \text{ ft} = 245 \cancel{\text{ft}} \times \frac{1 \text{ m}}{3.28 \cancel{\text{ft}}}$

$= \frac{245}{3.28} \text{ m}$

$= 74.70 \text{ m}$

15. Find the area of 1 tile:

$A = (8 \text{ in})(8 \text{ in}) = 64 \text{ in}^2$

Convert the area of the entryway to tile:

40 ft^2

$= 40 \cancel{\text{ft}^2} \times \frac{144 \cancel{\text{in}^2}}{1 \cancel{\text{ft}^2}} \times \frac{1 \text{ tile}}{64 \cancel{\text{in}^2}}$

$= \frac{40 \times 144}{64} \text{ tiles}$

$= 90 \text{ tiles}$

It will take 90 tiles to cover the pantry.

17. a. $5 \text{ hr } 30 \text{ min} = 5 \cancel{\text{hr}} \times \frac{60 \text{ min}}{1 \cancel{\text{hr}}} + 30 \text{ min}$

$= 300 \text{ min} + 30 \text{ min}$

$= 330 \text{ min}$

b. $5 \text{ hr } 30 \text{ min} = 5 \text{ hr} + 30 \cancel{\text{min}} \times \frac{1 \text{ hr}}{60 \cancel{\text{min}}}$

$= 5 \text{ hr} + 0.5 \text{ hr}$

$= 5.5 \text{ hr}$

19. $80 \text{ km per hour} = 80 \frac{\cancel{\text{km}}}{\text{hr}} \times \frac{1 \text{ mi}}{1.61 \cancel{\text{km}}}$

$= \frac{80}{1.61} \frac{\text{mi}}{\text{hr}}$

$= 50 \frac{\text{mi}}{\text{hr}}$

You could only out run it with a car.

Chapter 9 Preview

1. $3+4+2=(3+4)+2=7+2=9$

3. $9+(-12)=-3$

5. $5-8=5+(-8)=-3$

7. $-8-(-6)=-8+6=-2$

9. $2^6=2\cdot 2\cdot 2\cdot 2\cdot 2\cdot 2=64$

11. $8.64\times 10{,}000=86{,}400$

13.
$$\begin{array}{r} 1.3 \\ \times\ 5 \\ \hline 6.5 \end{array}$$

15. $-2x+3x=(-2+3)x=1x=x$

17. $7(2x-3)=7\cdot 2x-7\cdot 3=14x-21$

19.
$$\begin{aligned}(3xy)^3&=(3xy)(3xy)(3xy)\\&=3\cdot 3\cdot 3\cdot x\cdot x\cdot x\cdot y\cdot y\cdot y\\&=27x^3y^3\end{aligned}$$

Chapter 9 Pretest

1. $a^3 \cdot a^5 = a^{3+5} = a^8$

3. $\left(3ab^5\right)^3\left(2a^7b^2\right) = 3^3a^3\left(b^5\right)^3 \cdot 2a^7b^2$

$$= 27a^3b^{15} \cdot 2a^7b^2$$
$$= (27 \cdot 2)\left(a^3a^7\right)\left(b^{15}b^2\right)$$
$$= 54a^{3+7}b^{15+2}$$
$$= 54a^{10}b^{17}$$

5. $\left(a^3 - 3a^2 + a\right) + \left(2a^2 + 2a - 1\right)$

$$= a^3 + \left(-3a^2 + 2a^2\right) + (a + 2a) - 1$$
$$= a^3 - a^2 + 3a - 1$$

7. $\left(4x^3 + 6x^2 + 3\right) - \left(x^3 - 4x - 9\right)$

$$= 4x^3 + 6x^2 + 3 - x^3 + 4x + 9$$
$$= \left(4x^3 - x^3\right) + 6x^2 + 4x + (3 + 9)$$
$$= 3x^3 + 6x^2 + 4x + 12$$

9. $x^3\left(x^2 + x^4\right) = x^3x^2 + x^3x^4 = x^5 + x^7$

11. $(x+4)(3x-2)$

$$= (x+4)3x + (x+4)(-2)$$
$$= x \cdot 3x + 4 \cdot 3x + x(-2) + 4(-2)$$
$$= 3x^2 + 12x - 2x - 8$$
$$= 3x^2 + 10x - 8$$

13. $(-2)^{-3} = \dfrac{1}{(-2)^3} = -\dfrac{1}{8}$

15. $\dfrac{10^{-3}}{10^{-4}} = 10^{-3-(-4)} = 10^{-3+4} = 10^1 = 10$

17. $0.000000002 \text{ m} = 2 \times 10^{-9} \text{ m}$

19. $\left(5 \times 10^{12}\right)\left(2.4 \times 10^{-5}\right) = (5 \times 2.4)\left(10^{12} \times 10^{-5}\right)$

$$= 12 \times 10^{12+(-5)}$$
$$= 12 \times 10^7$$
$$= 1.2 \times 10^8$$

Section 9.1

1. $x^3 \cdot x^5 = x^{3+5}$
$= x^8$

3. $a^3 \cdot a^5 = a^{3+5}$
$= a^8$

5. $y^{10} \cdot y^5 = y^{10+5}$
$= y^{15}$

7. $x^4 \cdot x = x^4 x^1$
$= x^{4+1}$
$= x^5$

9. $x^2 \cdot x^3 \cdot x^4 = x^{2+3+4}$
$= x^9$

11. $a^6 \cdot a^4 \cdot a^2 = a^{6+4+2}$
$= a^{12}$

13. $y^3 \cdot y^2 \cdot y = y^{3+2+1}$
$= y^6$

15. $2x^4 \cdot 7x^5 = (2 \cdot 7)(x^4 \cdot x^5)$
$= (2 \cdot 7)(x^{4+5})$
$= 14x^9$

17. $3y^5 \cdot 8y^3 = (3 \cdot 8)(y^5 \cdot y^3)$
$= (3 \cdot 8)(y^{5+3})$
$= 24y^8$

19. $7x \cdot 3x = (7 \cdot 3)(x \cdot x)$
$= (7 \cdot 3)(x^{1+1})$
$= 21x^2$

21. $4x^3 \cdot 6x^5 \cdot 8x^7 = (4 \cdot 6 \cdot 8)(x^3 \cdot x^5 \cdot x^7)$
$= (4 \cdot 6 \cdot 8)(x^{3+5+7})$
$= 192x^{15}$

23. $2a \cdot 3a \cdot 4a = (2 \cdot 3 \cdot 4)(a \cdot a \cdot a)$
$= (2 \cdot 3 \cdot 4)(a^{1+1+1})$
$= 24a^3$

25. $5y^4 \cdot 5y^4 \cdot 5y^4 = (5 \cdot 5 \cdot 5)(y^4 \cdot y^4 \cdot y^4)$
$= (5 \cdot 5 \cdot 5)(y^{4+4+4})$
$= 125y^{12}$

27. $(3x^2y^7)(2x^3y) = (3 \cdot 2)(x^2 \cdot x^3)(y^7 \cdot y)$
$= (3 \cdot 2)(x^{2+3})(y^{7+1})$
$= 6x^5y^8$

29. $(7a^3b^2)(8ab^3) = (7 \cdot 8)(a^3 \cdot a)(b^2 \cdot b^3)$
$= (7 \cdot 8)(a^{3+1})(b^{2+3})$
$= 56a^4b^5$

31. $(2ab^3)(3a^2b^2)(4a^3b)$
$= (2 \cdot 3 \cdot 4)(a \cdot a^2 \cdot a^3)(b^3 \cdot b^2 \cdot b)$
$= (2 \cdot 3 \cdot 4)(a^{1+2+3})(b^{3+2+1})$
$= 24a^6b^6$

33. $(7a^3b^5)(4a^4b^4)(a^5b^3)$
$= (7 \cdot 4)(a^3 \cdot a^4 \cdot a^5)(b^5 \cdot b^4 \cdot b^3)$
$= (7 \cdot 4)(a^{3+4+5})(b^{5+4+3})$
$= 28a^{12}b^{12}$

35. $(x^2)^5 = x^{2 \cdot 5}$
$= x^{10}$

37. $\left(a^3\right)^7 = a^{3\cdot7}$
$= a^{21}$

39. $\left(5y^4\right)^3 = 5^3\left(y^4\right)^3$
$= 5^3 y^{4\cdot3}$
$= 125y^{12}$

41. $\left(9x^4\right)^2 = 9^2\left(x^4\right)^2$
$= 9^2 x^{4\cdot2}$
$= 81x^8$

43. $\left(2ab\right)^3 = 2^3 \cdot a^3 \cdot b^3$
$= 8a^3b^3$

45. $\left(5x^2y^3\right)^4 = 5^4\left(x^2\right)^4\left(y^3\right)^4$
$= 5^4 x^{2\cdot4} y^{3\cdot4}$
$= 625x^8y^{12}$

47. $\left(x^2\right)^3 \cdot \left(x^4\right)^2 = x^{2\cdot3} \cdot x^{4\cdot2}$
$= x^6 \cdot x^8$
$= x^{6+8}$
$= x^{14}$

49. $\left(a^3\right)^7 \cdot \left(a^4\right)^2 = a^{3\cdot7} \cdot a^{4\cdot2}$
$= a^{21} \cdot a^8$
$= a^{21+8}$
$= a^{29}$

51. $\left(3x^3y^5\right)^2\left(2x^2y^4\right)^3$
$= 3^2\left(x^3\right)^2\left(y^5\right)^2 \cdot 2^3\left(x^2\right)^3\left(y^4\right)^3$
$= 3^2 x^{3\cdot2} y^{5\cdot2} \cdot 2^3 x^{2\cdot3} y^{4\cdot3}$
$= 9x^6y^{10} \cdot 8x^6y^{12}$
$= \left(9\cdot8\right)\left(x^6x^6\right)\left(y^{10}y^{12}\right)$
$= 72x^{6+6}y^{10+12}$
$= 72x^{12}y^{22}$

53. $\left(5a^2b^4\right)^2\left(7a^8b\right) = 5^2\left(a^2\right)^2\left(b^4\right)^2 \cdot 7a^8b$
$= 25a^4b^8 \cdot 7a^8b$
$= \left(25\cdot7\right)\left(a^4a^8\right)\left(b^8b\right)$
$= 175a^{4+8}b^{8+1}$
$= 175a^{12}b^9$

55. $\left(a^2b^4\right)^3\left(ab^5\right)^4\left(a^4b\right)^2$
$= \left(a^2\right)^3\left(b^4\right)^3 \cdot a^4\left(b^5\right)^4 \cdot \left(a^4\right)^2 b^2$
$= a^6b^{12} \cdot a^4b^{20} \cdot a^8b^2$
$= \left(a^6a^4a^8\right)\left(b^{12}b^{20}b^2\right)$
$= a^{6+4+8}b^{12+20+2}$
$= a^{18}b^{34}$

57. The table is as follows.

Number x	Square x^2
1	$1^2 = 1$
2	$2^2 = 4$
3	$3^2 = 9$
4	$4^2 = 16$
5	$5^2 = 25$
6	$6^2 = 36$
7	$7^2 = 49$

59. The table is as follows.

Number x	Square x^2
-2.5	$(-2.5)^2 = (-2.5)(-2.5) = 6.25$
-1.5	$(-1.5)^2 = (-1.5)(-1.5) = 2.25$
-0.5	$(-0.5)^2 = (-0.5)(-0.5) = 0.25$
0	$0^2 = 0$
0.5	$(0.5)^2 = (0.5)(0.5) = 0.25$
1.5	$(1.5)^2 = (1.5)(1.5) = 2.25$
2.5	$(2.5)^2 = (2.5)(2.5) = 6.25$

Review Problems

61.
$$\begin{aligned} P &= 4s \\ &= 4(13 \text{ ft}) \\ &= 52 \text{ ft} \end{aligned}$$

63.
$$\begin{aligned} P &= \text{sum of all sides} \\ &= 24 \text{ ft} + 18 \text{ ft} + 24 \text{ ft} + 18 \text{ ft} \\ &= 84 \text{ ft} \end{aligned}$$

65.
$$\begin{aligned} P &= \text{sum of all sides} \\ &= 3 \text{ in} + 4 \text{ in} + 2 \text{ in} \\ &= 9 \text{ in} \end{aligned}$$

Section 9.2

1. $\left(3x^2+2x+5\right)+\left(3x^2+4x+3\right)$
$=\left(3x^2+3x^2\right)+\left(2x+4x\right)+\left(5+3\right)$
$=6x^2+6x+8$

or, in column form:

$$\begin{array}{r} 3x^2+2x+5 \\ +\ 3x^2+4x+3 \\ \hline 6x^2+6x+8 \end{array}$$

3. $\left(3a^2-4a+2\right)+\left(2a^2-5a+6\right)$
$=\left(3a^2+2a^2\right)+\left(-4a-5a\right)+\left(2+6\right)$
$=5a^2-9a+8$

or, in column form:

$$\begin{array}{r} 3a^2-4a+2 \\ +\ 2a^2-5a+6 \\ \hline 5a^2-9a+8 \end{array}$$

5. $\left(6x^3-5x^2+3x\right)+\left(9x^2-6x+2\right)$
$=6x^3+\left(-5x^2+9x^2\right)+\left(3x-6x\right)+2$
$=6x^3+4x^2-3x+2$

or, in column form:

$$\begin{array}{r} 6x^3-5x^2+3x+0 \\ +\qquad 9x^2-6x+2 \\ \hline 6x^3+4x^2-3x+2 \end{array}$$

7. $\left(7t^2-4t-5\right)+\left(6t^2+4t+3\right)$
$=\left(7t^2+6t^2\right)+\left(-4t+4t\right)+\left(-5+3\right)$
$=13t^2+0t-2$
$=13t^2-2$

or, in column form:

$$\begin{array}{r} 7t^2-4t-5 \\ +\ 6t^2+4t+3 \\ \hline 13t^2+0t-2=13t^2-2 \end{array}$$

9. $\left(20x^3-12x^2+x-5\right)+\left(13x^3+2x^2-5x+1\right)$
$=\left(20x^3+13x^3\right)+\left(-12x^2+2x^2\right)+\left(x-5x\right)$
$+\left(-5+1\right)$
$=33x^3-10x^2-4x-4$

or, in column form:

$$\begin{array}{r} 20x^3-12x^2+x-5 \\ +\ 13x^3+2x^2-5x+1 \\ \hline 33x^3-10x^2-4x-4 \end{array}$$

11. $\left(4x^2+3x+2\right)-\left(2x^2+5x+1\right)$
$=4x^2+3x+2-2x^2-5x-1$
$=\left(4x^2-2x^2\right)+\left(3x-5x\right)+\left(2-1\right)$
$=2x^2-2x+1$

13. $\left(9y^3-8y^2-4\right)-\left(5y^3-3y+8\right)$
$=9y^3-8y^2-4-5y^3+3y-8$
$=\left(9y^3-5y^3\right)-8y^2+3y+\left(-4-8\right)$
$=4y^3-8y^2+3y-12$

15. $(4x-5)-(3x+2)-(5x-3)$
$=4x-5-3x-2-5x+3$
$=(4x-3x-5x)+(-5-2+3)$
$=-4x+(-4)$
$=-4x-4$

17. $(11x^2-10x+14)-(10x^2+20x-50)$
$=11x^2-10x+14-10x^2-20x+50$
$=(11x^2-10x^2)+(-10x-20x)+(14+50)$
$=x^2-30x+64$

19. $(10y^2+10y+10)-(3y^2+7y-15)$
$=10y^2+10y+10-3y^2-7y+15$
$=(10y^2-3y^2)+(10y-7y)+(10+15)$
$=7y^2+3y+25$

21. Substitute 2 for x:
$$\begin{aligned} x^2-10x+25 &= (2)^2-10(2)+25 \\ &= 4-20+25 \\ &= 9 \end{aligned}$$

23. Substitute −3 for a:
$$\begin{aligned} a^2+6a+9 &= (-3)^2+6(-3)+9 \\ &= 9+(-18)+9 \\ &= 0 \end{aligned}$$
Note: Use of parenthesis when substituting is important here. For example, $-3^2=-9$, while $(-3)^2=9$.

25. Substitute −1 for y:
$$\begin{aligned} 4y^2+12y+9 &= 4(-1)^2+12(-1)+9 \\ &= 4(1)+12(-1)+9 \\ &= 4-12+9 \\ &= 1 \end{aligned}$$

27. The table is as follows:

n	$3n-1$
1	$3(1)-1=3-1=2$
2	$3(2)-1=6-1=5$
3	$3(3)-1=9-1=8$
4	$3(4)-1=12-1=11$

29. The table is as follows:

n	$(n+1)^2$
1	$(1+1)^2=2^2=4$
2	$(2+1)^2=3^2=9$
3	$(3+1)^2=4^2=16$
4	$(4+1)^2=5^2=25$

31. The table is as follows:

n	n^2+9
1	$(1)^2+9=1+9=10$
2	$(2)^2+9=4+9=13$
3	$(3)^2+9=9+9=18$
4	$(4)^2+9=16+9=25$

33. The table is as follows:

n	$4n$
1	$4\cdot1=4$
2	$4\cdot2=8$
3	$4\cdot3=12$
4	$4\cdot4=16$

35. The table is as follows:

n	$n-3$
1	$1-3=-2$
2	$2-3=-1$
3	$3-3=0$
4	$4-3=1$

Review Problems

37. $A = s^2$

$= (13 \text{ ft})^2$

$= 169 \text{ ft}^2$

39. $A = lw$

$= (21 \text{ in})(10 \text{ in})$

$= 210 \text{ in}^2$

41. $A = bh$

$= (32 \text{ yd})(15 \text{ yd})$

$= 480 \text{ yd}^2$

Section 9.3

1. $x^2(x^4+x^3)=x^2\cdot x^4+x^2\cdot x^3$
$=x^6+x^5$

3. $(x^4+x^3)x^2=x^4\cdot x^2+x^3\cdot x^2$
$=x^6+x^5$

5. $2x^4(3x^2-7)=2x^4\cdot 3x^2-2x^4\cdot 7$
$=(2\cdot 3)(x^4\cdot x^2)-(2\cdot 7)x^4$
$=6x^6-14x^4$

7. $4x^2y^3(3x^4+2y^3)$
$=(4x^2y^3)(3x^4)+(4x^2y^3)(2y^3)$
$=(4\cdot 3)(x^2\cdot x^4)y^3+(4\cdot 2)x^2(y^3\cdot y^3)$
$=12x^6y^3+8x^2y^6$

9. $(3x^2-7)2x^4=3x^2\cdot 2x^4-7\cdot 2x^4$
$=(3\cdot 2)(x^2\cdot x^4)-(7\cdot 2)x^4$
$=6x^6-14x^4$

11. $3xy(4xy-9)$
$=3xy\cdot 4xy-3xy\cdot 9$
$=(3\cdot 4)(x\cdot x)(y\cdot y)-(3\cdot 9)xy$
$=12x^2y^2-27xy$

13. $(4xy-9)3xy$
$=4xy\cdot 3xy-9\cdot 3xy$
$=(4\cdot 3)(x\cdot x)(y\cdot y)-(9\cdot 3)xy$
$=12x^2y^2-27xy$

15. $2x^4y^5(3xy^2+2x^2y+5)$
$=2x^4y^5\cdot 3xy^2+2x^4y^5\cdot 2x^2y+2x^4y^5\cdot 5$
$=(2\cdot 3)(x^4x)(y^5y^2)+(2\cdot 2)(x^4x^2)(y^5y)$
$+(2\cdot 5)x^4y^5$
$=6x^5y^7+4x^6y^6+10x^4y^5$

17. $(x+2)(x+3)=(x+2)x+(x+2)3$
$=x\cdot x+2\cdot x+x\cdot 3+2\cdot 3$
$=x^2+2x+3x+6$
$=x^2+5x+6$

19. $(x-2)(x+3)=(x-2)x+(x-2)3$
$=x\cdot x-2\cdot x+x\cdot 3-2\cdot 3$
$=x^2-2x+3x-6$
$=x^2+x-6$

21. $(2x+5)(3x+2)=(2x+5)3x+(2x+5)2$
$=2x\cdot 3x+5\cdot 3x+2x\cdot 2+5\cdot 2$
$=6x^2+15x+4x+10$
$=6x^2+19x+10$

23. $(3x-6)(x+4)=(3x-6)x+(3x-6)4$
$=3x\cdot x-6\cdot x+3x\cdot 4-6\cdot 4$
$=3x^2-6x+12x-24$
$=3x^2+6x-24$

25. $(4x-3)(4x+3)=(4x-3)4x+(4x-3)3$
$=4x\cdot 4x-3\cdot 4x+4x\cdot 3-3\cdot 3$
$=16x^2-12x+12x-9$
$=16x^2-9$

27. $(3x-1)(4x+1)=(3x-1)4x+(3x-1)1$
$=3x\cdot 4x-1\cdot 4x+3x\cdot 1-1\cdot 1$
$=12x^2-4x+3x-1$
$=12x^2-x-1$

29. $(2x-5)(3x-4)$

$=(2x-5)3x+(2x-5)(-4)$

$=2x\cdot 3x-5\cdot 3x+2x(-4)-5(-4)$

$=6x^2-15x-8x+20$

$=6x^2-23x+20$

31. Large rectangle: $l=x+4$, $w=x+2$

In the diagram, find the area of the 4 small rectangles:

	x	4
x	x^2	$4x$
2	$2x$	8

Area of large rectangle = sum of areas of small rectangles:

$(x+4)(x+2)=x^2+4x+2x+8$

$=x^2+6x+8$

33. Large rectangle: $l=2x+3$, $w=3x+2$

In the diagram, find the area of the 4 small rectangles:

	$2x$	3
$3x$	$6x^2$	$9x$
2	$4x$	6

Area of large rectangle = sum of areas of small rectangles:

$(2x+3)(3x+2)=6x^2+9x+4x+6$

$=6x^2+13x+6$

35. Large rectangle: $l=7x+2$, $w=3x+4$

In the diagram, find the area of the 4 small rectangles:

	$7x$	2
$3x$	$21x^2$	$6x$
4	$28x$	8

Area of large rectangle = sum of areas of small rectangles:

$(7x+2)(3x+4)=21x^2+6x+28x+8$

$=21x^2+34x+8$

37. $(x+3)^2=(x+3)(x+3)$

$=(x+3)x+(x+3)3$

$=x\cdot x+3\cdot x+x\cdot 3+3\cdot 3$

$=x^2+3x+3x+9$

$=x^2+6x+9$

39. $(x+6)^2=(x+6)(x+6)$

$=(x+6)x+(x+6)6$

$=x\cdot x+6\cdot x+x\cdot 6+6\cdot 6$

$=x^2+6x+6x+36$

$=x^2+12x+36$

41. $(3x+2)^2=(3x+2)(3x+2)$

$=(3x+2)3x+(3x+2)2$

$=3x\cdot 3x+2\cdot 3x+3x\cdot 2+2\cdot 2$

$=9x^2+6x+6x+4$

$=9x^2+12x+4$

43. $(2x+3)^2=(2x+3)(2x+3)$

$=(2x+3)2x+(2x+3)3$

$=2x\cdot 2x+3\cdot 2x+2x\cdot 3+3\cdot 3$

$=4x^2+6x+6x+9$

$=4x^2+12x+9$

45. $(7x+6)^2 = (7x+6)(7x+6)$
$= (7x+6)7x + (7x+6)6$
$= 7x \cdot 7x + 6 \cdot 7x + 7x \cdot 6 + 6 \cdot 6$
$= 49x^2 + 42x + 42x + 36$
$= 49x^2 + 84x + 36$

47. $(4x+1)^2 = (4x+1)(4x+1)$
$= (4x+1)4x + (4x+1)1$
$= 4x \cdot 4x + 1 \cdot 4x + 4x \cdot 1 + 1 \cdot 1$
$= 16x^2 + 4x + 4x + 1$
$= 16x^2 + 8x + 1$

49. $(x-5)^2 = (x-5)(x-5)$
$= (x-5)x + (x-5)(-5)$
$= x \cdot x - 5 \cdot x + x(-5) - 5(-5)$
$= x^2 - 5x - 5x + 25$
$= x^2 - 10x + 25$

51. $(2x-4)^2 = (2x-4)(2x-4)$
$= (2x-4)2x + (2x-4)(-4)$
$= 2x \cdot 2x - 4 \cdot 2x + 2x(-4) - 4(-4)$
$= 4x^2 - 8x - 8x + 16$
$= 4x^2 - 16x + 16$

53. The table is as follows:

x	$(x+3)^2$	x^2+9	x^2+6x+9
1	$(1+3)^2$ $=4^2$ $=16$	1^2+9 $=1+9$ $=10$	$1^2+6(1)+9$ $=1+6+9$ $=16$
2	$(2+3)^2$ $=5^2$ $=25$	2^2+9 $=4+9$ $=13$	$2^2+6(2)+9$ $=4+12+9$ $=25$
3	$(3+3)^2$ $=6^2$ $=36$	3^2+9 $=9+9$ $=18$	$3^2+6(3)+9$ $=9+18+9$ $=36$
4	$(4+3)^2$ $=7^2$ $=49$	4^2+9 $=16+9$ $=25$	$4^2+6(4)+9$ $=16+24+9$ $=49$

Note: From exercise #37, we know $(x+3)^2 = x^2 + 6x + 9$. From the values in the table, you can see that $(x+3)^2 \neq x^2 + 9$.

55. Large rectangle: $l = x+3$, $w = x+3$
In the diagram, find the area of the 4 small rectangles:

	x	3
x	x^2	$3x$
3	$3x$	9

Area of large rectangle = sum of areas of small rectangles:
$(x+3)^2 = (x+3)(x+3)$
$= x^2 + 3x + 3x + 9$
$= x^2 + 6x + 9$

57. Large rectangle: $l = x+5$, $w = x+5$
In the diagram, find the area of the 4 small rectangles:

	x	5
x	x^2	$5x$
5	$5x$	25

Area of large rectangle = sum of areas of small rectangles:

$$\begin{aligned}(x+5)^2 &= (x+5)(x+5)\\ &= x^2+5x+5x+25\\ &= x^2+10x+25\end{aligned}$$

59. Large rectangle: $l = 2x+3$, $w = 2x+3$
In the diagram, find the area of the 4 small rectangles:

	$2x$	3
$2x$	$4x^2$	$6x$
3	$6x$	9

Area of large rectangle = sum of areas of small rectangles:

$$\begin{aligned}(2x+3)^2 &= (2x+3)(2x+3)\\ &= 4x^2+6x+6x+9\\ &= 4x^2+12x+9\end{aligned}$$

Review Problems

61. $V = lwh$

$= (5\text{ in})(7\text{ in})(5\text{ in})$

$= 175\text{ in}^3$

$S = 2lw+2lh+2wh$

$= 2(7\text{in})(5\text{in})+2(7\text{in})(5\text{in})+2(5\text{in})(5\text{in})$

$= 70\text{ in}^2+70\text{ in}^2+50\text{ in}^2$

$= 190\text{ in}^2$

63. $V = lwh$

$= (13\text{ ft})(12\text{ ft})(15\text{ ft})$

$= 2,340\text{ ft}^3$

$S = 2lw+2lh+2wh$

$= 2\cdot 13\text{ft}\cdot 12\text{ft}+2\cdot 13\text{ft}\cdot 15\text{ft}+2\cdot 12\text{ft}\cdot 15\text{ft}$

$= 312\text{ ft}^2+390\text{ ft}^2+360\text{ ft}^2$

$= 1,062\text{ ft}^2$

Section 9.4

1. $2^{-5} = \frac{1}{2^5} = \frac{1}{32}$

3. $10^{-2} = \frac{1}{10^2} = \frac{1}{100}$

5. $x^{-3} = \frac{1}{x^3}$

7. $(-4)^{-2} = \frac{1}{(-4)^2} = \frac{1}{16}$

9. $(-5)^{-3} = \frac{1}{(-5)^3} = -\frac{1}{125}$

11. $2^6 \cdot 2^{-8} = 2^{6+(-8)} = 2^{-2} = \frac{1}{2^2} = \frac{1}{4}$

13. $10^{-2} \cdot 10^5 = 10^{-2+5} = 10^3 = 1{,}000$

15. $x^{-4} \cdot x^{-3} = x^{-4+(-3)} = x^{-7} = \frac{1}{x^7}$

17. $3^8 \cdot 3^{-6} = 3^{8+(-6)} = 3^2 = 9$

19. $2 \cdot 2^{-4} = 2^{1+(-4)} = 2^{-3} = \frac{1}{2^3} = \frac{1}{8}$

21. $10^{-3} \cdot 10^8 \cdot 10^{-2} = 10^{-3+8+(-2)} = 10^3 = 1{,}000$

23. $x^{-5} \cdot x^{-4} \cdot x^{-3} = x^{-5+(-4)+(-3)} = x^{-12} = \frac{1}{x^{12}}$

25. $2^{-3} \cdot 2 \cdot 2^3 = 2^{-3+1+3} = 2^1 = 2$

27. $\frac{2^9}{2^7} = 2^{9-7} = 2^2 = 4$

29. $\frac{2^7}{2^9} = 2^{7-9} = 2^{-2} = \frac{1}{2^2} = \frac{1}{4}$

31. $\frac{x^4}{x^3} = x^{4-3} = x^1 = x$

33. $\frac{x^3}{x^4} = x^{3-4} = x^{-1} = \frac{1}{x}$

35. $10^{-2} \div 10^5 = \frac{10^{-2}}{10^5} = 10^{-2-5} = 10^{-7} = \frac{1}{10^7}$

37. $10^{-5} \div 10^2 = \frac{10^{-5}}{10^2} = 10^{-5-2} = 10^{-7} = \frac{1}{10^7}$

39. $\frac{10^{-2}}{10^{-5}} = 10^{-2-(-5)} = 10^{-2+5} = 10^3 = 1{,}000$

41. $x^{12} \div x^{-12} = \frac{x^{12}}{x^{-12}} = x^{12-(-12)} = x^{12+12} = x^{24}$

43. $2^{12} \div 2^{10} = \frac{2^{12}}{2^{10}} = 2^{12-10} = 2^2 = 4$

45. $\frac{10}{10^{-3}} = 10^{1-(-3)} = 10^{1+3} = 10^4 = 10{,}000$

47. $3x^{-2} \cdot 5x^7 = (3 \cdot 5)(x^{-2} \cdot x^7) = 15x^{-2+7} = 15x^5$

49. $2x^{-2} \cdot 3x^{-3} = (2 \cdot 3)(x^{-2} \cdot x^{-3})$
$= 6x^{-2+(-3)}$
$= 6x^{-5}$
$= \frac{6}{x^5}$

51. $\left(4x^{-4}\right)\left(-3x^{-3}\right)=4(-3)\left(x^{-4}\cdot x^{-3}\right)$
$=-12x^{-4+(-3)}$
$=-12x^{-7}$
$=-\dfrac{12}{x^{7}}$

53. $7x\cdot 3x^{-4}\cdot 2x^{5}=(7\cdot 3\cdot 2)\left(x\cdot x^{-4}\cdot x^{5}\right)$
$=42x^{1+(-4)+5}$
$=42x^{2}$

55. $\dfrac{x^{-3}\cdot x^{7}}{x^{-1}}=\dfrac{x^{-3+7}}{x^{-1}}=\dfrac{x^{4}}{x^{-1}}=x^{4-(-1)}=x^{4+1}=x^{5}$

57. $\dfrac{x^{-2}\cdot x^{-8}}{x^{-12}}=\dfrac{x^{-2+(-8)}}{x^{-12}}$
$=\dfrac{x^{-10}}{x^{-12}}$
$=x^{-10-(-12)}$
$=x^{-10+12}$
$=x^{2}$

Review Problems

59. a. 6 min 45 sec = $6\cancel{\text{min}}\times\dfrac{60\text{ sec}}{1\cancel{\text{min}}}+45$ sec
= 360 sec + 45 sec
= 405 sec

b. 6 min 45 sec = 6 min + $45\cancel{\text{sec}}\times\dfrac{1\text{ min}}{60\cancel{\text{sec}}}$
= 6 min + 0.75 min
= 6.75 min

61. a. 3 lbs 8 oz = $3\cancel{\text{lb}}\times\dfrac{16\text{ oz}}{1\cancel{\text{lb}}}+8$ oz
= 48 oz + 8 oz
= 56 oz

b. 3 lb 8 oz = 3 lb + $8\cancel{\text{oz}}\times\dfrac{1\text{ lb}}{16\cancel{\text{oz}}}$
= 3 lb + 0.5 lb
= 3.5 lb

63. a. 5 ft 9 in = $5\cancel{\text{ft}}\times\dfrac{12\text{ in}}{1\cancel{\text{ft}}}+9$ in
= 60 in + 9 in
= 69 in

b. 5 ft 9 in = 5 ft + $9\cancel{\text{in}}\times\dfrac{1\text{ ft}}{12\cancel{\text{in}}}$
= 5 ft + 0.75 ft
= 5.75 ft

Section 9.5

1. $425{,}000 = 4.25 \times 10^5$

3. $6{,}780{,}000 = 6.78 \times 10^6$

5. $11{,}000 = 1.1 \times 10^4$

7. $89{,}000{,}000 = 8.9 \times 10^7$

9. $3.84 \times 10^4 = 3.84 \times 10{,}000 = 38{,}400$

11. $5.71 \times 10^7 = 5.71 \times 1{,}000{,}000 = 57{,}100{,}000$

13. $3.3 \times 10^3 = 3.3 \times 1{,}000 = 3{,}300$

15. $8.913 \times 10^7 = 8.913 \times 1{,}000{,}000$
$= 89{,}130{,}000$

17. $0.00035 = 3.5 \times 10^{-4}$

19. $0.0007 = 7.0 \times 10^{-4}$

21. $0.06035 = 6.035 \times 10^{-2}$

23. $0.1276 = 1.276 \times 10^{-1}$

25. $$\begin{aligned} 8.3 \times 10^{-4} &= 8.3 \times \frac{1}{10^4} \\ &= 8.3 \times \frac{1}{10{,}000} \\ &= \frac{8.3}{10{,}000} \\ &= 0.00083 \end{aligned}$$

27. $$\begin{aligned} 6.25 \times 10^{-2} &= 6.25 \times \frac{1}{10^2} \\ &= 6.25 \times \frac{1}{100} \\ &= \frac{6.25}{100} \\ &= 0.0625 \end{aligned}$$

29. $3.125 \times 10^{-1} = 3.125 \times \frac{1}{10} = \frac{3.125}{10} = 0.3125$

31. $$\begin{aligned} 5 \times 10^{-3} &= 5 \times \frac{1}{10^3} \\ &= 5 \times \frac{1}{1{,}000} \\ &= \frac{5}{1{,}000} \\ &= 0.005 \end{aligned}$$

Applying the Concepts

33. Ad cost: $\$42{,}000 = \$4.2 \times 10^4$
Viewers: $4{,}000{,}000 = 4.0 \times 10^6$

35. Ad cost: $\$575{,}000 = \$5.75 \times 10^5$
Viewers: $87{,}00{,}000 = 8.7 \times 10^7$

37. The table is as follows:

Moon	Period (sec)	
Io	153,000	1.53×10^5
Europa	307,000	3.07×10^5
Ganymede	618,000	6.18×10^5
Callisto	1,440,000	1.44×10^6

39. $$\begin{aligned} 1 \text{ kilobyte} &= 2^{10} \\ &= 1{,}024 \\ &= 1.024 \times 10^3 \text{ bytes} \end{aligned}$$

41. 1 Gigabyte $= 2^{30}$

$$= 1{,}074{,}000{,}000$$
$$= 1.074 \times 10^9 \text{ bytes}$$

Review Problems

43. $7x - 3x = (7-3)x = 4x$

45. $5x + 3x - 11x = (5+3-11)x = -3x$

47. $7x + 3x - x = (7+3-1)x = 9x$

49. $3a - 14a + 5a = (3-14+5)a = -6a$

Section 9.6

1. $(2\times10^4)(3\times10^6)=(2)(3)\times(10^4)(10^6)$
$=6\times10^{4+6}$
$=6\times10^{10}$

3. $(2.5\times10^7)(6\times10^3)=(2.5)(6)\times(10^7)(10^3)$
$=15\times10^{7+3}$
$=15\times10^{10}$
$=1.5\times10^{11}$

5. $(7.2\times10^3)(9.5\times10^{-6})$
$=(7.2)(9.5)\times(10^3)(10^{-6})$
$=68.4\times10^{3+(-6)}$
$=68.4\times10^{-3}$
$=6.84\times10^{-2}$

7. $(36{,}000)(450{,}000)=(3.6\times10^4)(4.5\times10^5)$
$=(3.6)(4.5)\times(10^4)(10^5)$
$=16.2\times10^9$
$=1.62\times10^{10}$

9. $(4{,}200)(0.00009)=(4.2\times10^3)(9\times10^{-5})$
$=(4.2)(9)\times(10^3)(10^{-5})$
$=37.8\times10^{-2}$
$=3.78\times10^{-1}$

11. $\dfrac{3.6\times10^5}{1.8\times10^2}=\dfrac{3.6}{1.8}\times\dfrac{10^5}{10^2}=2\times10^{5-2}=2\times10^3$

13. $\dfrac{8.4\times10^{-6}}{2.1\times10^3}=\dfrac{8.4}{2.1}\times\dfrac{10^{-6}}{10^3}$
$=4\times10^{-6-3}$
$=4\times10^{-9}$

15. $\dfrac{3.5\times10^5}{7.0\times10^{-10}}=\dfrac{3.5}{7.0}\times\dfrac{10^5}{10^{-10}}$
$=0.5\times10^{5-(-10)}$
$=0.5\times10^{15}$
$=5\times10^{14}$

17. $\dfrac{540{,}000}{9{,}000}=\dfrac{5.4\times10^5}{9\times10^3}$
$=\dfrac{5.4}{9}\times\dfrac{10^5}{10^3}$
$=0.6\times10^{5-3}$
$=0.6\times10^2$
$=6\times10^1$

19. $\dfrac{0.00092}{46{,}000}=\dfrac{9.2\times10^{-4}}{4.6\times10^4}$
$=\dfrac{9.2}{4.6}\times\dfrac{10^{-4}}{10^4}$
$=2\times10^{-4-4}$
$=2\times10^{-8}$

21. $\dfrac{(3\times10^7)(8\times10^4)}{6\times10^5}=\dfrac{(3)(8)}{6}\times\dfrac{(10^7)(10^4)}{10^5}$
$=4\times10^{7+4-5}$
$=4\times10^6$

23. $\dfrac{(2\times10^{-3})(6\times10^{-5})}{3\times10^{-4}}=\dfrac{(2)(6)}{3}\times\dfrac{(10^{-3})(10^{-5})}{10^{-4}}$
$=4\times10^{-3+(-5)-(-4)}$
$=4\times10^{-4}$

25. $\dfrac{(3.5\times10^{-4})(4.2\times10^5)}{7\times10^3}$
$=\dfrac{(3.5)(4.2)}{7}\times\dfrac{(10^{-4})(10^5)}{10^3}$
$=2.1\times10^{-4+5-3}$
$=2.1\times10^{-2}$

27. $\dfrac{(0.00087)(40{,}000)}{1{,}160{,}000} = \dfrac{(8.7\times10^{-4})(4\times10^{4})}{1.16\times10^{6}}$

$= \dfrac{(8.7)(4)}{1.16}\times\dfrac{(10^{-4})(10^{4})}{10^{6}}$

$= 30\times10^{-4+4-6}$

$= 30\times10^{-6}$

$= 3\times10^{-5}$

29. $\dfrac{(525)(0.0000032)}{0.0025} = \dfrac{(5.25\times10^{2})(3.2\times10^{-6})}{2.5\times10^{-3}}$

$= \dfrac{(5.25)(3.2)}{2.5}\times\dfrac{(10^{2})(10^{-6})}{10^{-3}}$

$= 6.72\times10^{2+(-6)-(-3)}$

$= 6.72\times10^{-1}$

Applying the Concepts

31. $\dfrac{\text{Cost}}{\text{Viewer}} = \dfrac{\$4.2\times10^{4}}{4.0\times10^{6}} = \dfrac{4.2}{4.0}\times\dfrac{10^{4}}{10^{6}}$

$= \$1.05\times10^{4-6}$

$= \$1.05\times10^{-2}$ per viewer

33. $\dfrac{\text{Cost}}{\text{Viewer}} = \dfrac{\$5.75\times10^{5}}{8.7\times10^{7}} = \dfrac{5.75}{8.7}\times\dfrac{10^{5}}{10^{7}}$

$\approx \$0.661\times10^{5-7}$

$= \$0.661\times10^{-2}$

$= \$6.61\times10^{-3}$ per viewer

35. Total stones =

$\dfrac{\text{Total weight}}{\text{Weight per stone}} = \dfrac{10^{10}\text{ kg}}{4{,}000\text{ kg}}$

$= \dfrac{1.0\times10^{10}}{4.0\times10^{3}}$

$= \dfrac{1.0}{4.0}\times\dfrac{10^{10}}{10^{3}}$

$= 0.25\times10^{10-3}$

$= 0.25\times10^{7}$

$= 2.5\times10^{6}$ stones

Review Problems

37. When $x = 4$:

$3x-5 = 3(4)-5 = 12-5 = 7$

39. When $x = 4$:

$-5x+6 = -5(4)+6 = -20+6 = -14$

41. When $x = -3$:

$3x-5 = 3(-3)-5 = -9-5 = -14$

43. When $x = -3$:

$-5x+6 = -5(-3)+6 = 15+6 = 21$

Chapter 9 Review

1. $3y^6 \cdot 5y^8 = (3\cdot 5)(y^6 \cdot y^8)$
$= (3\cdot 5)(y^{6+8})$
$= 15y^{14}$

3. $(9x^5y^2)(6x^3y^7) = (9\cdot 6)(x^5 \cdot x^3)(y^2 \cdot y^7)$
$= (9\cdot 6)(x^{5+3})(y^{2+7})$
$= 54x^8y^9$

5. $(x^4)^5 = x^{4\cdot 5}$
$= x^{20}$

7. $(5x^3y^2)^4 = 5^4(x^3)^4(y^2)^4$
$= 5^4x^{3\cdot 4}y^{2\cdot 4}$
$= 625x^{12}y^8$

9. $(3t^2 - 2t - 5) + (6t^2 + 5t - 2)$
$= 3t^2 - 2t - 5 + 6t^2 + 5t - 2$
$= (3t^2 + 6t^2) + (-2t + 5t) + (-5 - 2)$
$= 9t^2 + 3t - 7$

11. $(8x^2 - 3x + 6) - (5x^2 + 7x + 4)$
$= 8x^2 - 3x + 6 - 5x^2 - 7x - 4$
$= (8x^2 - 5x^2) + (-3x - 7x) + (6 - 4)$
$= 3x^2 - 10x + 2$

13. When $x = 2$:
$x^2 + 4x + 4 = (2)^2 + 4(2) + 4$
$= 4 + 8 + 4$
$= 16$

15. $2x^3(3x^4 - 8) = 2x^3 \cdot 3x^4 - 2x^3 \cdot 8$
$= (2\cdot 3)(x^3 \cdot x^4) - (2\cdot 8)x^3$
$= 6x^7 - 16x^3$

17. $(3x-5)(x+3) = (3x-5)x + (3x-5)3$
$= 3x\cdot x - 5\cdot x + 3x\cdot 3 - 5\cdot 3$
$= 3x^2 - 5x + 9x - 15$
$= 3x^2 + 4x - 15$

19. $(x-3)^2 = (x-3)(x-3)$
$= (x-3)x + (x-3)(-3)$
$= x\cdot x - 3\cdot x + x(-3) - 3(-3)$
$= x^2 - 3x - 3x + 9$
$= x^2 - 6x + 9$

21. Large rectangle: $l = x + 3$, $w = x + 5$
In the diagram, find the area of the 4 small rectangles:

	x	5
x	x^2	$5x$
3	$3x$	15

Area of large rectangle = sum of areas of small rectangles:
$(x+5)(x+3) = x^2 + 5x + 3x + 15$
$= x^2 + 8x + 15$

23. $5^{-3} = \frac{1}{5^3} = \frac{1}{125}$

25. $3^5 \cdot 3^{-7} = 3^{5+(-7)} = 3^{-2} = \frac{1}{3^2} = \frac{1}{9}$

27. $y^2 \cdot y^{-6} = y^{2+(-6)} = y^{-4} = \frac{1}{y^4}$

29. $\frac{x^7}{x^9} = x^{7-9} = x^{-2} = \frac{1}{x^2}$

31. $\frac{2^5}{2^8} = 2^{5-8} = 2^{-3} = \frac{1}{2^3} = \frac{1}{8}$

33. $(4x^{-3})(-5x^{-5}) = 4(-5)(x^{-3} \cdot x^{-5})$
$$= -20x^{-3+(-5)}$$
$$= -20x^{-8}$$
$$= -\frac{20}{x^8}$$

35. $73{,}800{,}000 = 7.38 \times 10^7$

37. $4.4 \times 10^9 = 4.4 \times 1{,}000{,}000{,}000$
$$= 4{,}400{,}000{,}000$$

39. $(4.8 \times 10^7)(3.6 \times 10^{-5})$
$$= (4.8)(3.6) \times (10^7)(10^{-5})$$
$$= 17.28 \times 10^{7+(-5)}$$
$$= 17.28 \times 10^2$$
$$= 1.728 \times 10^3$$

41. $\frac{6.8 \times 10^6}{3.4 \times 10^{-8}} = \frac{6.8}{3.4} \times \frac{10^6}{10^{-8}}$
$$= 2.0 \times 10^{6-(-8)}$$
$$= 2.0 \times 10^{14}$$

43. $\frac{(4.5 \times 10^{-6})(2.4 \times 10^5)}{1.2 \times 10^6}$
$$= \frac{(4.5)(2.4)}{1.2} \times \frac{(10^{-6})(10^5)}{10^6}$$
$$= 9.0 \times 10^{-6+5-6}$$
$$= 9 \times 10^{-7}$$

45. Total videos =
$$\frac{\text{Total bits}}{\text{bits per video}} = \frac{2.0 \times 10^6}{5.0 \times 10^5}$$
$$= \frac{2.0}{5.0} \times \frac{10^6}{10^5}$$
$$= 0.4 \times 10^{6-5}$$
$$= 0.4 \times 10^1$$
$$= 4 \text{ videos}$$

Chapters 1-9 Cumulative Review

1.
$$\begin{array}{r} \overset{1}{5{,}309} \\ +\ 687 \\ \hline 5{,}996 \end{array}$$

3.
$$\begin{array}{r} 11.090 \\ -\ 6.531 \\ \hline 4.559 \end{array}$$

5.
$$\begin{aligned} &(3x-5)(6x+7) \\ &=(3x-5)6x+(3x-5)7 \\ &=3x\cdot 6x-5\cdot 6x+3x\cdot 7-5\cdot 7 \\ &=18x^2-30x+21x-35 \\ &=18x^2-9x-35 \end{aligned}$$

7.
$$\begin{array}{r} 42 \\ 314\overline{)13188} \\ 1256\downarrow \\ \hline 628 \\ 628 \\ \hline 0 \end{array}$$

9. Locate the digit to the right of the ten thousands place: 43**5**,906 ; Since $5 \geq 5$, add 1 to the ten thousands place and change all digits to the right to 0: 440,000

11.
$$\begin{array}{r} 6 \\ 12\overline{)76} \\ \underline{72} \\ 4 \end{array} \qquad \frac{76}{12}=6\frac{4}{12}=6\frac{1}{3}$$

13. $0.8 = 0.8\times 100\% = 80\%$

15. $124\% = \dfrac{124}{100} = 1\dfrac{24}{100} = 1\dfrac{6}{25}$

17. $\sqrt{32xy^2} = \sqrt{16y^2\cdot 2x} = 4y\sqrt{2x}$

19. $-|-7| = -7$

21.
$$\begin{aligned} 19-5(7-4) &= 19-5(3) \\ &= 19-15 \\ &= 4 \end{aligned}$$

23.
$$\begin{aligned} \frac{3}{8}y &= 21 \\ \frac{8}{3}\cdot\frac{3}{8}y &= \frac{8}{3}\cdot 21 \\ y &= 56 \end{aligned}$$

25.
$$\begin{aligned} \frac{3.6}{4} &= \frac{4.5}{x} \\ 3.6x &= 4(4.5) \\ 3.6x &= 18 \\ \frac{3.6x}{3.6} &= \frac{18}{3.6} \\ x &= 5 \end{aligned}$$

27.
$$\begin{aligned} (6x-8)-(2x-9) &= 6x-8-2x+9 \\ &= (6x-2x)+(-8+9) \\ &= 4x+1 \end{aligned}$$

29.
$$\begin{aligned} S &= 2lw+2lh+2wh \\ &= 2(7\text{in}\cdot 3\text{in})+2(7\text{in}\cdot 2\text{in})+2(3\text{in}\cdot 2\text{in}) \\ &= 42\text{ in}^2+28\text{ in}^2+12\text{ in}^2 \\ &= 82\text{ in}^2 \end{aligned}$$

31. $\dfrac{432\text{ mi}}{27\text{ gal}} = \dfrac{16\text{ mi}}{1\text{ gal}} = 16\text{ mi/gal}$

33. $a^2 + b^2 = c^2$

$5^2 + 12^2 = c^2$

$25 + 144 = c^2$

$c^2 = 169$

$c = \sqrt{169}$

$c = 13$ m

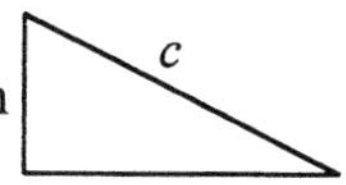

35. Note: For $I = prt$, t is in years. 90 days is

$$90\ \text{days} \cdot \frac{1\ \text{year}}{365\ \text{days}} = \frac{90}{365}\ \text{year}$$

$$\begin{aligned} I &= prt \\ &= \$1{,}400(0.06)\left(\frac{90}{365}\right) \\ &= \$20.71 \end{aligned}$$

37. Opening balance – checks = new balance ;

$$\begin{aligned} &\$469 - \$376 - \$138 \\ &= \$93 - \$138 \\ &= -\$45 \end{aligned}$$

39. Convert gallons to ounces, then ounces to glasses:

$$\begin{aligned} 15\ \text{gal} &= 15\ \text{gal} \times \frac{128\ \text{fl oz}}{1\ \text{gal}} \times \frac{1\ \text{glass}}{8\ \text{fl oz}} \\ &= \frac{15 \cdot 128}{8}\ \text{glasses} \\ &= 240\ \text{glasses} \end{aligned}$$

Chapter 9 Test

1. $3x^5 \cdot 4x^4 = (3\cdot 4)(x^5 \cdot x^4)$
$= (3\cdot 4)(x^{5+4})$
$= 12x^9$

3. $(a^4)^3 \cdot (a^7)^2 = a^{4\cdot 3} \cdot a^{7\cdot 2}$
$= a^{12} \cdot a^{14}$
$= a^{12+14}$
$= a^{26}$

5. $(4a^4b^3)^3(2a^2b)^2$
$= 4^3(a^4)^3(b^3)^3 \cdot 2^2(a^2)^2 b^2$
$= 64a^{12}b^9 \cdot 4a^4b^2$
$= (64\cdot 4)(a^{12}a^4)(b^9b^2)$
$= 256a^{12+4}b^{9+2}$
$= 256a^{16}b^{11}$

7. $(6a^2 - 7a - 4) - (2a^2 - 8a + 5)$
$= 6a^2 - 7a - 4 - 2a^2 + 8a - 5$
$= (6a^2 - 2a^2) + (-7a + 8a) + (-4 - 5)$
$= 4a^2 + a - 9$

9. Substitute 4 for x:
$2x^2 - 6x - 5 = 2(4)^2 - 6(4) - 5$
$= 2(16) - 24 - 5$
$= 32 - 24 - 5$
$= 3$

11. $(3x-5)(2x+6)$
$= (3x-5)2x + (3x-5)6$
$= 3x\cdot 2x - 5\cdot 2x + 3x\cdot 6 - 5\cdot 6$
$= 6x^2 - 10x + 18x - 30$
$= 6x^2 + 8x - 30$

13. $4^5 \cdot 4^{-7} = 4^{5+(-7)} = 4^{-2} = \dfrac{1}{4^2} = \dfrac{1}{16}$

15. $\dfrac{5^{-6}}{5^{-8}} = 5^{-6-(-8)} = 5^{-6+8} = 5^2 = 25$

17. $0.0000385 = 3.85\times 10^{-5}$

19. $(2.7\times 10^6)(6.3\times 10^{-4})$
$= (2.7)(6.3)\times(10^6)(10^{-4})$
$= 17.01\times 10^{6+(-4)}$
$= 17.01\times 10^2$
$= 1.701\times 10^3$

21. $\dfrac{6.8\times 10^{-8}}{1.7\times 10^{-4}} = \dfrac{6.8}{1.7}\times\dfrac{10^{-8}}{10^{-4}}$
$= 4.0\times 10^{-8-(-4)}$
$= 4.0\times 10^{-8+4}$
$= 4.0\times 10^{-4}$

23. Total videos =
$\dfrac{\text{Total bits}}{\text{bits per video}} = \dfrac{3.0\times 10^6}{2.5\times 10^5}$
$= \dfrac{3.0}{2.5}\times\dfrac{10^6}{10^5}$
$= 1.2\times 10^{6-5}$
$= 1.2\times 10^1$
$= 12$ videos